METHODS IN MOLECULAR BIOLOGY™

Series Editor
John M. Walker
School of Life Sciences
University of Hertfordshire
Hatfield, Hertfordshire, AL10 9AB, UK

For other titles published in this series, go to
www.springer.com/series/7651

Argonaute Proteins

Methods and Protocols

Edited by

Tom C. Hobman

Departments of Cell Biology and Medical Microbiology & Immunology, Li Ka Shing Institute of Virology, University of Alberta, Edmonton, AB, Canada

Thomas F. Duchaine

Department of Biochemistry, Goodman Cancer Research Centre, Division of Experimental Medicine, McGill University, Montreal, QC, Canada

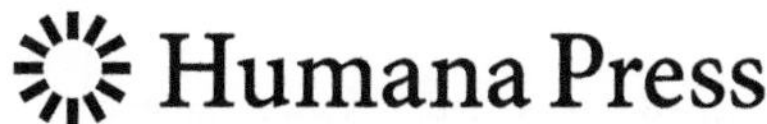

Editors
Tom C. Hobman
Departments of Cell Biology and Medical Microbiology & Immunology
Li Ka Shing Institute of Virology
University of Alberta
Edmonton, AB, Canada
tom.hobman@ualberta.ca

Thomas F. Duchaine
Department of Biochemistry,
Goodman Cancer Research Centre,
Division of Experimental Medicine,
McGill University, Montreal, QC, Canada
thomas.duchaine@mcgill.ca

ISSN 1064-3745 e-ISSN 1940-6029
ISBN 978-1-61779-045-4 e-ISBN 978-1-61779-046-1
DOI 10.1007/978-1-61779-046-1

Library of Congress Control Number: 2011926589

Printed on acid-free paper

Humana Press is part of Springer Science+Business Media (www.springer.com)

Preface

The discovery of RNA interference (RNAi) is one of the most important biomedical findings of the past 25 years. First reported in 1998 by Fire et al. (Nature 391:806–11), the pace of advancement in this research area has been nothing less than breath taking. A mere 6 years after its formal discovery, RNAi-based therapeutics were already in clinical trials in humans … with promising results. Two years later, the two scientists who were credited with the discovery of this gene-silencing mechanism, Craig Mello and Andrew Fire, were awarded the Nobel Prize in physiology or medicine.

The use of RNAi in academic and biotechnology research laboratories is now ubiquitous and, indeed, has revolutionized the study of eukaryotic gene function. Perhaps contributing to the wide spread incorporation of this technique into the toolbox of modern molecular biology is the fact that it offered a low-cost and fast-paced alternative to other reverse genetic technologies in a period that coincided with the dawn of the genomic era. Moreover, the idea that RNAi could serve as a molecular therapeutic for treating human disease had captured the attention of the biomedical and biotechnology communities. While the therapeutic potential for small RNA-based gene regulation is indeed exciting, it is important to remember that RNAi is mediated by an intricate gene-silencing apparatus that controls more than half of the human genes through a diversity of transcriptional and posttranscriptional mechanisms. Because of the importance of RNAi in the control of global gene expression, components of the RNAi machinery are undoubtedly subject to extensive regulation. Accordingly, a major challenge in the field now is to understand how this occurs.

Argonaute proteins are the central effectors of RNAi and are highly conserved among eukaryotes and some archaebacteria. These RNA-binding proteins use small guide RNAs to silence the expression of genes at the mRNA, chromatin, and DNA levels. By investigating how the activities of Argonaute proteins are regulated through *trans*-acting factors and associated regulatory RNAs, we will gain insight into how the RNAi apparatus modulates gene expression on a global level. The purpose of this book is to provide the reader with step-by-step methods to study Argonaute protein functions and interactions in a wide variety of cell types ranging from yeast to mammalian systems, as well as in vitro. The book is intended for researchers who have already acquired a working knowledge of Argonaute proteins as well as for scientists who are new to the field.

Edmonton, AB — *Tom C. Hobman*
Montreal, QC — *Thomas F. Duchaine*

Contents

Contributors

PEDRO J. BATISTA • *Program in Molecular Medicine, University of Massachusetts Medical School, Worcester, MA, USA; Gulbenkian PhD Programme in Biomedicine, Oeiras, Portugal*

SHANE M. BUKER • *Department of Cell Biology, Harvard Medical School, Boston, MA, USA*

EDWARD K. L. CHAN • *Department of Oral Biology, University of Florida, Gainesville, FL, USA*

JULIE M. CLAYCOMB • *Department of Molecular Genetics, University of Toronto, Toronto, Canada*

DARRYL CONTE • *Program in Molecular Medicine, University of Massachusetts Medical School, Worcester, MA, USA*

NABANITA DE • *Department of Molecular Biology, The Scripps Research Institute, La Jolla, CA, USA*

JESSICA DEMAIO • *Feinberg School of Medicine, Northwestern University, Chicago, IL, USA*

THOMAS F. DUCHAINE • *Department of Biochemistry, Goodman Cancer Research Centre, Division of Experimental Medicine, McGill University, Montreal, QC, Canada*

OLIVIA A. EBNER • *Max Delbrück Center for Molecular Medicine, Berlin, Germany*

STEPHANIE ECKHARDT • *EMBL International PhD Programme, European Molecular Biology Laboratory, Grenoble, France*

MARC R. FABIAN • *Department of Biochemistry, Goodman Cancer Research Center, McGill University, Montreal, QC, Canada*

MARVIN J. FRITZLER • *Department of Biochemistry and Molecular Biology, University of Calgary, Calgary, AB, Canada*

DERRICK GIBBINGS • *Department of Biology, Swiss Federal Institute of Technology (ETH-Z), Zurich, Switzerland*

WEIFENG GU • *Program in Molecular Medicine, University of Massachusetts Medical School, Worcester, MA, USA*

TOM C. HOBMAN • *Departments of Cell Biology and Medical Microbiology & Immunology, Li Ka Shing Institute of Virology, University of Alberta, Edmonton, AB, Canada*

GUILLAUME JANNOT • *Laval University Cancer Research Centre, Hôtel-Dieu de Québec (CHUQ), Quebec City, QC, Canada*

TOMOKO KAWAMATA • *Institute of Molecular and Cellular Biosciences, The University of Tokyo, Tokyo, Japan*

EUGENE KHANDROS • *Division of Neuropathology, Department of Pathology and Laboratory Medicine, University of Pennsylvania School of Medicine, Philadelphia, PA, USA*

YOHEI KIRINO • *Division of Neuropathology, Department of Pathology and Laboratory Medicine, University of Pennsylvania School of Medicine, Philadelphia, PA, USA*
PATRICIA LANDRY • *Centre de Recherche en Rhumatologie et Immunologie, CHUL Research Center/CHUQ, Quebec, QC, Canada*
JING LI • *Faculty of Medicine and Dentistry, Department of Cell Biology, University of Alberta, Edmonton, AB, Canada*
SONGQING LI • *Department of Oral Biology, University of Florida, Gainesville, FL, USA*
SHANG LI LIAN • *Department of Oral Biology, University of Florida, Gainesville, FL, USA*
WENDY LONG • *Faculty of Medicine and Dentistry, Department of Cell Biology, University of Alberta, Edmonton, AB, Canada*
JOAQUIN LOPEZ-OROZCO • *Department of Cell Biology, University of Alberta, Edmonton, AB, Canada*
IAN J. MACRAE • *Department of Molecular Biology, The Scripps Research Institute, La Jolla, CA, USA*
LYDIA V. MCCLURE • *Molecular Genetics & Microbiology, The University of Texas at Austin, Austin, TX, USA*
GUNTER MEISTER • *Center for Integrated Protein Science Munich (CIPSM), Laboratory for RNA Biology, Max-Planck-Institute of Biochemistry, Martinsried, Germany; University of Regensburg, Regensburg, Germany*
CRAIG C. MELLO • *Program in Molecular Medicine, University of Massachusetts Medical School, Worcester, MA, USA; Howard Hughes Medical Institute, Worcester, MA, USA*
KEITA MIYOSHI • *Department of Molecular Biology, Keio University School of Medicine, Tokyo, Japan*
MOHAMMAD R. MOTAMEDI • *Department of Cell Biology, Harvard Medical School, Boston, MA, USA*
ZISSIMOS MOURELATOS • *Division of Neuropathology, Department of Pathology and Laboratory Medicine, University of Pennsylvania School of Medicine, Philadelphia, PA, USA*
NIMA NAJAND • *Faculty of Medicine and Dentistry, Department of Cell Biology, University of Alberta, Edmonton, AB, Canada*
TOMOKO N. OKADA • *Department of Molecular Biology, Keio University School of Medicine, Tokyo, Japan*
JUSTIN M. PARE • *Department of Cell Biology, University of Alberta, Edmonton, AB, Canada*
MARJORIE P. PERRON • *Centre de Recherche en Rhumatologie et Immunologie, CHUL Research Center/CHUQ, Quebec, QC, Canada*
RAMESH PILLAI • *European Molecular Biology Laboratory, Grenoble, France*
ISABELLE PLANTE • *Centre de Recherche en Rhumatologie et Immunologie, CHUL Research Center/CHUQ, Quebec, QC, Canada*
PATRICK PROVOST • *CHUL Research Center/CHUQ and Faculty of Medicine, Université Laval, Quebec, QC, Canada*
MATTHIAS SELBACH • *Max Delbrück Center for Molecular Medicine, Berlin, Germany*

Gil Ju Seo • *Molecular Genetics & Microbiology, The University of Texas at Austin, Austin, TX, USA*
Martin J. Simard • *Laval University Cancer Research Centre, Hôtel-Dieu de Québec (CHUQ), Quebec City, QC, Canada*
Andrew Simmonds • *Faculty of Medicine and Dentistry, Department of Cell Biology, University of Alberta, Edmonton, AB, Canada*
Haruhiko Siomi • *Department of Molecular Biology, Keio University School of Medicine, Tokyo, Japan*
Mikiko C. Siomi • *Department of Molecular Biology, Keio University School of Medicine, Tokyo, Japan; Core Research for Evolutional Science and Technology (CREST), Japan Science and Technology Agency (JST), Saitama, Japan*
Nahum Sonenberg • *Department of Biochemistry, Goodman Cancer Research Center, McGill University, Montreal, QC, Canada*
Julia Stoehr • *Center for Integrated Protein Science Munich (CIPSM), Laboratory for RNA Biology, Max-Planck-Institute of Biochemistry, Martinsried, Germany*
Hong Su • *Department of Biochemistry, Northwestern University, Evanston, IL, USA*
Christopher S. Sullivan • *Molecular Genetics & Microbiology, The University of Texas at Austin, Austin, TX, USA*
Yuri V. Svitkin • *Department of Biochemistry, Goodman Cancer Research Center, McGill University, Montreal, QC, Canada*
Emilia Szostak • *European Molecular Biology Laboratory, Grenoble, France*
Yukihide Tomari • *Institute of Molecular and Cellular Biosciences, The University of Tokyo, Tokyo, Japan*
Alejandro Vasquez-Rifo • *Laval University Cancer Research Centre, Hôtel-Dieu de Québec (CHUQ), Quebec City, QC, Canada*
Thomas A. Volpe • *Feinberg School of Medicine, Northwestern University, Chicago IL, USA*
Anastassios Vourekas • *Division of Neuropathology, Department of Pathology and Laboratory Medicine, University of Pennsylvania School of Medicine, Philadelphia, PA, USA*
Xiaozhong Wang • *Department of Biochemistry, Northwestern University, Evanston, IL, USA*
Edlyn Wu • *Division of Experimental Medicine, Department of Biochemistry, Goodman Cancer Research Centre, McGill University, Montreal, QC, Canada*
Zhaolin Yang • *European Molecular Biology Laboratory, Grenoble, France*
Bing Yao • *Department of Oral Biology, University of Florida, Gainesville, FL, USA*

Chapter 1

Purification of Native Argonaute Complexes from the Fission Yeast *Schizosaccharomyces pombe*

Shane M. Buker and Mohammad R. Motamedi

Abstract

Small interfering (si) RNAs, produced by the RNA interference (RNAi)-mediated processing of long double-stranded (ds) RNAs, can inhibit gene expression by post-transcriptional or transcriptional gene silencing mechanisms. At the heart of all small RNA-mediated silencing lies the key RNAi effector protein Argonaute, which once loaded with small RNAs can recognize its target transcript by siRNA–RNA Watson–Crick base pairing interactions. In the fission yeast *Schizosaccharomyces pombe*, the formation of the epigenetically heritable centromeric heterochromatin requires RNAi proteins including the sole fission yeast Argonaute homolog, Ago1. Two distinct native Ago1 complexes have been purified and studied extensively, both of which are required for siRNA production and heterochromatin formation at the fission yeast centromeres. The purification and analysis of the Argonaute siRNA chaperone (ARC) complex and RNA-induced transcriptional silencing (RITS) complex have provided insight into the mechanism of siRNA-Ago1 loading and the *cis* recruitment of silencing complexes at fission yeast centromeres, respectively. These discoveries have been instrumental in shaping the current models of RNA-mediated epigenetic silencing in eukaryotes. Below, we describe the protocol used for affinity purification of the native Ago1 complexes from *S. pombe*.

Key words: Fission yeast, FLAG purifications, Argonaute (Ago1), Epitope tagging, Rapid silver staining, Lithium acetate transformation

1. Introduction

The catalytic engine of all small RNA-mediated gene silencing is the Argonaute family of eukaryotic proteins (recently reviewed in refs. 1, 2). When loaded with small RNAs, Argonautes are targeted to homologous transcripts and mediate sequence-specific repression by a variety of mechanisms, including endonucleolytic cleavage of target transcripts (3, 4), inhibition of protein translation (5), or *cis* recruitment of chromatin modifying/binding

Tom C. Hobman and Thomas F. Duchaine (eds.), *Argonaute Proteins: Methods and Protocols*, Methods in Molecular Biology, vol. 725, DOI 10.1007/978-1-61779-046-1_1, © Springer Science+Business Media, LLC 2011

complexes to the site of transcription (reviewed in ref. 6). Argonaute proteins contain an N-terminal RNA-binding domain called PAZ and a C-terminal catalytic domain called PIWI, which shares extensive structural similarity to the catalytic core of RNAse H enzymes – an ancient class of ribonucleases that bind to RNA/DNA duplexes and cleave RNA (1). These proteins can cleave ("slice") their target transcripts via their conserved Asp-Asp-His catalytic triad within the PIWI domain (4). Purification of Argonaute complexes from a variety of organisms has revealed a large network of Argonaute-interacting proteins and has suggested a regulatory role for these proteins in both targeting and regulating Argonaute activity (1). For example, the recent purification of several Argonaute-interacting proteins identified a common PIWI-binding domain called glycine-tryptophan (GW) rich or Ago hook motif, which may regulate PIWI-mediated slicing activity (7).

The fission yeast has only one copy of the Argonaute homolog, Ago1. Ago1 (8) and its associated slicing activity (9, 10) are required for heterochromatin formation at pericentromeric repeats. This protein is found in two distinct complexes: the Argonaute siRNA chaperone (ARC) complex (9) and the RNA-induced transcriptional silencing (RITS) complex (11). In addition to Ago1, ARC is composed of two auxiliary proteins called Arb1 and Arb2 and is primarily associated with double-stranded small RNAs. In RITS, Ago1 is in a complex with Chp1, a heterochromatin protein capable of binding to methylated histone H3 lysine 9 (H3K9me) (the hallmark of eukaryotic heterochromatin),

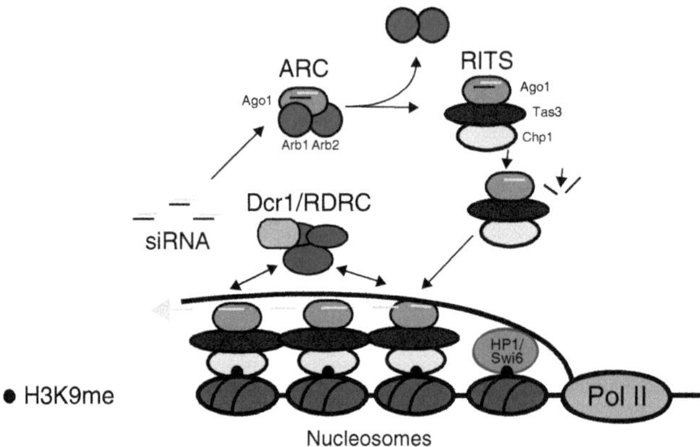

Fig. 1. The nascent transcript model for heterochromatin assembly at the fission yeast centromeres (13). Nascent transcripts, tethered to chromatin via HP1/Swi6 (14) recruit RNAi complexes RITS and RNA-dependent RNA polymerase complex, RDRC. RITS is tethered to chromatin via Chp1 binding to H3K9me and RITS bound single-stranded (ss) siRNAs base pairing with the nascent transcript. RDRC produces double-stranded (ds) RNA using the nascent transcript as a substrate, and with the help of ribonuclease enzyme Dicer (Dcr1) processes long dsRNA (*solid* and *dashed parallel lines*) into siRNAs. ds siRNAs are loaded into ARC and processed into ss siRNAs found in RITS. RNAi complexes also recruit chromatin-modifying proteins such as histone methyltransferase Clr4, leading to the spreading of heterochromatin to the surrounding chromosomal regions.

and with the GW protein, Tas3, and is predominately associated with single-stranded siRNAs. The discovery of the RITS complex provided a physical link between RNA interference (RNAi) and heterochromatin proteins (11), thus revealing the first clue about how small non-coding siRNAs regulate chromatin modifications in vivo (Fig. 1). Furthermore, the coimmunoprecipitation of single-stranded or double-stranded small RNAs in RITS and ARC complexes, respectively, suggests that Ago1-interacting proteins regulate Ago1 slicing activity in vivo. Here we describe the purification of the native Ago1 complexes using a functional FLAG-Ago1 construct.

2. Materials

2.1. Strain Construction, Transformation, Colony PCR Test, and FLAG Western Blot

1. PCR reaction buffer, 10× concentration (Roche).
2. Taq DNA polymerase with ready-to-use PCR-grade nucleotide mix (Roche).
3. PCR tubes (Denville Scientific, Inc.).
4. Agarose.
5. 50× TAE buffer: 121 g of Tris-HCl is dissolved in 28.6 ml of glacial acetic acid, and in 50 ml of 0.5 M ethylenediaminetetraacetic acid (EDTA) (pH 8.0) in approximately 300 ml of water. Adjust the volume with water to 500 ml.
6. DNA gel running apparatus (Owl Scientific).
7. Power supply.
8. 3 M Sodium acetate (NaOAc).
9. YES medium: 5 g/L yeast extract, 30 g/L dextrose and 0.225 g/L each of adenine, leucine, lysine, histidine, and uracil.
10. Lithium acetate (LiOAc)/TE transformation mix: 100 mM lithium acetate adjusted to pH 7.5 with diluted acetic acid, 10 mM Tris–HCl, pH 7.5, and 1 mM EDTA pH 7.5.
11. 10 mg/ml salmon sperm DNA (SIGMA).
12. 40% Polyethylene glycol (PEG) 4000 (Sigma). Water is added up to 75% of the final volume of the solution. PEG solution is heated while stirring with a magnetic bar, and once most of the PEG is dissolved, enough water is added to bring up to the final volume.
13. PEG mix: 0.8 ml of 40% PEG plus 100 μl of 10× TE plus 100 μl of 1 M LiOAc.
14. Dimethylsulfoxide (DMSO) (SIGMA).
15. YES plates are made with YES medium (see item 9 in Subheading 2.1 for recipe) plus 20 g of Agar, Bacteriological (USB) per liter. The mixture is autoclaved for 20 min, cooled

for 20 min at room temperature, and poured into 100 × 15 mm Petri plates (Corning).

16. Nourseothricin dihydrogen sulfate (Nat) (Werner Bioagent).
17. Phosphate-buffered saline (PBS) 10× stock: 137 mM sodium chloride (NaCl), 2.7 mM potassium chloride (KCl), 10 mM sodium phosphate (Na_2HPO_4), and 1.8 mM potassium phosphate (KH_2PO_4) adjust to pH 7.4 with hydrochloric acid (HCl) and autoclave.
18. Mini-Beadbeater-8 (BioSpec).
19. Microwave (Panasonic Inverter, Model NNH624BFR).
20. 50 ml Polypropylene centrifuge tubes (Corning).
21. Fast Prep FP120 bead beater (Qbiogene).
22. Laemmli sample buffer: 0.5 M Tris–HCl pH 6.8, 5% glycerol, 2% sodium dodecyl sulfate (SDS), 100 mM dithiothreitol (DTT).
23. NuPAGE Bis-Tris 4–12% acrylamide gel (Invitrogen) and XCell SureLock Mini-Cell gel Running apparatus (Invitrogen).
24. Nitrocellulose membrane (Bio-Rad).
25. Tween 20 (SIGMA).
26. FLAG M2 mouse monoclonal antibody (SIGMA).
27. ECL Anti-mouse IgG antibody (GE).
28. Novex® ECL Chemiluminescent Substrate Reagent Kit (Invitrogen).
29. KODAK™ X-OMAT™ Blue (XB) Film (Perkin Elmer).

2.2. Cell Growth and Lysis, Affinity Purification of Ago1 and TCA Precipitation

1. 1.5 L of YES (see item 9 in Subheading 2.1) medium in 2.8 L Erlenmeyer Flasks for inoculation.
2. Beckman Coulter Avanti J-20XP centrifuge with JLA 8.1000 rotor.
3. Sorvall RC 5C plus with SLA-1500 and SH-3000 rotors.
4. PBS (see item 17 in Subheading 2.1).
5. Lysis buffer: 50 mM HEPES (pH 7.6), 300 mM potassium acetate (CH_3COOK), 5 mM magnesium acetate ($Mg(CH_3COO)_2$), 20 mM β-glycerol phosphate, 1 mM ethylene glycol-bis(2-aminoethylether)-N,N,N',N'-tetraacetic acid (EGTA), 1 mM EDTA, 0.1% (v/v), 0.25% Nonidet P40 (NP-40) containing protease inhibitors added immediately prior to use: 1 mM Phenylmethanesulfonyl fluoride (PMSF) (from a 0.1 M stock, which is made fresh in 100% ethanol every time before purification), one Complete Protease Inhibitor Tablet, EDTA-free (Roche)/50 ml lysis buffer volume.
6. 50 ml polypropylene centrifuge tubes (Corning).
7. Coffee Grinder (Krups GX4100).

8. ANTI-FLAG M2 Affinity Agarose Gel (SIGMA).
9. Clay Adams Brand Nutator (Model 421105).
10. Polyprep columns (Bio-Rad).
11. 3XFLAG peptide (SIGMA), resuspended to 5 mg/ml in ddH_2O.
12. 100% Trichloroacetic acid (TCA).
13. 100% Acetone.

3. Methods

To purify the native Ago1 complexes from *Schizosaccharomyces pombe*, we used an N-terminally 3XFLAG-tagged protein, inserted using homologous recombination, as described in Subheading 3.1. This protein is functional, stable, and co-purifies with Arb1 and Arb2, components of the ARC complex, and Chp1 and Tas3, components of the RITS complex. Furthermore, repeat-associated siRNAs are coimmunoprecipitated and can be labeled and visualized using end-labeling or Northern blot analyses.

Roughly 5–10 g of logarithmically growing cells was used to purify enough protein for mass spectroscopy and visualization by rapid silver staining.

3.1. Strain Construction and Transformation

1. To obtain affinity-tagged fusion proteins expressed from their native promoter, a PCR-amplified double-stranded DNA is integrated at the desired locus via homologous recombination (12). This PCR fragment contains sequence coding for a selective drug resistance marker as well as the affinity tag. Long PCR primers are designed with flanking sequences homologous to the site of integration, so that the tag is inserted in register with the protein-coding sequence. While a tag can be inserted at the N- or C-termini of the gene of interest, we found that several tags inserted at the *S. pombe ago1* C-terminus resulted in a nonfunctional proteins. On the other hand, we found that 3XFLAG tag inserted at the N-terminus of Ago1 protein produces a functional protein in vivo (see Note 1 for description of the plasmid used for PCR amplification of the DNA fragment used for N-terminal tagging of Ago1.)
2. Several (3–6) identical PCR reactions are set up to amplify roughly 10–30 μg of the DNA used for transformation. PCR reactions (50 μl final volume) are set up using the following recipe (per reaction):

   Plasmid DNA template: 5 ng.

   Primers: 0.5 μl of 50 μM oligo solution.

10× dNTP: 5.0 μl of 2 mM stock solution per dNTP.

Buffer: 5.0 μl of a 10× stock.

Taq polymerase: 0.5 μl.

H_2O: Bring up to 50 μl final volume.

3. All PCR reactions are pooled, and ethanol precipitated using the following protocol: 3.5 volume of 100% ethanol is added to the pooled PCR reactions plus 0.5 volume of 3 M NaOAc pH 5.2. The sample is mixed, stored at −20°C for 1 h, and spun at 20,000 × *g* for 15 min. The DNA pellet is then washed with 200 μl of 70% ethanol, spun again, and air-dried for 2 h at room temperature. After drying, the DNA pellet is dissolved in 50 μl of TE and DNA concentration is determined by measuring UV_{260} absorbance. 1–2 μl of the concentrated DNA is run on a 1.2% agarose DNA gel at 100 V for 45 min. The gel is then stained with ethidium bromide and visualized under UV light to confirm the presence of the desired PCR product. A wild-type *S. pombe* is then transformed with this PCR product using the following steps.
4. A 20 ml culture is grown in YES to OD_{600} 0.6, washed once with ddH_2O and then with 1 ml of LiOAc/TE.
5. The cells are then resuspended in 100 μl of LiOAc/TE plus 2 μl of 10 mg/ml salmon sperm DNA and 10–30 μg of the PCR product. After vortexing, the mixture of cells and DNA is incubated for 10 min at room temperature.
6. 260 μl of PEG mix is then added, vortexed thoroughly, and incubated for 30–60 min at 30°C. Longer PEG incubation times are correlated with enhanced transformation efficiencies.
7. 43 μl of DMSO is then added, mixed thoroughly, and the cells are heat-shocked at 42°C for 5 min.
8. Cells are then washed once with room temperature water, resuspended in 200 μl of water, and the entire volume is spread onto one YES plate for each transformation. These plates are incubated for 18–24 h at 32°C.
9. The next day, a lawn of cells is visible which is replica plated onto YES plates containing 0.1 g/L Nat, and incubated for 3 days at 32°C.
10. Several resulting colonies are restreaked onto Nat plates for colony purification and PCR testing.

3.2. Colony PCR Test and FLAG Western Blot

1. To identify the true *3XFLAG-ago1* clones from among the false positives, a PCR-based screen is performed (see Note 2 for primer design).
2. Using an autoclaved P20 tip, a small portion of a colony from a candidate strain (following colony purification (step 10, Subheading 3.1)) is scraped at the bottom of a PCR tube.

The tubes, placed in a rack with open caps, are then put inside a microwave. The microwave is turned on at high setting for 3 min, after which the cell pellets at the bottom of the tube are mixed thoroughly with the PCR mix (see below) before thermocycling.

PCR reactions (12.5 μl final volume) for testing FLAG-Ago1 integration were set up using the following recipe (per reaction):

Primers: 0.125 μl of 50 μM oligo solution.

10× dNTP: 1.25 μl of 2 mM stock solution per dNTP.

Buffer: 1.25 μl of a 10× stock.

Taq: 0.0625 μl.

H_2O: bring up to 12.5 μl final volume.

3. Upon completion of the PCR, the entire PCR reaction is run on a 1.2% agarose gel for 45 min at 100 V.
4. The gel is then stained with ethidium bromide and visualized under UV light. Positive integrants should have a DNA band corresponding to the correct size in contrast to the no-tag negative control. Once a positive PCR clone is identified, a FLAG western blot is performed to confirm 3XFLAG-Ago1 expression.
5. 10 ml YES medium is inoculated with the candidate yeast strain(s) and grown overnight at 32°C with rotation at 225 rpm. Also, a no-tag negative control is grown in parallel for the FLAG western blot.
6. This culture is harvested by centrifugation using Sorvall RC 5C plus with SH-3000 rotor at 3,000 × *g* in or 10 min at room temperature, washed once with 1 ml PBS, transferred to a microcentrifuge tube, and centrifuged again.
7. The pellet is resuspended in 100 μl Laemmli sample buffer plus 1 mM PMSF added immediately prior to use, and an equivalent volume of glass beads is added.
8. The cells are lysed by bead-beating three times 30 s at the settings given in step 6 in the fastprep bead beater, with 5′ of rest on ice between cycles.
9. The lysate is then incubated at 95°C for 10 min and spun for 1 min at maximum speed in a microfuge.
10. 10 μl of the lysate is run on a 4–10% acrylamide gel in MOPS SDS running buffer, according to the manufacturer's instructions.
11. XCell SureLock Mini-Cell apparatus is assembled and filled with 1,000 ml 1× NuPAGE SDS Running Buffer.
12. Samples are loaded on gel and run at 200 V for approximately 50 min.

13. Proteins are then transferred from the gel onto a nitrocellulose membrane at 300 mAmp for 3 h at 4°C in a mini western blot apparatus according to the manufacturer's instructions.
14. The membrane is blocked with PBS + 0.5% Tween (PBS-T) + 5% milk for 1 h at room temperature.
15. The membrane is then incubated with PBS-T + 1 μg/ml FLAG M2 mouse monoclonal antibody.
16. The membrane is then washed three times with PBS-T for 5 min each wash.
17. The membrane is incubated with PBS-T + 1:10,000 ECL Anti-mouse IgG antibody.
18. After another set of three times 5 min washes with PBS-T, the membrane is incubated with equal volumes of ECL reagents A and B for 5 min.
19. The membrane is then exposed to film to visualize FLAG antibody reactive bands, thereby confirming the expression of FLAG-Ago1.

3.3. Cell Growth

1. 20 ml YES medium is inoculated with a single colony of *S. pombe* expressing 3XFLAG-Ago1 under the expression of the endogenous promoter (see Subheading 3.1). The culture is grown at 32°C with rotation at 225 rpm.
2. The next day, 1.5 L of YES (in 2.8 ml Erlenmeyer flasks) is inoculated with 2–5 ml of a saturated overnight culture and grown for ~20 h at 32°C, while shaking at 225 rpm, to an A_{600} optical density of 2–3.
3. For harvesting, cells are spun at 8,000 × *g* for 10 min at room temperature in Avanti J-20XP centrifuge. This yields approximately 7–10 g of wet cell pellet per liter of culture.
4. The cell pellets are washed once with 20 ml PBS, combined, and harvested again by centrifugation at 4,000 × *g* for 5 min in Sorvall RC 5C plus centrifuge using SLA-1500 rotor. The supernatant is removed and the pellet is resuspended in 0.25 volumes lysis buffer. Protease inhibitors – 1 mM PMSF, one tablet of Roche Complete Protease Inhibitor Tablets/50 ml lysis buffer volume – are added immediately prior to mixing with cells.
5. A small liquid nitrogen storage dewar is cleaned thoroughly before use, and a 50 ml polypropylene centrifuge tube is filled half way with liquid nitrogen for this step. The cell lysis buffer suspension is slowly added to the liquid nitrogen in the 50 ml tube by direct drop-wise addition. The frozen droplets can be stored in the 50 ml tube at −80°C indefinitely.

3.4. Affinity Purification of Ago1

1. Thirty minutes prior to starting, a clean 500 ml beaker is placed at −20°C. Cells are lysed by grinding the frozen cell/lysis buffer droplets in a coffee grinder filled with ~20 g dry ice for two times 5 min, with a 5 min rest in between each grinding.
2. The dry ice cell mixture powder is transferred into the −20°C 500 ml beaker with a magnetic stirring rod at the bottom. This is then placed on a stirring platform in the cold room and the mixture is stirred gently for roughly 30–45 min until all the dry ice has sublimed. All subsequent steps of this protocol are performed in the cold room, using pre-cooled (4°C) buffers, tubes, and pipette tips.
3. After sublimation, one volume of cold (4°C) lysis buffer (with protease inhibitors added immediately prior to use) is added to the extract, and the mixture is transferred to one or two pre-cooled 50 ml polypropylene centrifuge tubes and spun at 4,000×*g* in Sorvall RC 5C plus centrifuge using SH-3000 rotor for 15 min at 4°C.
4. During the spin, 100 μl of dry ANTI-FLAG M2 Affinity Agarose Gel is measured for every 5 g of cell pellet. The beads are washed three times with 10 volumes of cold lysis buffer, and centrifuged for 3 min at 500×*g* at 4°C between each wash.
5. Equilibrated beads and the cleared extracts are added to a new cold 50 ml polypropylene centrifuge tube and incubated for 3 h on a Clay Adams Brand Nutator (or suitable mixing device) at 4°C. The beads are then recovered by centrifugation *g* in Sorvall RC 5C plus centrifuge using SH-3000 rotor at 500×*g* at 4°C for 5 min.
6. The extract is discarded and the beads are batched-washed three times with 10 volumes of cold lysis buffer and spun at 500×*g* between each wash. The beads are transferred to a cold disposable 10-ml PolyPrep column (see Note 3). The beads are washed again with three times with 10 ml of lysis buffer, allowing the beads to settle to the bottom of the column after each wash (see Note 4).
7. After washes, the yellow cap supplied with the column is used to block flow from the bottom of the column. To elute the bound protein, the beads are resuspended in 1.5× bead volume of cold lysis buffer containing 200 μg/ml 3XFLAG peptide and incubated for 20 min at 4°C. The eluate is collected by draining the column into a fresh pre-cooled Eppendorf tube. Elution is repeated once more with another 1.5 volume of cold lysis buffer containing 200 μg/ml 3XFLAG peptide and combined with the first eluate.
8. If the purified protein is not going to be used immediately, glycerol is added up to 5%, and the purified mixture is flash-frozen in liquid nitrogen, and stored at −80°C.

3.5. TCA Precipitation

1. Half of the FLAG eluate is combined with enough 100% TCA to make a 20% TCA final solution. The sample is vortexed thoroughly and stored for 20 min on ice.
2. The sample is spun at 20,000 × *g* for 20 min at 4°C and the pellet washed once with 0.5 ml of −20°C 100% acetone. The sample is air dried for roughly 30 min at room temperature and the protein precipitate can be stored at −20°C for 2 weeks before analysis by mass spectrometry.
3. For mass spectrometry, whole protein mixtures are analyzed after in-solution digestion with trypsin.
4. Peptide matches were filtered to 0.5% false positives using a target-decoy database strategy. Final lists of Ago1-interacting proteins were obtained by subtracting protein matches that were also found in an untagged control sample (see Fig. 2).

3.6. SDS-PAGE and Silver Stain

1. The remaining half of the final eluate is TCA precipitated as above and resuspended in Laemmli SDS-PAGE sample buffer.
2. The sample is run on a 4–12% acrylamide gel in MOPS SDS running buffer, according to the manufacturer's instructions (see steps 11 and 12 in Subheading 3.2.)
3. Samples, which are resuspended in 10–20 μl of Laemmli sample buffer, are heated for 10 min at 95°C.
4. Samples are then loaded on the gel and run at a constant voltage (200 V) for approximately 50 min.

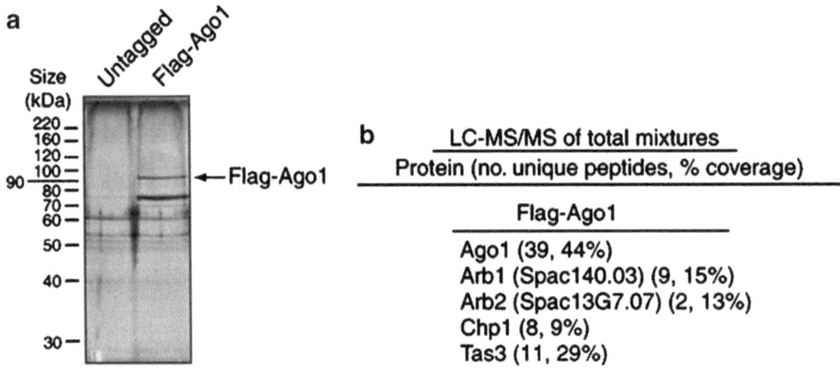

Fig. 2. (**a**) Silver stained SDS-PAGE gel of FLAG-Ago1 purification. The protein that migrates at ~75 kDa was determined by mass spectrometry to be a heat shock protein that often co-purifies with FLAG epitope-tagged proteins. (**b**) Results of mixture tandem MS sequencing (LC-MS/MS) of FLAG-Ago1 purification. Proteins are indicated as the number of unique peptides and percent of total number of amino acid residues covered.

5. The two plates are carefully detached and the bottom lip and top wells of the gel are removed.
6. The gel is briefly washed with water and incubated in 50% methanol with gentle shaking for 10 min.
7. The gel is then incubated in 5% methanol with gentle shaking for 10 min.
8. The gel is then incubated in 2.5 mM DTT with gentle shaking for 10 min after which it is washed three times with 100 ml of ddH_2O.
9. The gel is then incubated with freshly made 0.1% $AgNO_3$ dissolved in ddH_2O for 10 min and briefly washed three times with 10–20 ml of ddH_2O.
10. The gel is developed using freshly made 7.5 g sodium carbonate and 125 μl 37% formaldehyde in 250 ml ddH_2O. Formaldehyde is added immediately prior to use.
11. Just before silver-stained protein bands reach the desired intensity, anhydrous citric acid is sprinkled over the gel until the solution no longer bubbles and no additional citric acid can be dissolved. Incubate for 5 min.
12. The gel is then washed two times 5 min with 100 ml of ddH_2O and immediately photographed, scanned, or mounted to dry.

4. Notes

1. The pFA6 plasmid cassette for N-terminal integrations, described in (11), was modified such that a 3XFLAG tag was inserted in the place of MYC or HA tags, and the *nmtl* promoter was replaced with a 600 bp DNA fragment upstream of *ago1* locus. The resulting plasmid encoding the Nourseothricin (Nat) drug resistance marker (*NatR*), *ago1* promoter, and 3XFLAG sequence was amplified using PCR with primers containing 80 nucleotides flanking the *ago1* start codon.
2. Short (20 nt) PCR primers are designed such that the resulting PCR fragment will span the *NatR* gene, *ago1* upstream sequence, 3XFLAG tag, and the N-terminal portion of *ago1* gene. This product will be of a specified size (depending on choice of primers) and should only appear in clones that have the correct integration of the PCR fragment in the upstream region of the *ago1* locus. This diagnostic PCR strategy is used to quickly screen tens of *NatR* candidates. A no-tag negative control clone must always be included in parallel in these analyses.

3. To transfer the beads, they are first suspended in 1 ml of cold lysis buffer and mixed thoroughly using a P1000 Pipette tip with 1 cm of the its tip cut off. The tip is cut off to increase the P1000 tip circumference allowing for the efficient transfer of FLAG agarose beads. The 50 ml polypropylene centrifuge tube is washed twice more with 1 ml aliquots of lysis buffer. For each wash, 1 ml of lysis buffer is added using another P1000 pipette, and the same cut off P1000 tip is used to transfer the beads to increase the transfer efficiency of beads to the column.
4. Using a Pipette Aid and a pre-cooled disposable pipet, 10 ml of cold lysis buffer is released onto the beads at the bottom of the column with enough pressure to dislodge the beads from the bottom of the column, creating a lysis buffer/bead slurry. This helps in reducing the coimmunoprecipitation of background proteins.

Acknowledgments

The authors would like to thank the Howard Hughes Medical Professor in Cell Biology at Harvard Medical School, Dr. Danesh Moazed, for his advice, guidance, and encouragement. This work was supported by an NIH grant (RO1 (GM72805)) to Danesh Moazed.

References

1. Jinek, M., and Doudna, J.A. (2009). A three-dimensional view of the molecular machinery of RNA interference. *Nature* **457**, 405–412.
2. Nowotny, M., and Yang, W. (2009). Structural and functional modules in RNA interference. *Curr. Opin. Struct. Biol.* **19**, 286–293.
3. Liu, J., Carmell, M.A., Rivas, F.V., Marsden, C.G., Thomson, J.M., Song, J.J., et al (2004). Argonaute 2 is the catalytic engine of mammalian RNAi. *Science* **305**, 1437–41.
4. Rivas, F.V., Tolia, N.H., Song, J.J., Aragon, J.P., Liu, J., Hannon, G.J., et al (2005). Purified Argonaute2 and an siRNA form recombinant human RISC. *Nat. Struct. Mol. Biol* **12**, 340–349.
5. Mathonnet, G., Fabian, M.R., Svitkin, Y.V., Parsyan, A., Huck, L., Murata, T., et al. (2007). MicroRNA inhibition of translation in vitro by targeting the cap binding complex eIF4F. *Science* **317**, 1764–1767.
6. Moazed, D. (2009). Small RNAs in transcriptional gene silencing and genome defense. *Nature* **457**, 413–420.
7. Till, S., Thermann, R., Bortfeld, M., Hothorn, M., Enderle, D., Heinrich, C., et al. (2007). A conserved motif in Argonaute-interacting proteins mediates functional interactions through the Argonaute PIWI domain. *Nat. Struct. Mol. Biol* **14**, 897–903.
8. Volpe, T.A., Kidner, C., Hall, I.M., Teng, G., Grewal, S.I., and Martienssen, R.A. (2002). Regulation of heterochromatic silencing and histone H3 lysine-9 methylation by RNAi. *Science* **297**, 1833–1837.
9. Buker, M.S., Iida, T., Buhler, M., Villen, J., Gygi, S.P., Nakayama, J.I., et al. (2006). Two different Argonaute complexes are required for siRNA generation and heterochromatin assembly in fission yeast. *Nat. Struct. Mol. Biol.* **14**, 200–207.

10. Irvine, D.V., Zaratiegui, M., Tolia, N.H., Goto, D.B., Chitwood, D.H., Vaughn M.W., et al. (2006). Argonaute slicing is required for heterochromatic silencing and spreading. *Science* **313**, 1134–1137.
11. Verdel, A., Jia, S., Gerber, S., Sugiyama, T., Gygi, S., Grewal, S.I., et al. (2004). RNAi-mediated targeting of heterochromatin by the RITS complex. *Science* **303**, 672–676.
12. Bahler, J., Wu, J.Q., Longtine, M.S., Shah, N.G., McKenzie, A.3rd., Steever, A.B., et al. (1998). Heterologous modules for efficient and versatile PCR-based gene targeting in *Schizosaccharomyces pombe. Yeast* **14**, 943–51.
13. Motamedi, M.R., Verdel, A., Colmenares, S.U., Gerber, S.A., Gygi, S.P., and Moazed, D. (2004). Two RNAi complexes, RITS and RDRC, physically interact and localize to noncoding centromeric RNAs. *Cell* **119**, 789–802.
14. Motamedi, M. R., Hong, E.J., Li X., Gerber, S., Denison, C., Gygi, S., and Moazed, D. (2008). HP1 proteins form distinct complexes and mediate heterochromatic gene silencing by nonoverelapping mechanisms. *Mol. Cell* **32**, 778–790.

Chapter 2

Chromatin Immunoprecipitation in Fission Yeast

Thomas A. Volpe and Jessica DeMaio

Abstract

A tremendous amount of information regarding the nature and regulation of heterochromatin has emerged in the past 10 years. This rapid progress is largely due to the development of techniques such as chromatin immunoprecipitation or "ChIP," which allow analysis of chromatin structure. Further technological advances such as microarray analysis and, more recently, deep sequencing technologies, have made ChIP an even more powerful tool. ChIP allows the investigator to identify protein interactions and/or the presence of various chromatin modifications at specific genomic loci.

Key words: ChIP, Heterochromatin, Immunoprecipitation, RNAi, Histone, Epigenetics, Fission yeast, *S. pombe*

1. Introduction

The term *heterochromatin* was first used by Emil Heitz in the late 1920s to describe densely staining material observed in interphase nuclei (1). More recent work has implicated heterochromatin to be involved in a wide range of cellular functions including the regulation of gene expression and maintenance of genome integrity (2). In addition, the role of heterochromatin in regulation of epigenetic changes in gene expression (heritable changes in gene expression that do not result from altered nucleotide sequence) has been recognized in a broad range of eukaryotic species (3, 4).

Nucleosomes, the basic building blocks of heterochromatin, consist of approximately two turns of DNA wrapped around core histone octamers containing two molecules each of histone H2A, H2B, H3, and H4. Observations that specific modifications of core histone proteins occur within heterochromatin domains led to the idea that these histone modifications could function as a

Tom C. Hobman and Thomas F. Duchaine (eds.), *Argonaute Proteins: Methods and Protocols*,
Methods in Molecular Biology, vol. 725, DOI 10.1007/978-1-61779-046-1_2, © Springer Science+Business Media, LLC 2011

histone code that can be interpreted by adapter proteins that dictate the transcriptional activity of chromosomal regions (5, 6). For example, heterochromatic centromere repeats in the fission yeast (*Schizosaccharomyces pombe*) are enriched with histone H3 methylated at lysine 9 (H3mK9). This histone modification recruits downstream effector molecules such as Swi6 protein that bind H3mK9 via its amino-terminal chromodomain and is required for heterochromatin assembly (7, 8).

Studies in *S. pombe* have revealed a role for RNA interference (RNAi) in targeting histone modifications to centromere repeats. Transcription of centromere repeat sequences results in double-stranded RNA (dsRNA) formation. This dsRNA is rapidly processed by the RNAi apparatus resulting in siRNA production followed by sequence-specific targeting of heterochromatin to cognate sequences (9). Several components of the RNAi machinery including the RNA-dependent RNA polymerase, Rdp1, and Argonaute, Ago1, have also been found to be associated with heterochromatin regions in fission yeast (9, 10). Interestingly, RNAi-mediated heterochromatin assembly has been observed in other eukaryotes including plants, trypanosomes, ciliates, flies, and humans (11–15).

2. Materials

1. Yeast extract medium [supplemented with adenine (YEA)]: 0.5% yeast extract, 3% glucose, supplemented with 75 mg adenine per liter.
2. 30% Paraformaldehyde solution (pFA): 30% (w/v) pFA (Sigma, P 6148), 0.25 M NaOH prepared in YEA liquid medium. pFA is very toxic and should be handled with care (see Note 1). Incubate solution in a 50 ml conical tube(s) at 65°C for 5–10 min or until solution is clear (shaking every few minutes will aid pFA dissolution). Once pFA is completely dissolved cool solution to room temperature. Always prepare fresh 30% pFA solution (do not store for subsequent use).
3. PBS (phosphate-buffered saline, pH = 7.4): 137 mM NaCl, 2.7 mM KCl, 10 mM Na_2HPO_4, and 2 mM KH_2PO_4. Store at 4°C.
4. ChIP lysis buffer: 50 mM HEPES–KOH (pH = 7.5), 140 mM NaCl, 1 mM EDTA (pH = 8.0), 1% Triton X-100, and 0.1% DOC (sodium deoxycholate monohydrate, Sigma D 5670). Just prior to use, add one tablet (per 10 ml of lysis buffer) of Complete Protease Inhibitor Cocktail (Roche) and PMSF (phenylmethylsulfonyl fluoride) to a final concentration of

1 mM. PMSF is very toxic and should be handled with care. Store at 4°C.

5. High salt lysis buffer: 50 mM HEPES–KOH (pH = 7.5), 0.5 M NaCl, 1 mM EDTA (pH = 8.0), 1% Triton X-100, 0.1% DOC. Just prior to use add Complete Protease Inhibitor Cocktail Tablet (Roche, one tablet per 10 ml of lysis buffer) and PMSF to a final concentration of 1 mM. PMSF is very toxic and should be handled with care. Store at 4°C.
6. Wash Buffer: 10 mM Tris–HCl (pH = 8.0), 0.25 M LiCl, 0.5% IGEPAL® CA-630 (Sigma), 0.5% DOC, and 1 mM EDTA (pH = 8.0). Store at 4°C.
7. TES: 50 mM Tris–HCl (pH = 8.0), 10 mM EDTA, 1% SDS (sodium dodecyl sulfate). Store at room temperature (to avoid precipitation of SDS).
8. TE: 100 mM Tris–HCl (pH = 8.0) and 10 mM EDTA. Store at 4°C.
9. Phenol–Chloroform: Combine 25 ml Tris-equilibrated phenol (pH = 8.0) with 24 ml chloroform and 1 ml Isoamyl alcohol. Mix and allow to settle (avoid aqueous phase). Phenol–chloroform is very toxic and should be handled with care (use only in fume hood and avoid contact with skin).
10. 0.5 μm glass beads (Sigma G8772).
11. BeadBeater type homogenizer (BioSpec).
12. Protein A- or G-Sepharose.
13. Antibodies specific for protein of interest or antibodies raised against a specific epitope for use with epitope-tagged proteins. Because some commercially available antibodies may not work well for ChIP applications, it is best to use antibodies that have previously been tested for use in ChIP. For example, we have used mouse anti-HA to isolate chromatin that is associated with HA-Ago1 in fission yeast.

3. Methods

To obtain reliable data from any chromatin immunoprecipitation, extreme care must be taken at several critical steps in the ChIP protocol. A flow chart outlining key steps in this protocol is shown in Fig. 1. Those that are most vulnerable to human error include the fixation step and the immunoprecipitation step. It is essential that the fixation times for all samples are equal. Over fixation can lead to high levels of background enrichment, while interactions may go undetected with too little fixation. The success of any immunoprecipitation is dependent on the quality of antibodies

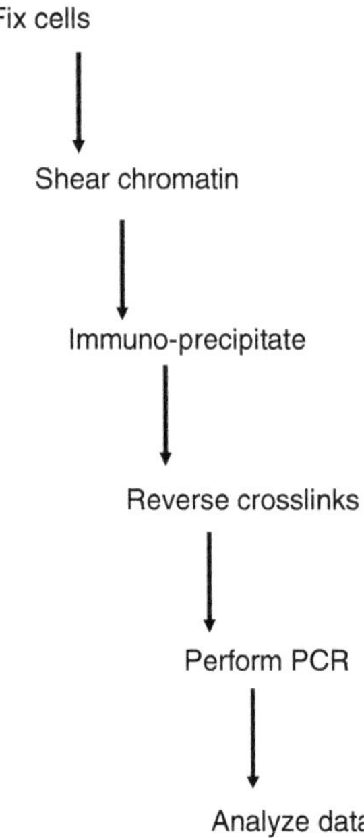

Fig. 1. Flow chart of the ChIP procedure. Cells are lysed after fixation with formaldehyde. Lysates are then cleared and sonicated to shear chromatin to ~500 bp fragments. Antibodies are then used to immunoprecipitate-specific proteins along with any chromatin fragments they are associated with. Crosslinks are then reversed to prepare for chromatin analysis. PCR reactions with primers specific for a specific genomic region of interest are prepared and analyzed using real-time PCR.

used, however, there is also room for considerable experimental error due to inaccurate pipetting and removal of wash solutions from agarose beads.

The ChIP protocol described below is modified from previous methods (16, 17) and can be modified for use with any protein for which an antibody suitable for ChIP is available as well as for any genomic sequence of interest.

3.1. Preparation on the Day Before Experiment

1. The morning before performing the ChIP experiment, prepare starter cultures for the desired strains to be tested by inoculating 10 ml of YEA followed by incubation of cultures at 33°C with shaking.
2. Once starter cultures reach log phase growth [$0.5–1 \times 10^7$ cells/ml ($A_{595} = 0.25–0.5$)] dilute cells in 50 ml of YEA (use 125 ml flasks) so that they will grow to a cell density of 1×10^7 cells/ml ($A_{595} = 0.5$) by the next morning. Incubate cultures at 33°C with shaking.
3. Make sure to prepare an 18°C shaker/water bath for use on the day of the ChIP experiment (see Note 2).
4. Prepare all solutions ahead of time (except 30% pFA solution, which is made fresh for each experiment).

3.2. Growth and Fixation of Yeast Cells

1. Once cell cultures have reached the desired cell density (0.5–1 × 10^7 cells/ml (A_{595} = 0.25–0.5)) transfer culture flasks to an 18°C water bath to incubate for an additional 2 h with gentle shaking (~150 rpm).
2. While cells are incubating at 18°C, begin preparation of fresh 30% pFA. Pre-chill ChIP lysis buffer, High salt lysis buffer, Wash Buffer, PBS, and TE on ice for later use.
3. Once cell cultures have incubated at 18°C for 2 h, begin fixation by adding 30% pFA to each sample to a final concentration of 3% (5.6 ml of 30% solution). Continue shaking.
4. Fix cells at 18°C for a maximum of 30 min (fixation time is critical).
5. To stop fixation, add glycine to each sample to a final concentration of 0.125 M (3 ml of 2.5 M solution) (see Note 3). Place flasks on ice.
6. Transfer cells to 50 ml conical and centrifuge at ~2,060 × g for 5 min at 4°C in a tabletop centrifuge.
7. Wash cell pellet by resuspending cells in 20 ml ice-cold PBS followed by centrifugation at 3,000 rpm for 5 min at 4°C in a tabletop centrifuge.
8. Resuspend cell pellet in 1 ml of ice-cold PBS and transfer to pre-chilled 2 ml screw cap microcentrifuge tubes on ice.
9. Centrifuge cells at ~17,500 × g in a microfuge for 1 min at 4°C and remove supernatant.
10. Wash cells by resuspending pellet in 500 μl of ice-cold lysis buffer (without protease inhibitors) followed by centrifugation at 15,000 rpm in a microfuge at 4°C for 1 min. If desired, cell pellets can be stored at −80°C.
11. Resuspend cell pellet in 400 μl of ice-cold lysis buffer with protease inhibitors.
12. Carefully add ice-cold 0.5 μm glass beads to each sample up to the meniscus of the lysis buffer. Make sure caps are screwed on tightly.
13. Lyse cells at 4°C using a bead beater homogenizer at 4°C at maximum power with three 5 min pulses. Cool samples for 10 min in ice/water slush between each 5 min pulse.
14. Examine an aliquot of cell lysate using a microscope to ensure >95% cell breakage (lysed cells will appear dark when viewing with phase/contrast). Perform additional 5 min pulses with bead beater if necessary.
15. After last pulse with bead beater, place tubes in ice/water slush for additional 10 min.
16. Set up 15 ml conical tubes on ice with 1.5 ml microcentrifuge tubes at the bottom of each (remove cap of microfuge tube

(if present) with scissors so that the microcentrifuge tube rests at the bottom of the 15 ml conical tube). Place tubes on ice. These tubes will be used to collect cell lysates in the next step.

17. With caps screwed on tightly, gently tap sample tubes on bench top upside down until contents are away from bottom of tubes. Using a 25 gauge syringe needle, poke a small hole in the bottom of the tube (one at a time) and then place tube, hole side down, in chilled 15 ml collection tube assembly on ice (one for each sample) (see Note 4).
18. Centrifuge collection tubes for 1 min at ~2,060 × *g* at 4°C in a table top centrifuge to collect lysate.
19. Remove tubes containing beads (see Note 5). Transfer the cell lysates to fresh (chilled) 2 ml screw cap microfuge tubes containing 1.1 ml of ice cold lysis buffer and place on ice (volume should now be ~1.5 ml).
20. Sonicate lysates on ice using 15 s pulses (four times) with 1 min incubations on ice/water slush between each pulse until chromatin is sheared to ~500 bp fragments (Fig. 2). Also see Note 6.
21. Centrifuge lysates at ~17,500 × *g* at 4°C for 5 min in a microfuge to remove cell debris.
22. Transfer supernatants to new (cold) 1.5 ml microcentrifuge tubes on ice.
23. Centrifuge lysates at 15,000 rpm at 4°C for 10 min in a microfuge.
24. Transfer supernatants to new (cold) 1.5 ml microcentrifuge tubes on ice.
25. For each sample, save 50 μl of lysate in a new 2 ml screw cap microcentrifuge tube (on ice). These samples will be used later as total input controls and should be stored on ice or at 4°C until step 33.
26. Determine the number of antibodies (up to four antibodies can be tested for each lysate) to be used adding one for a no

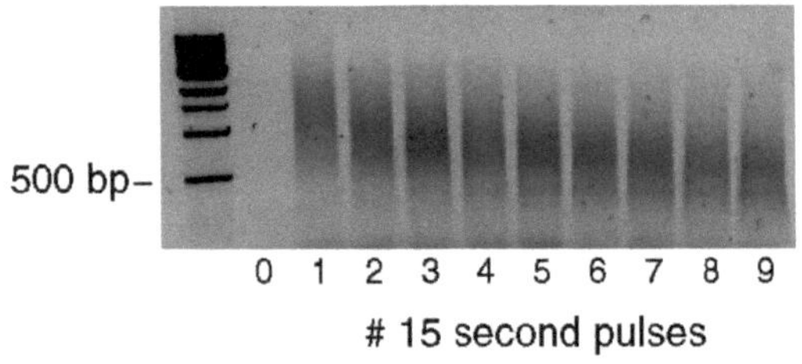

Fig. 2. Sheared chromatin samples. Lysates were sonicated on ice using 15 s pulses (four times with 1 min incubations in ice/water slush between pulses) at 40% duty cycle with output set at 2.5. DNA was extracted and resolved on a 0.8% agarose gel.

antibody control. Divide lysate equally into pre-chilled 1.5 ml microcentrifuge tubes. Bring volume of each sample up to 500 µl with ice-cold lysis buffer and place on ice.

27. Add antibody to lysates (except for the no antibody control) (see Note 7).
28. Incubate samples (including no antibody control sample) at 4°C for 4 h on orbital shaker or overnight on ice.
29. Determine the amount of 1:1 Protein A- or G-Sepharose bead slurry (see Note 8) needed by multiplying the number of samples × 50 µl per sample. It is best to wash 10% more slurry than needed to insure availability of 50 µl for each sample. Centrifuge beads at 1,000 × *g* for 1 min at 4°C and remove supernatant. Equilibrate beads by washing four times with ice-cold lysis buffer without protease inhibitors (trim tip of p200 pipet tip with scissors to enlarge opening and prevent clogging while dispensing beads). After last wash dilute beads in an equal volume of ice-cold lysis buffer and resuspend (see Note 9).
30. Add 50 µl of washed 1:1 agarose bead slurry to each sample (to avoid settling, gently draw slurry up and down with pipet just prior to dispensing beads into each tube).
31. Incubate on orbital shaker for 1 h at 4°C.
32. Centrifuge samples at 1,000 × *g* for 1 min.
33. Wash beads with (see Note 9):

    1 ml ice-cold lysis buffer (two times)

    1 ml ice-cold high salt lysis buffer

    1 ml ice-cold wash buffer

    1 ml ice-cold TE.
34. After removing supernatant from last wash, add 125 µl room temperature TES to beads and incubate at 65°C for 10 min (add 200 µl TES to input controls (from step 26) and incubate along with IP samples).
35. Spin samples with beads (not input controls) at ~2,000 × *g* in a microfuge for 1 min and save supernatants to screw cap tubes at room temperature.
36. Add an additional 125 µl TES to beads and incubate at 65°C for 10 min.
37. Spin at 5,000 rpm in a microfuge for 1 min and save supernatants (pool samples in tubes from step 32).
38. Incubate all samples (including input control, there should be ~250 µl for each sample) for at least 6 h at 65°C or overnight.
39. Cool samples to RT

40. Add 12.5 μl of 20 μg/ml Proteinase K and 237.5 μl of dH_2O to each sample.
41. Incubate samples at 37°C for 2 h.
42. Extract once with equal volume of phenol–chloroform.
43. For each sample, divide aqueous phase into two new tubes (~250 μl each) and add 75 μl 1 M NaOAc (pH = 5.2), 0.5 μl 20 mg/ml glycogen, and 815 μl 100% ethanol.
44. Incubate at −20°C for at least 2 h (or overnight).
45. Centrifuge samples at ~17,500 × *g* for 15 min at 4°C in a microfuge.
46. Carefully remove and discard supernatants.
47. Wash pellets with 1 ml 70% ethanol.
48. Carefully remove supernatants and air-dry pellets for 5–10 min.
49. Resuspend pellets in 40 μl TE + 10 mg/ml RNase A.
50. Incubate at 37°C for 30 min.
51. Samples are now ready for analysis by PCR but can be stored at −20°C for future analysis.
52. An example of a typical qPCR protocol to analyze ChIP samples can be found below (see Note 10).

4. Notes

1. Paraformaldehyde is toxic and should be weighed out carefully in a fume hood to avoid the inhalation of powder. Other precautions, such as wearing a dusk mask while working with pFA powder, are also suggested. Any solutions containing paraformaldehyde should be handled with extreme care since it is very toxic if inhaled, ingested, and can be absorbed through skin.
2. We use a temperature controlled water bath set at 18°C that is placed in a cold room (or large refrigerator). Make sure to prepare water bath the night before performing the ChIP experiment to allow sufficient time for the water to warm to 18°C.
3. Addition of glycine quelches the fixation reaction so this step should occur at precisely 30 min after the addition of pFA.
4. Retrieval of lysate is simplified by allowing lysate to drain through a small hole at the bottom of the tube using centrifugation. Use caution when using syringe needle. Replace syringe needle when tip becomes bent.

5. There are several ways to remove the tube containing beads. We typically use a long syringe needle or scalpel to carefully pierce the top of the tube containing glass beads and slowly lift it from the 15 ml conical tube.
6. Sonication efficiency will vary depending on several factors. These include fixation time, the sonicator being used as well as the technique of the individual performing the sonication. We use a Branson Sonifier cell disrupter (model S-450A) and typically sonicate lysates on ice using 15 s pulses (four times with 1 min incubations in ice/water slush between pulses) at 40% duty cycle with output set at 2.5. It is suggested that a preliminary experiment be performed with fixed cells to determine how many pulses are required to shear chromatin to ~500 bp fragments.
7. The amount of antibody used for each immunoprecipitation will vary depending on how abundant the protein of interest is. It is important that the amount of antibody added is in excess over the amount of protein being immunoprecipitated. This should be determined empirically for each antibody used.
8. Which agarose beads to use for immunoprecipitation, either Protein A agarose or Protein G-agarose, will depend on the antibody being used (see manufacturer specifications).
9. One of the most time consuming aspects of immunoprecipitation experiments is the removal of wash buffers from Protein-A beads. Since each wash is an opportunity for loss of beads it is extremely important to perform the washes carefully but at the same time not make the procedure too time consuming. We typically set up an aspiration apparatus (see Fig. 3) consisting of a vacuum trap attached to a flexible tube with a custom made pasture pipet attached to end. For washes, use lower gauge syringe (18 gauge) being careful to leave ~50 μl to insure no loss of beads. For the final wash, use a 27 gauge syringe needle to remove remaining wash buffer (use of the high gauge syringe needle will prevent significant bead loss).
10. There are many variables to consider when analyzing ChIP samples by quantitative real-time PCR (qPCR). Guidelines describing the minimum information necessary for evaluating qPCR experiments can be found within Bustin et al. (18). We perform qPCR using an MJ Research/BioRad Chromo4 thermocycler using Opticon 3.0 software (BioRad). Data are then imported into an Excel spreadsheet for further analysis. There are several different methods for normalizing qPCR data for ChIP analysis. These include background subtraction, percent of input, fold enrichment, normalization relative to a

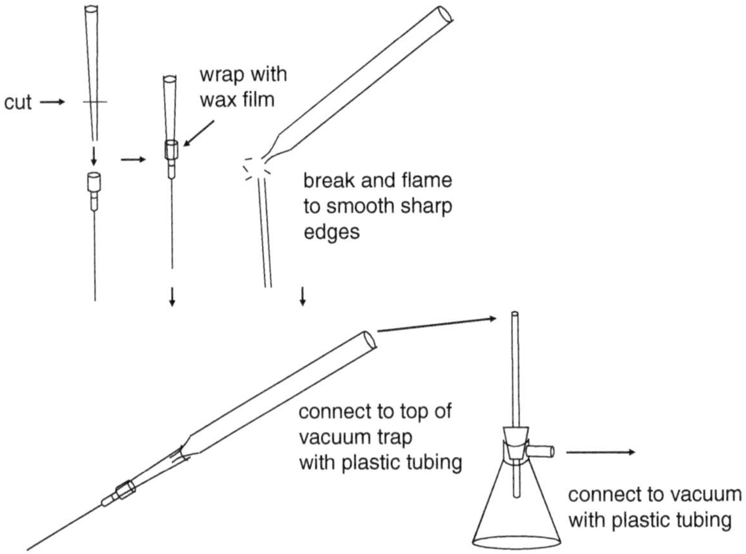

Fig. 3. Assembly of aspirator used for rapid removal of wash buffer from beads. Prepare syringe needles (both 27 and 18 gauge) for aspirator by cutting a p200 tip (non-filtered) with scissors so that the tip snaps into the plastic end of a syringe needle. Once the tip has a snug fit secure with several wraps of wax film so that the seal is air and water tight. Prepare a Pasteur pipet by breaking the end so that the tip fits snug inside a p200 tip. Keeping a bit of length on the end will help to steady the tip while aspirating. There is no need to wrap wax film around the Pasteur pipet and the p200 tip, the suction from the vacuum will hold the syringe needle on while aspirating (this allows easy changing of syringe needle). The Pasteur pipette should be attached to a vacuum trap using flexible plastic tubing.

control sequence, and normalization relative to nucleosome density (reviewed in ref. 19). One method involving background subtraction is described below and is modified from Mutskov and Felsenfeld (20). The following formula would be used for each primer set where IP/In represents the fold change in enrichment in the immunoprecipitated fraction compared with the level of input chromatin:

$$\frac{\mathrm{IP}}{\mathrm{In}} = 2^{-\Delta\mathrm{Ct}} = 2^{-(\mathrm{Ct(IP)}-\mathrm{Ct(In)})}$$

This formula makes the assumption that primer pairs will amplify with 100% efficiency by PCR. Methods to correct for primer efficiency have been developed (21).

Although it may be best to compare levels of background obtained from no antibody controls to target enrichment side by side, the following formula can be used to subtract background signal determined from the no antibody control if desired:

$$\left(\frac{\mathrm{IP}}{\mathrm{In}}\right)^{\mathrm{t}} - \left(\frac{\mathrm{IP}}{\mathrm{In}}\right)^{\mathrm{t0}}$$

Target enrichments can also be normalized to signal obtained from a reference gene, if appropriate, using the following formula:

$$\frac{(\mathrm{IP} / \mathrm{In})^{\mathrm{t}} - \mathrm{IP} / \mathrm{In})^{\mathrm{t0}}}{(\mathrm{IP} / \mathrm{In})^{\mathrm{c}} - (\mathrm{IP} / \mathrm{In})^{\mathrm{c0}}}.$$

DNA sample from input control (In)

DNA sample from immunoprecipitation using specific antibody (IP)

Signal from IP sample compared with In (IP/In)

Difference in threshold cycle (Ct) value (ΔCt)

Target signal from immunoprecipitated sample ($(\mathrm{IP}/\mathrm{In})^{\mathrm{t}}$)

Target signal from no antibody control ($(\mathrm{IP}/\mathrm{In})^{\mathrm{t0}}$)

Reference signal from immunoprecipitated sample ($(\mathrm{IP}/\mathrm{In})^{\mathrm{c}}$)

Reference signal from no antibody control ($(\mathrm{IP}/\mathrm{In})^{\mathrm{c0}}$)

Epitope-tagged Ago1 is typically used for ChIP analysis of chromatin bound Argonaute protein in fission yeast, although commercially available antibodies raised against *S. pombe* Ago1 have also be used successfully in the past (22–24).

qPCR primers used for previous ChIP analysis are included (see Table 1 and Fig. 4) (25).

Table 1
List of qPCR primers

D25 AATAGTAAGTCGAATTGAGATGTAAACG D26 AGAGAAGTCTATATCTTGAACAGAAGG	Imr
D23 CGTAACCGATACATAATTTAGG D24 TTAATGTGTTTGCCATCTTAC	Imr
D1 AGAGCATGGTGGTGGTTATGG D2 TTTGGCGACTAAACCGAAAGC	dg
C98 CAGGAAGATGATACACAATG C99 TTTGGACAGAATGGATGG	dg
C9 CCGCAGTTGGGAGTACATCATTC C10 ACAGCACTCAACAACAGTCTTGG	dg
C7 TGTGCCTCGTCAAATTATCATCCATCC C8 ACTTGGAATCGAATTGAGAACTTGTTATGC	dg
C5 TTTCCCGCCCAGTGGATGCTTC C6 TCAGTGCCAAACAACCGCTTTAACC	dg
C80 CTTGGCTTGTCTTCTGTATG C81 TTGACGAGTCTTGGAACC	dg
C78 GTGCGTCTAGGTATCCTTAGC C79 CGGTTCAATCACAATTCATAAGC	dg

(continued)

Table 1 (continued)

Primers	
C1 GTCAGTATAGGCATCAACATCATCACC C2 AGTGGTAAGTGGAAGTGGTAAGTGG	dg
C76 GGAATATCAGACAACATATACG C77 AATCTCATCACTATATTATACTTGG	dg
C74 GAGTGCTCATGCGACATTTGG C75 ACACTTCTTACTCAATCTCACATAACC	dg
C72 GTTTGGAAGAACAAGAACTTTGAAGG C73 CTTAGTGAAGTAATGGTGACTGAGC	dg
C68 TTGTGGTGGTGTGGTAATACG C69 TGAGCACTAAATAGACGGAACC	dg
C66 AAGTGAATGAGTAGTAGAAGG C67 ATCCAATAATAGCGACTTCC	dg
D39 CACTACTATCATACAGTTTCTTCTCC D40 GCTTATTCATAAACAATCCAATTTCG	dh
D37 AACGTAGCATGAAGAATCC D38 TTACAATAGTCAATTAAGATATTCG	dh
D35 TATTCAACAGCAGATACACC D36 TTACCGCAGAACTCTAGC	dh
D33 GACCAGGAACAAATCAGGAAACC D34 GATTATATTCTTCCATTCATGTCGTAGATG	dh
D31 TCGGTAGGTATGAGTGAAATCTTCC D32 AAATTGCGAACCTGAAACTGAAATATC	dh
C90 AATTATTCAAGTGCTCAATGTTATTTAG C91 TTCTAACAGTCTAAAGTAGAGATTGG	dh
C92 ATCACTATCATTCTTCCAAAGTAAATAC C93 AAACACGGCGATAAGAAATGG	dh
C94 TTATTCAAGAGAAGATTCATCC C95 GAGTAGGTGTAGGAGTAGG	dh
C96 CTGTGTCGATGTAGTTCTCTATACC C97 GCCCATTCATCAAGCGAGTC	dh
C15 ATTAACTGTCAGGATGTGTTGTCGTTCTTG C16 CGCATCTACCTCAGCAGTCCTTGG	dh
C17 GGAATAGCATACCGTCAAGTCGTTAGTTG C18 GGTCAACGCACGCCTAAACTAGC	dh
D29 ATTGCCTTGTTCTTGAGTAC D30 AGGGAGTAACTTCTTCACC	dh
D27 AGAGCAAATGTGAATAAAATGATAACG D28 AATTACACATGATATTCTTACTGATAACC	dh
C48 TCAAGTGGTCTGCCTCTGG C49 CACCGACGACGAACATGG	gpd3

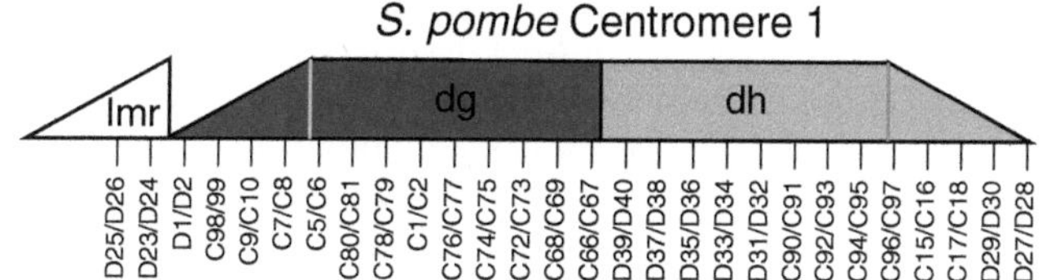

Fig. 4. Schematic representation of one arm of *Schizosaccharomyces pombe* Centromere 1 including innermost (lmr), dg and dh centromere repeats as well as locations of qPCR primers listed in Table 1.

Acknowledgments

The authors would like to thank Michele McDonough for helpful comments on the manuscript. T.V. and J.D. are supported by the NIH (R01 GM074986) and the generous support of the Robert H. Lurie Comprehensive Cancer Center.

References

1. Heitz, E. (1928) Das heterochromatin der Moose. *Jahrbuecher Wiss Botanik*, **69**, 762–818.
2. Lippman, Z. and Martienssen, R. (2004) The role of RNA interference in heterochromatic silencing. *Nature*, **431**, 364–370.
3. Reik, W. and Walter, J. (2001) Genomic imprinting: parental influence on the genome. *Nat Rev Genet*, **2**, 21–32.
4. Wolffe, A.P. and Matzke, M.A. (1999) Epigenetics: regulation through repression. *Science*, **286**, 481–486.
5. Hake, S.B., Xiao, A. and Allis, C.D. (2007) Linking the epigenetic 'language' of covalent histone modifications to cancer. *Br J Cancer*, **96 Suppl**, R31–39.
6. Strahl, B.D. and Allis, C.D. (2000) The language of covalent histone modifications. *Nature*, **403**, 41–45.
7. Bannister, A.J., Zegerman, P., Partridge, J.F., Miska, E.A., Thomas, J.O., Allshire, R.C. and Kouzarides, T. (2001) Selective recognition of methylated lysine 9 on histone H3 by the HP1 chromo domain. *Nature*, **410**, 120–124.
8. Lachner, M., O'Carroll, D., Rea, S., Mechtler, K. and Jenuwein, T. (2001) Methylation of histone H3 lysine 9 creates a binding site for HP1 proteins. *Nature*, **410**, 116–120.
9. Volpe, T.A., Kidner, C., Hall, I.M., Teng, G., Grewal, S.I. and Martienssen, R.A. (2002) Regulation of heterochromatic silencing and histone H3 lysine-9 methylation by RNAi. *Science*, **297**, 1833–1837.
10. Verdel, A., Jia, S., Gerber, S., Sugiyama, T., Gygi, S., Grewal, S.I. and Moazed, D. (2004) RNAi-mediated targeting of heterochromatin by the RITS complex. *Science*, **303**, 672–676.
11. Durand-Dubief, M. and Bastin, P. (2003) TbAGO1, an argonaute protein required for RNA interference, is involved in mitosis and chromosome segregation in Trypanosoma brucei. *BMC Biol*, **1**, 2.
12. Mochizuki, K., Fine, N.A., Fujisawa, T. and Gorovsky, M.A. (2002) Analysis of a piwi-related gene implicates small RNAs in genome rearrangement in tetrahymena. *Cell*, **110**, 689–699.
13. Morris, K.V., Chan, S.W., Jacobsen, S.E. and Looney, D.J. (2004) Small interfering RNA-induced transcriptional gene silencing in human cells. *Science*, **305**, 1289–1292.
14. Pal-Bhadra, M., Leibovitch, B.A., Gandhi, S.G., Rao, M., Bhadra, U., Birchler, J.A. and Elgin, S.C. (2004) Heterochromatic silencing and HP1 localization in Drosophila are dependent on the RNAi machinery. *Science*, **303**, 669–672.
15. Zilberman, D., Gehring, M., Tran, R.K., Ballinger, T. and Henikoff, S. (2007) Genome-wide analysis of Arabidopsis thaliana DNA methylation uncovers an interdependence between methylation and transcription. *Nat Genet*, **39**, 61–69.

16. Ekwall K, P.J. (1999) In W, B. (ed.), *Chromosome Structural Analysis: A Practical Approach* Oxford University Press.
17. Nakayama, J., Klar, A.J. and Grewal, S.I. (2000) A chromodomain protein, Swi6, performs imprinting functions in fission yeast during mitosis and meiosis. *Cell*, **101**, 307–317.
18. Bustin, S.A., Benes, V., Garson, J.A., Hellemans, J., Huggett, J., Kubista, M., Mueller, R., Nolan, T., Pfaffl, M.W., Shipley, G.L. *et al.* (2009) The MIQE guidelines: minimum information for publication of quantitative real-time PCR experiments. *Clin Chem*, **55**, 611–622.
19. Haring, M., Offermann, S., Danker, T., Horst, I., Peterhansel, C. and Stam, M. (2007) Chromatin immunoprecipitation: optimization, quantitative analysis and data normalization. *Plant Methods*, **3**, 11.
20. Mutskov, V. and Felsenfeld, G. (2004) Silencing of transgene transcription precedes methylation of promoter DNA and histone H3 lysine 9. *EMBO J*, **23**, 138–149.
21. Pfaffl, M.W. (2001) A new mathematical model for relative quantification in real-time RT-PCR. *Nucleic Acids Res*, **29**, e45.
22. Buhler, M., Verdel, A. and Moazed, D. (2006) Tethering RITS to a nascent transcript initiates RNAi- and heterochromatin-dependent gene silencing. *Cell*, **125**, 873–886.
23. Irvine, D.V., Zaratiegui, M., Tolia, N.H., Goto, D.B., Chitwood, D.H., Vaughn, M.W., Joshua-Tor, L. and Martienssen, R.A. (2006) Argonaute slicing is required for heterochromatic silencing and spreading. *Science*, **313**, 1134–1137.
24. Partridge, J.F., DeBeauchamp, J.L., Kosinski, A.M., Ulrich, D.L., Hadler, M.J. and Noffsinger, V.J. (2007) Functional separation of the requirements for establishment and maintenance of centromeric heterochromatin. *Mol Cell*, **26**, 593–602.
25. Lawrence, R.J. and Volpe, T.A. (2009) Msc1 links dynamic Swi6/HP1 binding to cell fate determination. *Proc Natl Acad Sci USA*, **106**, 1163–1168.

Chapter 3

Biochemical Analyzes of Endogenous Argonaute Complexes Immunopurified with Anti-Argonaute Monoclonal Antibodies

Keita Miyoshi, Tomoko N. Okada, Haruhiko Siomi, and Mikiko C. Siomi

Abstract

Argonaute proteins are key factors in RNA silencing. After association with small RNAs of 20–30 nucleotides, Argonaute proteins are targeted to homologous RNA molecules that are to be silenced. To understand the functional contributions of Argonaute proteins to RNA silencing at a biochemical level, immunoisolation of Argonaute proteins from living cells of various organisms has been performed. This has enabled the analysis of Argonaute-associated proteins and RNAs. Identifying the small RNAs that associate with individual Argonaute proteins, for instance, could help to elucidate the silencing pathways in which particular Argonaute proteins are involved. However, it is also necessary to note that the results obtained through such biochemical analyzes are greatly affected by the quality and properties of the antibodies used, as well as by the immunoprecipitation conditions employed, including buffer contents and/or salt concentration. In this chapter, we describe fundamental methods for immunoprecipitating Argonaute proteins using monoclonal antibodies as well as for detecting associated proteins and small RNAs. Furthermore, we will also explain how various parameters, such as antibody properties and buffer conditions, can alter the production and interpretation of experimental data.

Key words: RNA silencing, Small RNA, Argonaute, Monoclonal antibody, *Drosophila*

1. Introduction

In RNA silencing, Argonaute proteins associate with tiny noncoding RNAs [20–30 nucleotides (nt) long], such as small-interfering RNAs (siRNAs) and microRNAs (miRNAs), to negatively regulate the expression of genes targeted by the RNA–protein (RNP) complexes (1). Many genes targeted by Argonaute proteins are involved in fundamental processes, such as development, differentiation, metabolism, and controlled cell death. Indeed, the loss of Argonaute function leads to severe damage at the cell, organ, and whole organism levels, in both plants and animals; thus, Argonaute proteins are essential for life in most organisms (2–4).

Tom C. Hobman and Thomas F. Duchaine (eds.), *Argonaute Proteins: Methods and Protocols*, Methods in Molecular Biology, vol. 725, DOI 10.1007/978-1-61779-046-1_3, © Springer Science+Business Media, LLC 2011

Argonaute proteins are broadly conserved in species ranging from unicellular eukaryotes (i.e., yeast) to fungi, plants, invertebrates, and higher vertebrates, including humans (reviewed in ref. 5). The number of Argonaute proteins expressed in a species differs; for example, *Schizosaccharomyces pombe* has only one Argonaute, while humans possess eight Argonaute proteins (5). Members of the "Argonaute family of proteins" are defined by the presence of two characteristic domains, the PAZ and PIWI domains. The PAZ domain was shown to associate with 3′ end of small RNAs (6–8). The PIWI domain was shown to fold into a structure that resembles that of Ribonuclease H (RNase H) (9, 10). Indeed, most Argonaute proteins have been shown to contain the essential residues for an RNase H-like activity (in particular a DDH triad), and cleave RNAs that are targeted by the Argonaute–small RNA complexes (9–11). The endonucleolytic activity that Argonaute proteins exhibit through the PIWI domain is referred to as Slicer activity (9–11).

Phylogenetic analysis showed that Argonaute proteins are divided into two subgroups: the Argonaute (AGO) and PIWI subfamilies (12). In *Drosophila*, five genes encode Argonaute proteins; of these, AGO1 and AGO2 belong to the AGO subfamily, while AGO3, Aub (Aubergine) and Piwi (P-element insertion wimpy testis) belong to the PIWI subfamily. AGO1 and AGO2 are similar at the peptide sequence level, but their mode of action in RNA silencing is different. AGO1 associates mostly with miRNAs to act as a translational inhibitor, or mRNA destabilizer of target genes, whereas AGO2 associates mostly with siRNAs and acts through its Slicer activity in the destruction of target RNAs (13, 14). However, it is noted that in some cases, the miRNAs that are associated with AGO1 can also downregulate the expression of specific genes by cleaving the target mRNAs. This selection of "cleaving" and "noncleaving" depends on the complementarity between miRNAs and their targets (15).

The origins of siRNA and miRNAs differ from one another. Mature miRNAs are processed from primary transcripts that arise from miRNA-coding genes, through two consecutive steps operated by distinct processing complexes, the Drosha–Pasha (16) and the Dicer1–Loquacious (Loqs) complexes (17, 18). In contrast, siRNAs are processed by the Dicer2–R2D2 complex from long, nearly completely complementary double-stranded RNAs (dsRNAs) that are introduced into, or expressed in living cells (19). Recent studies have demonstrated that siRNAs can also be derived from endogenously expressed dsRNAs that originate from intergenic repetitive regions of the genome, including transposons (reviewed in ref. 20). These latter siRNAs are specifically called endogenous siRNAs (esiRNAs or endo-siRNAs) to distinguish them from the above-mentioned siRNAs, which have been lately referred to as exo-siRNAs. Protein factors involved in esiRNA processing from transposon transcripts or other hairpin-shaped RNA molecules are the siRNA factor, Dicer2, and the miRNA factor, Loqs (20). Requirement of Loqs in miRNA and esiRNA production systems was recently investigated and we now know that two Loqs isoforms, Loqs-PA and Loqs-PB, derived from alternative splicing of the *loqs* gene transcript, are necessary for the miRNA production pathway, while the Loqs-PD isoform functions specifically in the esiRNA production pathway (21–23). esiRNAs processed by the Dicer2–Loqs-PD complex are loaded onto AGO2 and target this protein to homologous RNAs (20–23).

In this chapter, we describe detailed methods for the immunoprecipitation of AGO2 and AGO1 from *Drosophila* S2 cells using monoclonal antibodies as well as for the detection of their associated RNAs and proteins. Using these methods, we demonstrate that the location of the epitope recognized by the antibody, and the composition of the buffer chosen for immunoprecipitation will drastically affect the detection of Argonaute-interacting proteins and their association with small RNAs.

Outline of the methods described in this chapter:

- Immunoprecipitation of AGO2 from S2 cells using anti-AGO2 antibodies in high and low salt concentrations.
- Visualization of small RNAs (esiRNAs) associated with AGO2 using 5′ end ^{32}P-labeling.
- Detection of AGO1-associated proteins in S2 cells using a strong detergent (Empigen).
- Detection of AGO1-associated miRNAs in S2 cells by northern blotting.

2. Materials

2.1. Immunoprecipitation of AGO2 from S2 Cells

1. Monoclonal anti-AGO2 antibody (9D6) (15).
2. Monoclonal anti-AGO2 antibody (4D2) (13).
3. GammaBind G Sepharose (GE Healthcare Bio-Sciences, Piscataway, NJ, USA).
4. Phosphate-buffered saline (PBS).
5. 30-Gauge needles.
6. IP-NaCl-150 buffer: 30 mM HEPES–KOH, pH 7.4, 150 mM NaCl, 2 mM Mg(OAc)$_2$, 5 mM dithiothreitol (DTT), 0.1% Nonidet P-40 (NP-40), 2 μg/mL Pepstatin, 2 μg/mL Leupeptin, and 0.5% Aprotinin.
7. IP-NaCl-800 buffer: 30 mM HEPES–KOH, pH 7.4, 800 mM NaCl, 2 mM Mg(OAc)$_2$, 5 mM DTT, 0.1% NP-40, 2 μg/mL Pepstatin, 2 μg/mL Leupeptin, and 0.5% Aprotinin.

8. 5 M NaCl.
9. Sample buffer-DTT (2×): 20% glycerol, 100 mM Tris–HCl (pH 6.8), 4% sodium dodecyl sulfate (SDS), 0.12% Bromophenol blue.

2.2. Visualization of Small RNAs Associated with AGO2 Using ^{32}P-ATP

1. Nuclease-free water.
2. RNA extraction reagent (ISOGEN-LS, Nippon gene, Toyama, Japan, or an equivalent such as Trizol or tri-reagent).
3. Phenol/chloroform/isoamyl alcohol (25:24:1).
4. Chloroform.
5. Isopropanol.
6. RNA carrier: Pellet Paint Co-Precipitant (EMD Bioscience, Darmstadt, Germany).
7. Ethanol (70%).
8. Calf Intestinal Alkaline Phosphatase (CIP) (New England BioLabs, MA, USA).
9. 3 M NaOAc.
10. ^{32}P-γ-ATP (259 TBq/mmol) (Institute of Isotopes, Budapest, Hungary).
11. T4 Polynucleotide Kinase (PNK) (New England BioLabs, MA, USA).
12. Gel-filtration columns: Micro Bio-Spin Columns P-30 Tris, RNase-Free (Bio-Rad, Hercules, CA, USA).
13. Gel loading buffer II: 95% formamide, 18 mM ethylenediaminetetraacetic acid (EDTA) and 0.025% each of SDS, Xylene cyanol, and Bromophenol blue (Applied Biosystems/Ambion, Austin, TX, USA).
14. TBE (10×): 890 mM Tris–borate and 2 mM EDTA.
15. 40% Acrylamide/bisacrylamide (19:1).
16. Ammonium persulfate (APS) solution: 10% (w/v) in distilled water, prepared fresh.
17. 12% Acrylamide/bisacrylamide denaturing gel: 1× TBE buffer (89 mM Tris–borate, 2 mM EDTA), 12% acrylamide/bisacrylamide (19:1), 6 M urea, tetramethylethylenediamine (TEMED), APS solution.
18. Image plates: BAS-MS2040 (Fujifilm, Tokyo, Japan).
19. BAS-2500 imaging system (Fujifilm, Tokyo, Japan).

2.3. Detection of AGO1-Associated Proteins in S2 Cells

1. Monoclonal anti-AGO1 antibody (1B8): raised against the N-terminal 300 amino acids of *Drosophila* AGO1 (15).
2. Empigen (e.g., Empigen BB Detergent (35%), Sigma–Aldrich, MO, USA).

3. Empigen buffer: 1% Empigen, 1 mM EDTA, and 100 μM DTT, 2 μg/mL Pepstatin, 2 μg/mL Leupeptin and 0.5% Aprotinin with PBS.
4. Sample buffer-DTT (2×): 20% glycerol, 100 mM Tris–HCl (pH 6.8), 4% SDS, and 0.12% Bromophenol blue.
5. Silver stain kit: SilverQuest Silver Staining Kit (Invitrogen, Carlsbad, CA, USA).

2.4. Detection of AGO1-Associated miRNAs in S2 Cells

1. 12% Acrylamide/bisacrylamide denaturing gel.
2. Gel loading buffer II.
3. Hybond-N^+ (GE Healthcare Bio-Sciences, Piscataway, NJ, USA).
4. 0.5× TBE buffer.
5. UV crosslinker (e.g., Stratalinker, Stratagene, La Jolla, CA, USA).
6. Hybridization buffer: 200 mM sodium phosphate (pH 7.2), 7% SDS, and 1 mM EDTA.
7. Hybridization bags.
8. DNA oligonucleotide probe for miR-bantam: 5′-CAGCTTT CAAAATGATCTCAC-3′.
9. DNA oligonucleotide probe for miR-2b: 5′-GCTCCTCAA AGCTGGCTGTGATA-3′.
10. 2× SSC+0.1% SDS: 300 mM NaCl, 30 mM sodium citrate, 0.1% SDS.
11. Stripping buffer: 0.1× SSC and 0.5% SDS.

3. Methods

An important factor that should be considered before performing immunoprecipitation experiments is the choice of antibodies that recognize the protein of interest. The selected antibodies should have been tested for their specificity and cross-reactivity. Also, of great importance is the location of the epitope(s) to which an antibody reacts because this can affect the experimental results. To illustrate this point, we describe here two anti-*Drosophila* AGO2 monoclonal antibodies, 9D6 and 4D2, which we developed in mice using the N-terminal (409 amino acids) and C-terminal (300 amino acids) regions of AGO2, respectively. Figure 1 depicts the mapping of the recognition sites for 4D2 within AGO2. Using western blot analysis, we mapped the recognition site of 4D2 to 16 amino acid residues (1145–1160) located within the catalytic center of the AGO2 PIWI domain (close to His1173 which is part of the DDH motif).

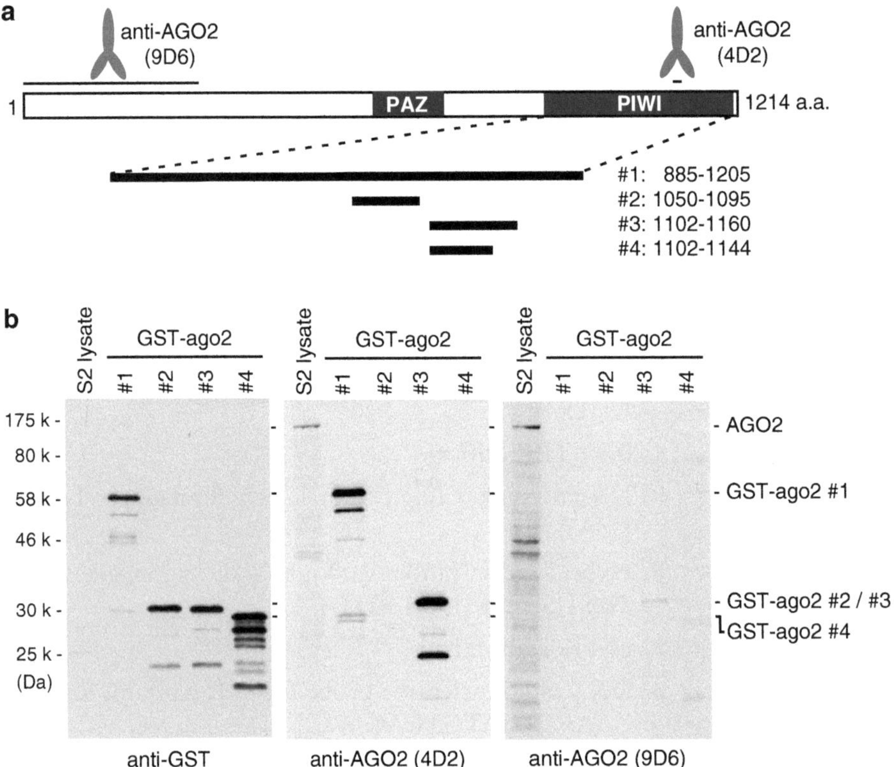

Fig. 1. Identification of the recognition site for anti-AGO2 antibody (4D2). (**a**) A schematic diagram of GST-tagged full-length AGO2 and recombinant fragments encoding part of the PIWI domain (#1 to #4). The numbers refer to amino acid positions. (**b**) Western blotting analyzes. While the anti-GST antibody recognized all GST-fusion proteins to similar extents, 4D2-recognized GST-ago2 #1 and GST-ago2 #3, but not GST-ago2 #2 or GST-ago2 #4. Therefore, 4D2 most likely recognizes the 16 amino acids, 1145–1160, as the epitope. The catalytic center of AGO2 consists of D965, D1037, and H1173 (11, 24); this means that the epitope of 4D2 resides within the DDH catalytic center of AGO2. Anti-AGO2 (9D6) monoclonal antibody only recognized AGO2 in the S2 lysate.

3.1. Immunoprecipitation of AGO2 from S2 Cells Using Anti-AGO2 Antibodies

Because its binding site maps near, or within the catalytic center of AGO2, we speculated that 4D2 may not be able to immunoprecipitate AGO2 while it is bound to siRNAs (either exo-siRNAs or esiRNAs). In our earlier studies, we found that both 4D2 and 9D6 antibodies poorly immunoprecipitated AGO2 under mild conditions, i.e., in buffer containing 150 mM NaCl. These results suggest that the epitopes for the antibodies may associate with some yet-to-be-identified molecules in vivo and were, therefore, not accessible for antibody binding. Therefore, we performed immunoprecipitation in a buffer containing 800 mM NaCl, which should disrupt most protein–protein interactions occurring in vivo. In such conditions, we found that both 9D6 and 4D2 were able to precipitate AGO2 from S2 cells to similar extents (Fig. 2a). The details of the immunoprecipitation protocol in high salt are as follows.

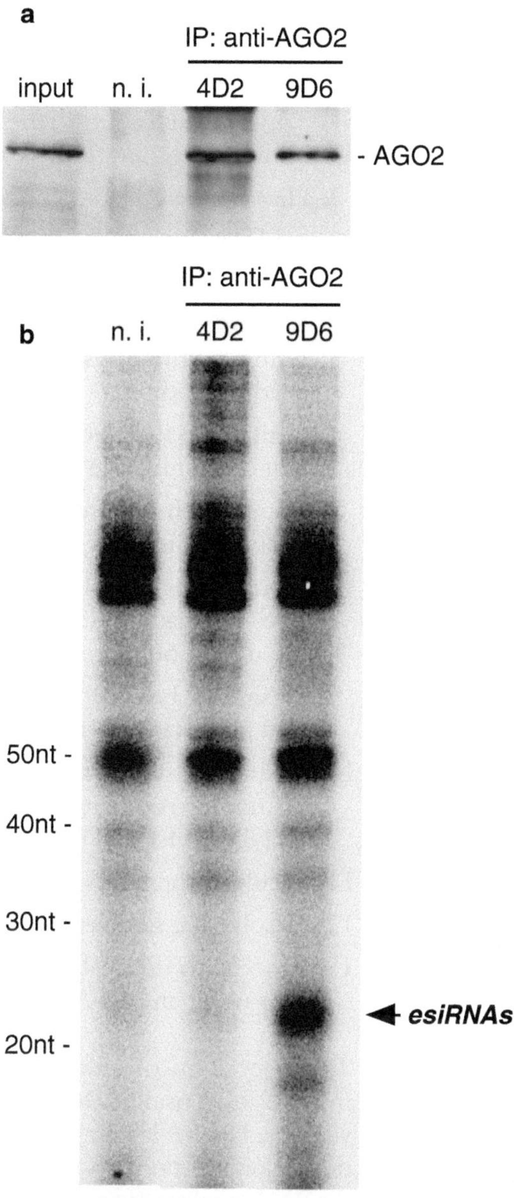

Fig. 2. Visualization of small RNAs associated with AGO2. (**a**) Immunoprecipitates of anti-AGO2 antibodies (4D2 and 9D6) analyzed by western blot analysis with anti-Ago2 antibody (4D2). Non-immune antibody (n.i.) was used as a negative control. (**b**) Small RNAs associated with AGO2 antibodies (4D2 and 9D6) in S2 cells visualized by ^{32}P-labeling. Anti-AGO2 antibody (9D6) immunoprecipitates are associated with small RNAs ranging from 21 to 23 nt. In contrast, anti-AGO2 antibody (4D2) immunoprecipitates are not associated with small RNAs.

3.1.1. Preparation of Anti-AGO2 Antibody (Either 4D2 or 9D6) Immobilized on GammaBind G Sepharose Beads

1. Wash the GammaBind G Sepharose beads (30 μL) with PBS.
2. Incubate the beads with 0.7 mL PBS and 3 μg of anti-AGO2 antibody (4D2). Alternatively, incubate the beads with 0.5 mL of the supernatant of 9D6 (anti-Ago2 antibody) hybridoma cell culture (see Note 1).

3. Rock the mixtures at 4°C for at least 30 min.
4. Wash the beads twice with IP-NaCl-800 buffer.

3.1.2. Immunoprecipitation

1. Approximately 5×10^7 S2 cells are needed per immunoprecipitation.
2. Harvest the cells by centrifugation for 5 min at 400 × *g*. Wash the cells twice with PBS buffer.
3. Suspend the cells in 500 μL of IP-NaCl-150 buffer.
4. Incubate on ice for 10 min.
5. Using a syringe, lyse the cells by passing through a 30-gauge needle five times.
6. Centrifuge at 20,000 × *g* for 20 min and recover the supernatant containing the cytoplasmic lysate (>5 mg/mL proteins).
7. Prior to immunoprecipitation, add 5 M NaCl to the lysates to final concentration of 800 mM.
8. Incubate the cytoplasmic lysate with Bead-bound anti-AGO2 (see Subheading 3.1.1). Rock the reaction mixtures at 4°C for at least 60 min.
9. Wash the beads four times with IP-NaCl-800 containing 800 mM NaCl buffer.
10. Add 20 μL of 2× sample buffer-DTT, mix and incubate for 10 min at room temperature, recover the supernatant, and add 5 μL of 1 M DTT.
11. Incubate the sample at 95°C for 5 min.
12. Resolve the sample by SDS–PAGE.
13. Proceed to the analysis of the immunoprecipitates by western blot using an anti-AGO2 antibody (e.g., 4D2) (Fig. 2a).

3.2. Visualization of Small RNAs Associated with AGO2 Using ^{32}P-Labeling

The association of esiRNAs with AGO2, when immunopurified from S2 cells using 9D6 and 4D2 anti-AGO2 antibodies, can be visualized by ^{32}P-labeling. Small RNAs are isolated from the immunoprecipitates prior to end-labeling. Although multiple methods for RNA labeling are available, the protocols utilizing ^{32}P-γ-ATP or ^{32}P-pCp are commonly used to visualize Argonaute-associating small RNAs. Earlier studies reported that esiRNAs in *Drosophila* have 2′-*O*-methyl groups at their 3′ ends as do miRNAs in plants (but not in *Drosophila*) and that a *Drosophila* homolog of *Arabidopsis* methyltransferase HEN1, dHEN1 (alternatively called Pimet), is responsible for the 2′-*O*-methyl modification for *Drosophila* esiRNAs (25, 26). Thus, the ^{32}P-pCp-labeling method, whose efficiency of labeling seems to be greatly affected by the availability of free 2′ and 3′ ends of small RNAs, is unsuitable for esiRNA detection. We, therefore, used the protocol involving ^{32}P-γ-ATP.

The protocol for the ^{32}P-γ-ATP-labeling of coimmunoprecipitated RNAs is as follows.

1. Purify the RNA from the AGO2-immunoprecipitates using an RNA extraction reagent (ISOGEN-LS or an equivalent reagent). Alternatively, one may also use phenol/chloroform/isoamyl alcohol (25:24:1). For this, add 200 μL of nuclease-free water to the AGO2-immunoprecipitates (step 9 in Subheading 3.1.2).
2. Add 600 μL of ISOGEN-LS and mix thoroughly.
3. Add 160 μL of chloroform and mix thoroughly.
4. Centrifuge at 20,000 × *g* for 10 min.
5. Recover the aqueous phase and transfer to a new tube.
6. Precipitate the RNA by adding 0.8 volumes of isopropanol and 1 μL of Pellet Paint Co-precipitant then mix well.
7. Chill the mixture for at least 20 min at –80°C.
8. Centrifuge at 4°C for 20 min at 20,000 × *g* to pellet the RNA.
9. Wash pellet with 70% ethanol.
10. Dissolve the purified RNA in 10 μL of nuclease-free water.
11. For a 50 μL reaction, assemble on ice the 2 μL of RNA solution, 5 μL of 10× NEB buffer 3, and 1 μL of CIP (see Note 2).
12. Incubate the reaction at 37°C for 30 min.
13. Add 15 μL of 3 M NaOAc, 85 μL of nuclease-free water, and 150 μL of phenol/chloroform/isoamyl alcohol (25:24:1) and mix thoroughly. Recover the aqueous phase and transfer to a new tube.
14. Precipitate the RNA by adding 1 volume of isopropanol and 1 μL of Pellet Paint Co-Precipitant, and then mix well.
15. Chill the mixture for at least 20 min at –80°C, and then centrifuge at 4°C for 20 min at 20,000 × *g* to pellet the RNA.
16. Wash the pellet with 70% ethanol.
17. Dissolve the purified RNA in 10 μL of nuclease-free water.
18. For a 30 μL reaction, assemble on ice the 2 μL of RNA solution, 3 μL of 10× T4 PNK buffer, 0.25 μL of ^{32}P-γ-ATP, and 1 μL of T4 PNK.
19. Incubate the reaction at 37°C for 60 min.
20. Prepare a Micro Bio-Spin 30 Column, and remove the RNase-free Tris buffer from the matrix by centrifuging for 2 min in a microcentrifuge tube at 1,000 × *g* (follow the supplier's instructions).
21. Place the column in a clean 1.5 mL tube.

22. For the removal of the unincorporated ^{32}P-γ-ATP, load the reaction onto the Micro Bio-Spin 30 column.
23. Centrifuge for 4 min in a new 1.5 mL tube at 1,000 × *g*. The flow-through (20 μL) contains the purified RNA sample.
24. Add 15 μL of 3 M NaOAc, 105 μL of nuclease-free water, and 150 μL of phenol/chloroform/isoamyl alcohol (25:24:1) and then mix thoroughly. Centrifuge, recover the aqueous phase and transfer to a new tube.
25. Precipitate the RNA by adding 1 volume of isopropanol, 1 μL of Pellet Paint Co-Precipitant, and mix well.
26. Chill the mixture for at least 20 min at –80°C.
27. Centrifuge at 4°C for 20 min at 20,000 × *g* to pellet the RNA.
28. Wash pellet with 70% ethanol.
29. Dissolve the purified RNA in 3 μL of nuclease-free water.
30. Add 12 μL of loading dye (e.g., Gel loading buffer II).
31. Incubate at 95°C for 3 min, and then on ice.
32. Separate on a 12% acrylamide/bisacrylamide denaturing gel.
33. Dry the gel in a gel-dryer.
34. Expose the gel to an imaging plate and visualize the signals using the BAS-2500 system.

Using this procedure, we found that AGO2 immunopurified with 9D6 (which recognizes the N-terminal region of AGO2) was associated with esiRNAs, whereas AGO2 immunopurified with 4D2 (which recognizes the C-terminal region of AGO2) was not (Fig. 2b). These data indicate that esiRNA-bound AGO2 is not recognized by or immunoprecipitated with 4D2. In theory, this means that 4D2 may allow the isolation of a pool of AGO2 that is not bound to small RNAs.

3.3. Detection of AGO1-Associated Proteins in S2 Cells

Several buffer compositions may be used for immunoprecipitation under mild conditions. However, one has to carefully choose the appropriate buffer depending on the aim of the studies. For example, to obtain an antigen with a high purity from crude cell lysates by immunoprecipitation, a buffer containing a high salt or a strong detergent, such as Empigen that will disrupt most protein–protein interactions in vivo, should be chosen. It should be noted, however, that even in a buffer containing Empigen, the antigen–antibody association is often maintained. Accordingly, there is a higher chance that extremely tight binding factors will remain associated with the antigens. In this section, we demonstrate that the content of the complexes immunoprecipitated from S2 cells with the anti-AGO1 antibody is affected by the type of buffer that is used (see Fig. 3 and Note 3).

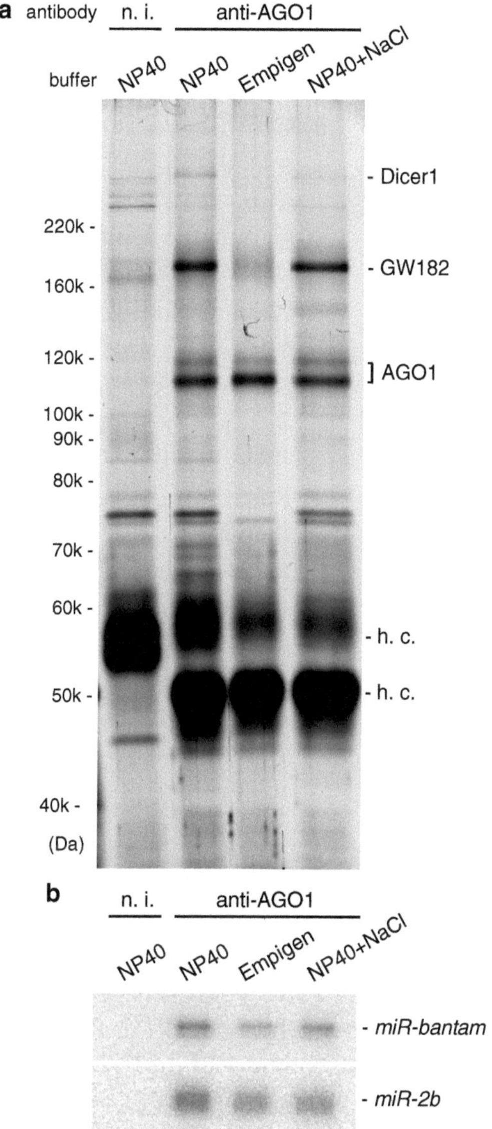

Fig. 3. Characterization of anti-AGO1 immunoprecipitates from S2 cells under different conditions. (**a**) Purification of Ago1-containing complex from S2 cell lysates by immunoprecipitation with anti-AGO1 monoclonal antibody (1B8) or a control non-immune antibody (n.i.). Immunoprecipitates were resolved by SDS–PAGE and visualized by silver staining. (**b**) miRNAs (*miR-bantam* and *miR-2b*) associated with AGO1 in S2 cell lysates prepared in a buffer containing NP-40, Empigen, or NP-40 + 800 mM NaCl, visualized by Northern blot analysis.

A detailed protocol for immunoprecipitation in an Empigen buffer is as follows.

1. Harvest S2 cells (>5×10^7 cells per immunoprecipitation) and wash twice with PBS.
2. Resuspend the cells in 0.5 mL of Empigen buffer.

3. Incubate on ice for 10 min.
4. Using a syringe, lyse the cells by passing five times through a 30-gauge needle.
5. Centrifuge at 20,000 × g at 4°C for 20 min in 1.5 mL tube.
6. Recover and incubate the supernatant (>5 mg/mL proteins) with the anti-AGO1 (1B8) antibody immobilized on GammaBind G Sepharose beads. Rock the reaction mixtures at 4°C for at least 1 h.
7. Wash the beads five times with 700 μL Empigen buffer at 4°C.
8. For silver staining of the AGO1 immunoprecipitates, add 20 μL of 2× sample buffer-DTT, mix, and incubate for 10 min at room temperature.
9. Recover the supernatant, and add 5 μL of 1 M DTT. Incubate the sample at 95°C for 5 min.
10. Resolve the sample by SDS–PAGE.
11. Visualize the AGO1-associating proteins using a silver stain kit (e.g., SilverQuest, follow the supplier's instructions).

3.4. Detection of AGO1-Associated miRNAs

3.4.1. Detection of AGO1-Associated miRNAs in S2 Cells by Northern Blotting

1. It is first necessary to purify the RNAs from the AGO1 immunoprecipitates. For this, add 200 μL of nuclease-free water to the AGO1 immunoprecipitates after wash (step 7 in Subheading 3.3).
2. Add 600 μL of ISOGEN-LS and mix thoroughly.
3. Add 160 μL of chloroform and mix thoroughly.
4. Centrifuge at 20,000 × g for 10 min.
5. Recover the aqueous phase and transfer to new tube.
6. Precipitate the RNA by adding 0.8 volumes of isopropanol and 1 μL of Pellet Paint Co-Precipitant, and then mix well.
7. Chill the mixture for at least 15 min at –80°C.
8. Centrifuge at 4°C for 20 min at 20,000 × g to pellet the RNA.
9. Wash the pellet with 70% ethanol.
10. Dissolve the purified RNA in 3 μL of nuclease-free water.
11. Add 12 μL of Gel loading buffer II
12. Incubate the solution at 95°C for 3 min.
13. Incubate on ice for 3 min.
14. Separate on a 12% acrylamide/bisacrylamide denaturing gel.
15. Transfer the gel to a nylon membrane (Hybond N^+) at constant current (1 mA/cm^2) for 60 min in 0.5× TBE buffer on a semi-dry apparatus.

16. After transfer, UV crosslink with 1,200 × 100 μJ energy.
17. Soak the membrane with hybridization buffer in a hybridization bag.
18. Incubate at 42°C for 2–4 h in a hybridization oven.
19. Incubate the probe (see Subheading 3.4.2) at 95°C for 3 min and then add it to the hybridization bag.
20. For hybridization, incubate with shaking at 42°C overnight.
21. Wash the membrane with 2× SSC + 0.1% SDS buffer while shaking at room temperature for 10 min. Repeat the wash step three times.
22. Place the membrane in a piece of Saran wrap.
23. Expose the membrane to an imaging plate and visualize the signals using the BAS-2500 system or an equivalent (see Notes 4 and 5).

3.4.2. Probe Labeling

1. For a 30 μL reaction, assemble on ice 23.5 μL of water, 3 μL of 10× T4 PNK buffer, 1.5 μL of 1 μM DNA oligo (~22 nt), 1 μL of ^{32}P-γ-ATP, and 1 μL of T4 PNK.
2. Incubate the reaction at 37°C for 1 h.
3. For the removal of unincorporated ^{32}P-γ-ATP, load the reaction on a Micro Bio-Spin column P-30 Tris RNase-Free (follow the manufacturer's instructions).

As shown by silver staining (Fig. 3), similar amounts of AGO1 protein were immunoisolated under all three buffer conditions. However, one of the proteins that is expected to associate with AGO1, Dicer1, was only evident when using a buffer containing 150 mM NaCl. Another AGO1-interacting protein, GW182, appeared in both low and high (800 mM) NaCl immunoprecipitation buffers. These data indicate that the interaction of GW182 with AGO1 may be very stable in vivo. As indicated in Fig. 3b, AGO1 still remains associated with miRNAs, *miR-2b* and *miR-bantam*, in an Empigen-containing buffer. These results demonstrate that GW182 and Dicer1 are dispensable for AGO1 to associate with miRNAs in S2 cells and that the binding of miRNAs with AGO1 is resistant to Empigen and high salt buffers.

4. Notes

1. This is a rather unusual experience, but we found that the purification of anti-AGO2 monoclonal antibody (9D6) through Thiophilic-Superflow resin (BD Biosciences Clontech, Palo Alto) does not work well for immunoprecipitation. Therefore, we use the supernatant of hybridoma cell culture.

2. esiRNAs and miRNAs contain 5′-phosphate terminus. Therefore, both small RNAs have to be dephosphorylated by CIP prior to the ^{32}P-labeling at their 5′ ends.
3. We already described above (Subheading 3.1.2) how to perform the immunoprecipitation in a buffer containing 800 mM NaCl. To perform immunoprecipitation under milder conditions, such as in IP-NaCl-150 buffer, just omit the addition of NaCl at step 7 in Subheading 3.1.2.
4. The membrane may be stripped and rehybridized with other probes. For this, incubate the membrane in stripping buffer at 95°C for 20 min in a hybridization oven. Repeat the wash step three times. This approach has worked well for stripping most DNA probes. Check the efficacy of the stripping using the BAS-2500 system. For the next hybridization, proceed from the prehybridization step (step 17 in Subheading 3.4.1).
5. When the signal for an RNA of interest is low using this Northern blotting method, the signal can be enhanced by a chemical cross-linking method (27). The transferred molecules are usually cross-linked to the membrane using UV irradiation to reduce the loss of the sample RNA during the subsequent hybridization and washing steps. A 1-ethyl-3-(3-dimethylaminopropyl) carbodiimide (EDC)-mediated, chemical cross-linking step can enhance the detection of small RNAs by up to 50-fold. Although chemical cross-linking takes longer than UV cross-linking, improved sensitivity means shorter periods of exposure are required to detect signal after hybridization.

Acknowledgments

We thank Asuka Azuma-Mukai for technical assistance and other members of the Siomi laboratory for comments and for critical reading of the manuscript. This work was supported by Mochida Memorial Foundation for Medical and Pharmaceutical Research grants to K.M., MEXT grants to H.S. and NEDO (New Energy and Industrial Technology Development Organization) grants to M.C.S. M.C.S. is supported by CREST from JST.

References

1. Siomi, H., and Siomi, M. C. (2009). On the road to reading the RNA-interference code. *Nature* **457**, 396–404.
2. Bartel, D. P. (2009). MicroRNAs: target recognition and regulatory functions. *Cell* **136**, 215–233.
3. Malone, C. D., and Hannon, G. J. (2009). Small RNAs as guardians of the genome. *Cell* **136**, 656–668.
4. Voinnet, O. (2009). Origin, biogenesis, and activity of plant microRNAs. *Cell* **136**, 669–687.

5. Hutvagner, G., and Simard, M. J. (2008). Argonaute proteins: key players in RNA silencing. *Nat. Rev. Mol. Cell Biol.* **9**, 22–32.
6. Song, J. J., Liu, J., Tolia, N. H., Schneiderman, J., Smith, S. K., Martienssen, R. A., Hannon, G. J., and Joshua-Tor, L. (2003). The crystal structure of the Argonaute2 PAZ domain reveals an RNA binding motif in RNAi effector complexes. *Nat. Struct. Biol.* **10**, 1026–1032.
7. Ma, J. B., Ye, K., and Patel, D. J. (2004). Structural basis for overhang-specific small interfering RNA recognition by the PAZ domain. *Nature* **429**, 318–22.
8. Lingel, A., Simon, B., Izaurralde, E., and Sattler, M. (2004). Nucleic acid 3′-end recognition by the Argonaute2 PAZ domain. *Nat. Struct. Mol. Biol.* **11**, 576–577.
9. Song, J. J., Smith, S. K., Hannon, G. J., and Joshua-Tor, L. (2004). Crystal structure of Argonaute and its implications for RISC slicer activity. *Science* **305**, 1434–1437.
10. Parker, J. S., Roe, S. M., and Barford, D. (2004) Crystal structure of a PIWI protein suggests mechanisms for siRNA recognition and slicer activity. *EMBO J.* **23**, 4727–4737.
11. Liu, J., Carmell, M. A., Rivas, F. V., Marsden, C. G., Thomson, J. M., Song, J. J., Hammond, S. M., Joshua-Tor, L., and Hannon, G. J. (2004). Argonaute2 is the catalytic engine of mammalian RNAi. *Science* **305**, 1437–1441.
12. Farazi, T. A., Juranek, S. A., and Tuschl, T. (2008). The growing catalog of small RNAs and their association with distinct Argonaute/Piwi family members. *Development* **135**, 1201–1214.
13. Okamura, K., Ishizuka, A., Siomi, H., and Siomi, M. C. (2004). Distinct roles for Argonaute proteins in small RNA-directed RNA cleavage pathways. *Genes Dev.* **18**, 1655–1666.
14. Förstemann, K., Horwich, M. D., Wee, L., Tomari, Y., and Zamore, P. D. (2007). *Drosophila* microRNAs are sorted into functionally distinct argonaute complexes after production by dicer-1. *Cell* **130**, 287–297.
15. Miyoshi, K., Tsukumo, H., Nagami, T., Siomi, H., and Siomi, M. C. (2005). Slicer function of *Drosophila* Argonautes and its involvement in RISC formation. *Genes Dev.* **19**, 2837–2848.
16. Denli, A. M., Tops, B. B., Plasterk, R. H., Ketting, R. F., and Hannon, G. J. (2004). Processing of primary microRNAs by the Microprocessor complex. *Nature* **432**, 231–235.
17. Saito, K., Ishizuka, A., Siomi, H., and Siomi, M. C. (2005). Processing of pre-microRNAs by the Dicer-1-Loquacious complex in *Drosophila* cells. *PLoS Biol.* **3**, e235.
18. Förstemann, K., Tomari, Y., Du, T., Vagin, V. V., Denli, A. M., Bratu, D. P., Klattenhoff, C., Theurkauf, W. E., and Zamore, P. D. (2005). Normal microRNA maturation and germ-line stem cell maintenance requires Loquacious, a double-stranded RNA-binding domain protein. *PLoS Biol.* **3**, e236.
19. Liu, Q., Rand, T. A., Kalidas, S., Du, F., Kim, H. E., Smith, D. P., and Wang, X. (2003). R2D2, a bridge between the initiation and effector steps of the *Drosophila* RNAi pathway. *Science* **301**, 1921–1925.
20. Kim, V. N., Han, J., and Siomi, M. C. (2009). Biogenesis of small RNAs in animals. *Nat. Rev. Mol. Cell Biol.* **10**, 126–139.
21. Zhou, R., Czech, B., Brennecke, J., Sachidanandam, R., Wohlschlegel, J. A., Perrimon, N., and Hannon, G. J. (2009). Processing of *Drosophila* endo-siRNAs depends on a specific Loquacious isoform. *RNA* **15**, 1886–1895.
22. Hartig, J. V., Esslinger, S., Böttcher, R., Saito, K., and Förstemann, K. (2009). Endo-siRNAs depend on a new isoform of loquacious and target artificially introduced, high-copy sequences. *EMBO J.* **28**, 2932–2944.
23. Miyoshi, K., Miyoshi, T., Hartig, J. V., Siomi, H., and Siomi, M. C. (2010). Molecular mechanisms that funnel RNA precursors into endogenous small-interfering RNA and micro RNA biogenesis pathways in *Drosophila*. *RNA*, **16**, 506–515.
24. Tolia, N. H., and Joshua-Tor, L. (2007). Slicer and the argonautes. *Nat. Chem. Biol.* **3**, 36–43.
25. Saito, K., Sakaguchi, Y., Suzuki, T., Suzuki, T., Siomi, H., and Siomi, M.C. (2007). Pimet, the *Drosophila* homolog of HEN1, mediates 2′-*O*-methylation of Piwi- interacting RNAs at their 3′ ends. *Genes Dev.,* **21**, 1603–1608.
26. Kawamura, Y., Saito, K., Kin, T., Ono, Y., Asai, K., Sunohara, T., Okada, T. N., Siomi, M. C., and Siomi, H. (2008). *Drosophila* endogenous small RNAs bind to Argonaute 2 in somatic cells. *Nature* **453**, 793–797.
27. Pall, G. S., and Hamilton, A. J. (2008). Improved northern blot method for enhanced detection of small RNA. *Nat. Protoc.* **3**, 1077–1084.

Chapter 4

Mapping of Ago2–GW182 Functional Interactions

Bing Yao, Songqing Li, Shang Li Lian, Marvin J. Fritzler, and Edward K.L. Chan

Abstract

MicroRNA (miRNA)-mediated posttranscriptional regulation of gene expression has become a major focus in understanding fine-tuning controls in many biological processes. Argonaute 2 protein (Ago2), a core component of RNA-induced silencing complex, directly binds miRNA and functions in both RNAi and miRNA pathways. GW182 is a marker protein of GW bodies (GWB, also known as mammalian P-bodies) and is known to bind the Ago2 protein. This Ago2–GW182 interaction is crucial for Ago2–miRNA-mediated translational silencing as well as the recruitment of Ago2 into GWB. Translational silencing of tethered Ago2 to a 3′UTR reporter requires GW182 for function, whereas tethered GW182 exerts a stronger repression than tethered Ago2 and does not apparently require Ago2. This chapter describes in detail the methods used in mapping Ago2–GW182 interactions.

Key words: Ago2, GW182, MicroRNA, GW bodies, RNA-induced silencing complex, Gene silencing, Translational repression

1. Introduction

MicroRNAs (miRNA), composed of 20–25 nucleotides, are largely derived from endogenous transcription of independent miRNA genes or gene clusters and play many key roles in a variety of normal and pathological cellular processes (1). miRNAs are incorporated into RNA-induced silencing complex (RISC) to induce translational repression or RNA degradation of their target mRNAs (2). At the core of RISC complexes are members of the Argonaute protein family; a group of RNA-binding proteins that is conserved across different species (3). There are four homologs in mammals, Ago1–Ago4, which are all involved in miRNA-mediated translational silencing, whereas only Ago2 harbors RNase H-type activity in its C-terminal P-element-induced wimpy

Tom C. Hobman and Thomas F. Duchaine (eds.), *Argonaute Proteins: Methods and Protocols*,
Methods in Molecular Biology, vol. 725, DOI 10.1007/978-1-61779-046-1_4, © Springer Science+Business Media, LLC 2011

testis (PIWI) domain. The endonuclease activity of Ago2 is critical for its functions in small-interfering RNA (siRNA)-mediated cleavage of mRNA targets (4, 5).

GW182 protein was first identified and characterized by our laboratory in 2002 as a novel protein using an autoimmune serum from a patient with motor and sensory neuropathy (6). It is a 182 kDa protein characterized by multiple glycine (G) and tryptophan (W) repeats. GW182 likely serves as an essential component of GW bodies because knockdown of this protein leads to the disassembly of these foci (7). It also plays a pivotal role as a strong repressor in RISC complex for silencing the microRNA-targeted mRNA translation (8–10).

In this chapter, methods used in the study of the interaction between these two proteins, including in vivo GST pull-downs and immunofluorescence, are described. The data indicate that Ago2 interacts with GW182 in multiple regions and this interaction is important for correct colocalization of Ago2 into GW bodies (11–14). To dissect the individual role of Ago2 and GW182 in its translational silencing of the mRNAs, the dual luciferase tethering assay is described using a transfected luciferase reporter harboring the 5BoxB secondary structures in the 3′UTR region (15). The 5BoxB element is bound by an N-terminal λN-hemagglutinin (NHA) polypeptide tag. Therefore, Ago2 or GW182 fused to this NHA-tag can be physically tethered to the reporter, which bypasses the requirement of miRNA–Ago2 guidance. A firefly luciferase reporter without the 5BoxB structures serves as internal transfection control. In tethering experiments, both Ago2 and GW182 can trigger posttranscriptional silencing of the reporter when tethered to the 3′UTR. However, siRNA knockdown of either endogenous Ago2 or GW182 in the tethering assay shows that GW182 can serve as a downstream direct repressor independent of Ago2 (9, 12). Thus, the current working model is that Ago2–miRNA complexes are responsible for target mRNA recognition and recruitment of GW182, which initiate translational repression and/or degradation of the targeted mRNA.

2. Materials

2.1. Plasmids Cloning Reagents and siRNA

1. PrimeSTAR high-fidelity DNA polymerase (Takara, Madison, WI).
2. Gateway BP and LR Clonase II Enzyme Mix (Invitrogen, Carlsbad, CA).
3. One Shot OmniMAX 2 T1 phage-resistant competent cells (Invitrogen, Carlsbad, CA).

4. QIAprep Spin Miniprep kit and Endo-free Plasmid Maxiprep kit (Qiagen, Valencia, CA).
5. Agarose (molecular biology grade).
6. QIAquick Gel Extraction Kit (Qiagen, Valencia, CA).
7. Five nanomoles of each siGENOME SMARTpool siRNA for TNRC6A (GW182, NM_014494) and EIF2C2 (human Ago2, NM_012154) (Dharmacon RNA Technologies, Lafayette, CO). 20 μM stocks are dissolved in molecular grade water (Fisher Scientific, Pittsburgh, PA) and stored in aliquots at −80°C before use.

2.2. Cell Culture, Transfection and Whole Cell Lysate Preparation

1. Dulbecco's Modified Eagle's Medium (DMEM) (Cellgro, Manassas, VA) supplemented with 10% fetal bovine serum (FBS, Cellgro, Manassas, VA).
2. Solution of trypsin (0.25%) and ethylenediamine tetraacetic acid (EDTA, 1 mM) (Cellgro, Manassas, VA).
3. Penicillin/streptomycin solution (Cellgro, Manassas, VA).
4. Lipofectamine 2000 (Invitrogen, Carlsbad, CA).
5. Opti-MEM I Reduced-Serum Medium (Gibco/BRL, Bethesda, MD).
6. Cell lysis buffer for GST pull-down: NET/NP40 buffer (150 mM NaCl, 5 mM EDTA, 50 mM Tris-HCl, pH 7.4, 0.3% NP40) with Complete Protease Cocktail Inhibitor (Roche Diagnostics, Indianapolis, IN).
7. Cell passive lysis buffer (5×) from Dual Luciferase Reporter Assay System (Promega, Madison, WI).
8. Tissue culture flasks, cell scrapers, 6-well and 24-well cell culture plates (Fisher Scientific, Pittsburgh, PA).

2.3. GST Pull-Down Assays and SDS–Polyacrylamide Gel Electrophoresis

1. Glutathione Sepharose 4B (GST) beads (GE Healthcare, Piscataway, NJ).
2. Mini-PROTEIN 3 cell SDS–PAGE gel system (Bio-Rad, Hercules, CA).
3. Separating buffer: 1.0 M Tris–HCl, pH 8.8, store at room temperature.
4. Stacking buffer: 1.0 M Tris–HCl, pH 6.8, store at room temperature.
5. Thirty percent acrylamide and 2% bis-acrylamide solution (Bio-Rad, Hercules, CA).
6. Ammonium persulphate (APS). Dissolve 0.1 g APS into 1 mL dH_2O, vortex briefly, and store at 4°C for no more than 1 week.
7. *N*,*N*,*N*,*N*′-tetramethyl-ethylenediamine (TEMED). Store at 4°C.

8. Laemmli loading buffer (2×): 0.125 M Tris–HCl, pH 6.8, 4% SDS, 0.004% bromophenol blue powder (Sigma, St. Louis, MO), 20% glycerol, 10% 2-mercaptoethanol, store at −20°C.
9. Water-saturated butanol: mix equal amount of butanol and water in a glass bottle to let them separate. Use the top layer and store at room temperature.
10. Gel running buffer (10×): 1.92 M glycine, 0.25 M Tris base, and 1% SDS. Store at room temperature.
11. Protein molecular weight markers: Kaleidoscope markers (Bio-Rad, Hercules, CA).

2.4. Western Blot Analysis of Ago2–GW182 Interactions

1. Mini Trans-Blot Electrophoretic Transfer Cell (Bio-Rad, Hercules, CA).
2. Transfer buffer (10×): 1.92 M glycine and 0.25 M Tris base. Store at room temperature.
3. Nitrocellulose membrane (GE Healthcare, Piscataway, NJ).
4. Filter papers (3MM, Bio-Rad, Hercules, CA).
5. PBS (10×): 2.8 M NaCl, 5 mM thimerosal, 140 mM Na_2HPO_4, and 46 mM NaH_2PO_4, adjust the pH to 7.4. Store at room temperature.
6. PBS buffer with 0.05% Tween (PBS-T): 400 mL 10× PBS plus 2 mL Tween-20 (Sigma, St. Louis, MO), bring the volume to 4 L with dH_2O, store at room temperature.
7. Blocking buffer: 5% (w/v) nonfat dry milk in PBS-T.
8. Primary antibody dilution buffer: PBS-T supplemented with 5% (w/v) nonfat dry milk.
9. Primary antibodies: mouse monoclonal anti-GST (MBL International, Woburn, MA), rabbit polyclonal anti-GFP (Invitrogen, Carlsbad, CA), and mouse monoclonal anti-FLAG (Sigma, St. Louis, MO).
10. Secondary antibody: Anti-mouse/rabbit IgG-conjugated to horse radish peroxidase (Southwestern Biotech, Birmingham, AL).
11. Enhanced Luminol-based chemiluminescent (ECL) reagents (Thermo Scientific, Rockford, IL). Store at room temperature.

2.5. Indirect Immunofluorescence for Subcellular Localization of Ago2 and GW182 Proteins

1. Microscope coverslips (22 × 40 × 0.15 mm) (Fisher Scientific, Pittsburgh, PA).
2. Eight-well chamber slides (BD Biosciences, San Jose, CA).
3. Paraformaldehyde stock (16%, Sigma, St. Louis, MO). Prepare a 3% working solution by diluting in PBS freshly upon use. Store at 4°C.
4. Permeabilization solution: 0.5% (v/v) Triton X-100 (Sigma, St. Louis, MO) in PBS.

5. Antibody dilution buffer: PBS. Primary antibody: mouse monoclonal anti-FLAG M2 antibody (Sigma, St. Louis, MO).
6. Secondary antibody: Alexa-Fluor-568 (Invitrogen, Carlsbad, CA).
7. Vectashield mounting media with DAPI counterstain for nuclei (Vector Laboratories, Inc., Burlingame, CA).

2.6. Dual Luciferase Assays

1. Dual-Glo Luciferase Reporter Assay System (Promega, Madison, WI).
2. Opaque cell culture plate 96-well (BD Biosciences, San Jose, CA).
3. FLUOstar Omega microplate reader (BMG Biotech, Durham, NC).

2.7. Reverse Transcriptase and Quantitative Real-Time PCR

1. mirVana miRNA isolation kit (Ambion, Austin, TX).
2. High Capacity cDNA Reverse Transcription Kit (Applied Biosystems, Foster City, CA).
3. TaqMan Fast Universal Master Mix (Applied Biosystems, Foster City, CA).
4. StepOne Real-time PCR system (Applied Biosystems, Foster City, CA).
5. iCYCLER thermocycler system (Bio-Rad, Hercules, CA).

3. Methods

The interaction between human GW182 and Ago2 was first reported by us and others in 2005 (7, 16). To map the exact interaction domain(s) and study their functional significance, a series of deletion constructs of GW182 were generated by PCR based on the distribution of known domains and subcloned into different destination vectors using the Invitrogen Gateway Cloning system (Fig. 1). Each of these constructs was subsequently co-transfected into HeLa cells with full-length Ago2, its N-terminal PAZ domain or C-terminal PIWI domain (Fig. 1). Co-immunoprecipitation and western blots indicated that the N-terminal, middle, and C-terminal region of GW182 bind the C-terminal PIWI domain of Ago2. A representative example of the results is shown in Fig. 2. Immunofluorescence experiments demonstrated that GW182–Ago2 interaction is crucial for the localization of Ago2 into the cytoplasmic foci (Fig. 3). Tethering and dual luciferase experiments showed Ago2 and its C-terminal PIWI domain, as well as other Ago family proteins, can trigger reporter silencing by tethering them to the 3′UTR of the reporter independent of endogenous miRNA and this repression ability is dependent on their

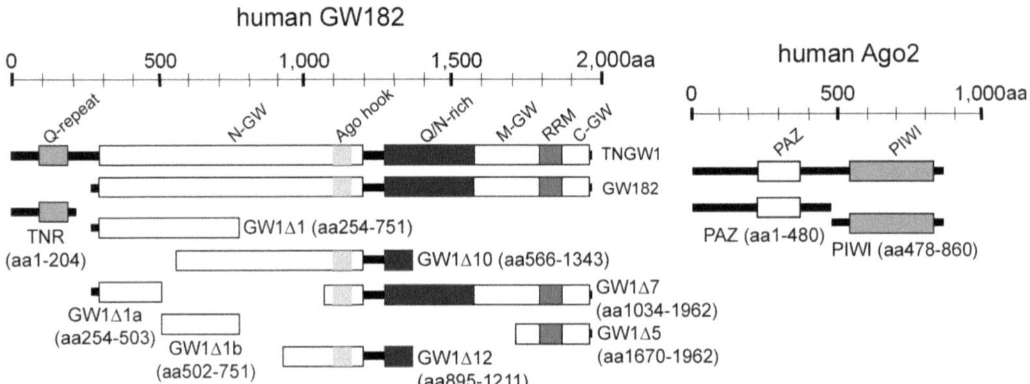

Fig. 1. Human GW182, Ago2 and the corresponding deletion constructs used in this study. TNGW1, the longer isoform of GW182 (NM_014494.2); TNR, trinucleotide repeat; Q-repeat, glutamine repeat; Q/N-rich, glutamine/asparagine-rich region; RRM, RNA recognition motif; GW-rich, glycine/tryptophan-rich region; N-GW, N-terminal GW-rich region; M-GW, middle GW-rich region; and C-GW, C-terminal GW-rich region. Human Ago2 protein contains two conserved domains, the P-element-induced wimpy testis (PIWI) domain and PIWI Argonaute Zwille (PAZ) domain. Modified from ref. 12.

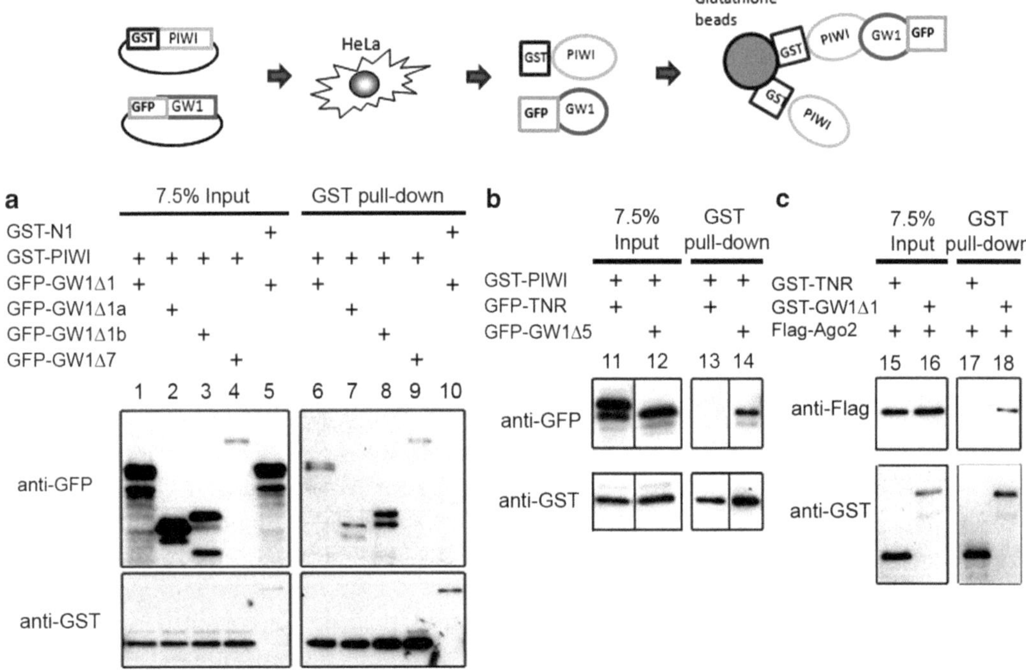

Fig. 2. Agonaute protein bound to multiple nonoverlapping GW-rich regions of GW182. (**a**) GW182 fragments interact with C-terminal half of Ago2. GST-PIWI (aa478–860) co-transfected with GFP-tagged GW1Δ1 (positive control shown interaction previously) (*lane 1*), GW1Δ1a (*lane 2*), GW1Δ1b (*lane 3*), GW1Δ7 (*lane 4*) into HeLa cells. GST-tagged N1, N-terminal fragment from an unrelated protein hZW10 served as a negative control. All GW182 fragments representing different regions were detected in GST-PIWI precipitates. (**b**) GW182 fragment TNR (*lane 11*), or GW1Δ5 (*lane 12*) co-precipitated with Ago2. (**c**) Flag-Ago2 was co-transfected with GST-tagged TNR (*lane 15*) or with GST-GW1Δ1 (*lane 16*). Flag-Ago2 was co-precipitated with GST-tagged GW1Δ1 but not TNR. Modified from ref. 12.

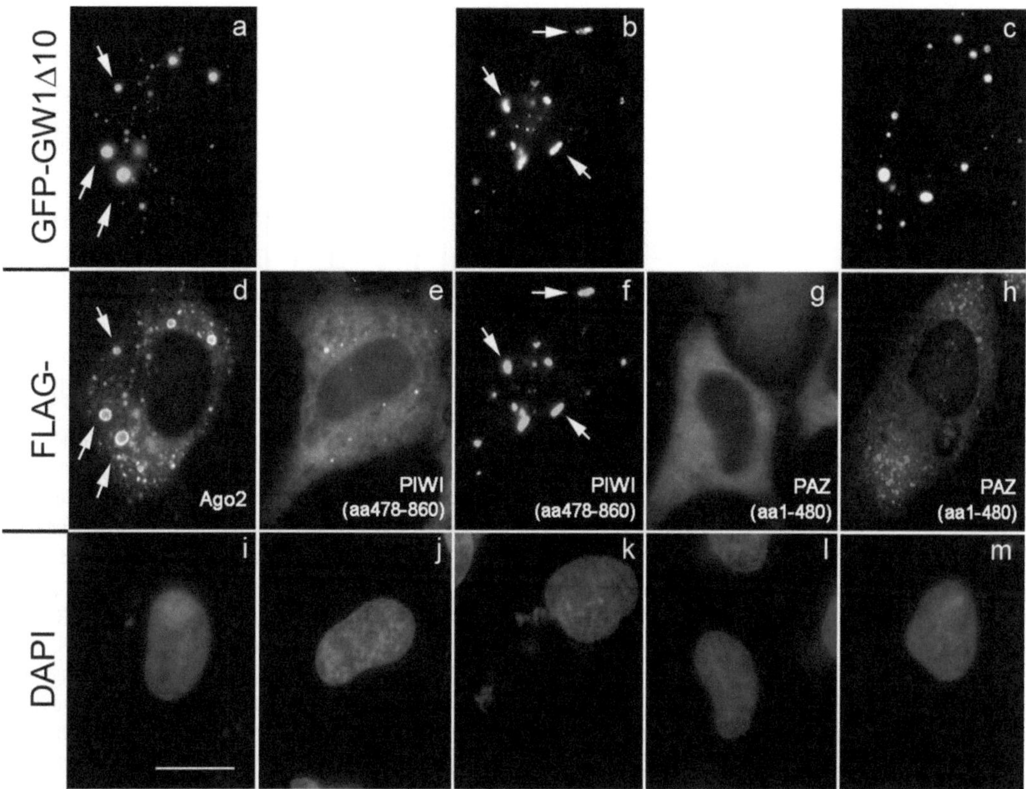

Fig. 3. GW182 fragment GW1Δ10 (aa566–1343) interacted with the C-terminus of Ago2 and recruited it to cytoplasmic foci. GFP-GW1Δ10 (**a–c**) was co-transfected with Flag-Ago2 (**d**), PIWI (aa478–860, **f**), or PAZ (aa1–480, **h**) into HeLa cells. As controls, cells were singly transfected with Flag-PIWI (**e**) and Flag-PAZ (**g**). The cells were stained with anti-FLAG antibody (**d–h**). *Arrows* show the cytoplasmic foci containing GFP-MGW and FLAG-Ago2, PIWI but not PAZ (**a**, **b**, **d**, **f**). *Panels* in the *bottom row* are the merged images (**i–m**). Nuclei were counterstained with DAPI. Scale bar, 10 μm. Modified from ref. 12.

interaction with GW182 (Fig. 4). SiRNA knockdown of GW182 confirmed its important role in Ago2 tethering experiments. Knockdown of Ago2 further implied the repression induced by tethering GW182 is independent of Ago2 (Fig. 5).

3.1. Plasmid Cloning

1. Full-length GW182 cDNA was previously amplified by PCR from a human testis cDNA library (9).
2. PCR amplification of GW182-truncated fragments utilized PrimeSTAR high-fidelity DNA polymerase (5× buffer and dNTP mix are provided). To generate fragments to accommodate pDONR entry clones, a two-step PCR was required to add the full recombination sequences onto the flank region of each amplicon (see Note 1).
3. For the first PCR, forward primer sequence is 5′-AA AAA GCA GGC TNN – template-specific sequence – 3′. Reverse

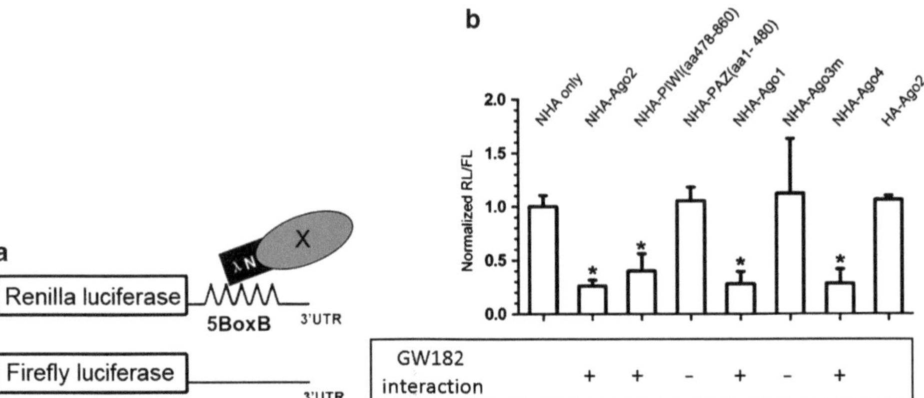

Fig. 4. Translational repression triggered by tethered Ago2 and other Ago family proteins required their interaction with GW182. Thus this translational repression reporter assay bypassed the normal requirement for miRNA-guided targeting. (**a**) *Schematic chart* illustrates the plasmid design for the tethering assay. A protein (X) fused with an N-terminal λN-hemagglutinin (NHA) polypeptide tag bound specifically to the 5BoxB hairpin structure in the 3′UTR of the Renilla luciferase (RL) reporter. When co-transfected with Firefly luciferase (FL) reporter plasmid as an internal control for transfection efficiency, the RL activity, normalized to FL level, was the readout for the repression activity of the NHA-tagged protein (or NHA-tagged protein domain) in inhibiting translation when tethered to the 3′UTR of the RL reporter. (**b**) The C-terminal PIWI domain of Ago2 repressed reporter expression comparable to the full-length Ago2 and other Ago proteins. HeLa cells were transfected with constructs expressing the RL-5BoxB reporter, control FL reporter, and indicated NHA-tagged proteins. The relative RL/FL values of each NHA-tagged protein were normalized to the NHA only control. NHA-tagged Ago2, PIWI, Ago1, and Ago4 showed significantly repression (*) when tethered to 3′UTR. In contrast, no repression was observed for NHA-tagged PAZ, the N-terminal domain of Ago2, and NHA-tagged Ago3m, a splicing variant of Ago3 missing amino acids 757–823, the C-terminal 66 amino acids of the PIWI domain. Both PAZ and Ago3m did not interact with GW182. HA-tagged Ago2 served as an additional control showing that the expression of tagged Ago had no effect when it was not tethered to the UTR reporter. Note that full-length Ago3 shows comparable repression activity (data not shown).

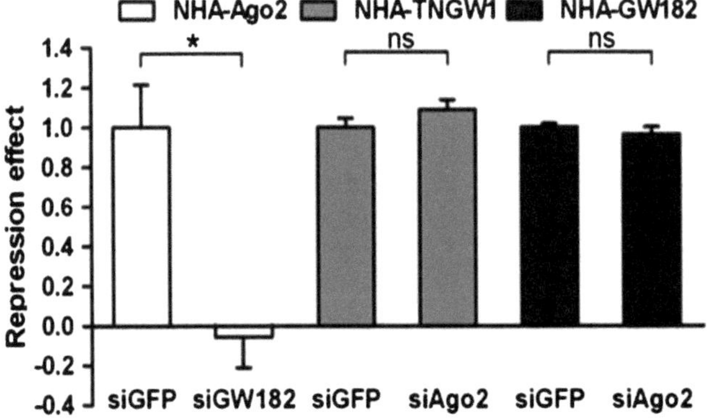

Fig. 5. Reporter repression of tethered Ago2 required endogenous GW182 but not vice versa. HeLa cells were transfected with different siRNAs 24 h prior to the transfection of NHA-tagged constructs and RL-5BoxB/FL reporters. The repression effects induced by tethered Ago2 were abolished when GW182 was knocked down, whereas knockdown Ago2 affected neither TNGW1 nor GW182 repression.

primer sequence is 5′-A GAA AGC TGG GTN (TCA) – template-specific sequences – 3′. TCA is an optional in-frame stop codon.

4. For the second PCR, forward primer sequence is 5′-G GGG ACA AGT TTG TAC AAA AAA GCA GGC T-3′. Reverse primer sequence is 5′-GGG GAC CAC TTT GTA CAA GAA AGC TGG GT-3′.
5. PCR is carried out in 50 μL volume with 10 μL 5× PCR Buffer, 4 μL 2.5 mM dNTP mixture, 0.2 μM (final conc.) forward and reverse primers, 0.5 μL PrimeSTAR polymerase (2.5 U/μL), 500 ng template, and add molecular grade water to 50 μL. Thermocycler conditions are set as 98°C 10 s, 55°C 5 s (T_m > 55°C) or 15 s (T_m < 55°C), 72°C 1 min/kb. Repeat 25–30 cycles to achieve the desired yield.
6. PCR products are loaded onto 1% agarose gels and run at 100 V about 30 min. Fragments are purified using the QIAquick Gel Extraction Kit.
7. Gel-purified PCR fragments are then cloned into pDONR207 donor vector using the Gateway BP clonase II Enzyme Mix by combining 1–7 μL PCR products, 150 ng Donor vector, 2 μL BP Clonase II Enzyme mix, and fill TE buffer (pH 8.0) to 10 μL. Incubate reactions at 25°C for 1 h and then add 1 μL of the Proteinase K solution to each sample to terminate the reaction, vortex briefly, and then incubate at 37°C for 10 min.
8. One microliter BP reaction is added to 50 μL One Shot OmniMAX competent cells and incubated on ice for 30 min. Set water bath to 42°C and heat shock each reaction for exactly 30 s. Rest the samples on ice for 1–2 min and then add 250 μL SOC medium to each sample, and then shake in a 37°C incubator for 1 h. Plate 50–100 μL of bacteria on ampicillin-infused LB agar plates and incubate at 37°C overnight at which time ampicillin-resistant colonies should be formed.
9. Pick two to five colonies for each construct and purify the plasmid DNA using a QIAprep Spin Miniprep kit following the manufacturer's instructions. Submit all plasmids to direct DNA sequencing for confirmation of expected sequences.
10. Subclone each construct from pDONR207 vectors to different destination vectors using Gateway LR clonase II Enzyme mix by mixing 50–150 ng pDONR207 clone, 150 ng Destination vector, 2 μL LR Clonase II Enzyme mix, and bring the volume to 10 μL with TE buffer (pH 8.0). Incubate at 25°C for 1 h and then add 1 μL of the Proteinase K solution to each sample to terminate the reaction. Vortex briefly and incubate at 37°C for 10 min.

11. Transform the LR reaction as in step 8 and perform small- and large-scale plasmid preparations using QIAprep Spin Miniprep and Endo-free Plasmid Maxiprep kits, respectively. Submit destination plasmids to direct DNA sequencing for confirmation if needed.

3.2. Cell Culture, Plasmid Transfection for In Vivo GST Pull-Down Assays and Preparation of Whole Cell Lysates

1. HeLa cells are obtained from ATCC and cultured in DMEM containing 10% fetal bovine serum (Thermo Fisher Scientific, Pittsburgh, PA) and penicillin/streptomycin at 37°C under 5% CO_2. One day before transfection of plasmids, the cells are trypsinized and seeded at $4–5 \times 10^5$ cells/well into 6-well plates with 2 mL medium (see Note 2).
2. On the day of transfection, mix 2 μg plasmid encoding GST-tagged fusion protein and 2 μg plasmid encoding other tags with 250 μL Opti-MEM transfection medium in tube A. In the meantime, mix 10 μL Lipofectamine 2000 with 250 μL Opti-MEM transfection medium in tube B. Mix well by pipetting and incubate at room temperature for 5 min.
3. Add 250 μL Lipofactamine-Opti-MEM tube B mixture to the plasmid-Opti-MEM tube A mixture, mix well by pipetting and incubate at room temperature for 20 min.
4. During step 3, wash cells in 6-well plate once with Opti-MEM, remove, and add 2 mL Opti-MEM supplement with 10% FBS. These cells are ready to receive the Lipo-DNA mixture from step 3.
5. After 20 min incubation, add total 500 μL transfection mixture into each well and incubate for an additional 4–6 h.
6. After 4–6 h, change the medium to regular DMEM containing 10% FBS but without penicillin/streptomycin. Incubate for 24 h before harvesting the cells (see Note 3).
7. To harvest the cells, remove the culture medium and wash with PBS at room temperature. Then add 300 μL GST pull-down buffer with complete Protease Cocktail Inhibitor to each well, scrape the cells, and collect the cell lysate into 1.5 mL tubes.

3.3. GST Pull-Down Assays and SDS–PAGE

1. Aliquot 35 μL of glutathione Sepharose 4B beads per sample and wash three times with complete GST pull-down buffer. Pellet the beads at 10,000 rpm ($9,300 \times g$) in a microfuge for 8–10 s at room temperature and discard the supernatant. Incubate the beads in 1 mL complete GST pull-down buffer and leave on the rotator at 4°C until required (see Note 4).
2. The lysates are sonicated with three 10 s pulses at 20% amplitude on ice and then centrifuged at 13,200 rpm ($16,100 \times g$) in a microfuge for 5 min at 4°C. The resulting supernatants (~200 μL) are incubated with 35 μL GST beads in a total of 1 mL GST pull-down buffer for 2 h at 4°C. Mix 15 μL supernatants with 15 μL Laemmli loading buffer (2× stock) and save as inputs.

3. After 2 h incubation, beads are collected by centrifugation at 10,000 rpm in a desktop microcentrifuge for 8–10 s at 4°C. The supernatant is removed and beads are washed four times with GST pull-down buffer for 5 min each at 4°C. Beads are collected after each wash by spinning down at 10,000 rpm in a microcentrifuge for 8–10 s at 4°C.
4. Add 40–50 μL Laemmli loading buffer (2× stock) to the beads and then heat the samples, including the beads, at 95°C for 5 min. After centrifugation at 13,200 rpm in a microcentrifuge for 30 s, 10–20 μL samples are loaded onto SDS–PAGE gels.

3.4. SDS–PAGE Gel

1. The Mini-PROTEIN 3 SDS–PAGE gel system is used but the following directions also apply to other SDS–PAGE gel systems. The glass plates should be washed with detergent after each use, rinsed with water, and allowed to air dry. Before each use, wipe the glass and spacer plates with 70% ethanol then let the ethanol evaporate.
2. Assemble the apparatus following the manufacturer's instruction. For each 1.5 mm 10% separating gel, mix 2.5 mL 40% acrylamide, 1.4 mL bis-acrylamide, 3.75 mL 1 M Tris (pH 8.7), 50 μL 20% SDS, 2.3 mL dH_2O, 50 μL freshly prepared 10% APS, and 5 μL TEMED.
3. Pour about 8 mL separating gel in between of the glass and leave about 2 cm space for stacking gel. Slowly overlay 1 mL butanol on top of the gel by pipetting and let the gel polymerize for 45 min or longer.
4. When the gel shows a clear layer separated from butanol, pour off the butanol, and rinse two times with distilled water. Prepare the 5% stacking gel by mixing 0.65 mL 40% acrylamide, 0.35 mL bis-acrylamide, 0.63 mL 1 M Tris (pH 6.8), 25 μL 20% SDS, 3.34 mL dH_2O, 25 μL fresh made 10% APS, and 12.5 μL TEMED.
5. Pour approximately 2 mL stacking gel on top of the separating gel; insert the comb gently to avoid any bubbles. Stacking gel should polymerize in about 15 min (see Note 5).
6. When the stacking gel is solidified, carefully remove the comb without disturbing each divider. Set up the gel running tank following the manufacturer's instruction and then load the samples (including the Kaleidoscope protein molecular weight markers) into the wells.
7. Run the gel at 50 V until the samples enter the separating gel and then increase the voltage to 100 V until favorable sample separation is achieved. Check the gel occasionally to ensure that target proteins are not running off the bottom of the gel. When the run is over, turn off the power and prepare to transfer the proteins to membranes as described below.

3.5. Western Blot to Identify the GW182–Ago2 Interaction

1. A mini Trans-Blot Electrophoretic Transfer Cell and wet transfer system are used. The proteins are transferred from the SDS–PAGE gel to nitrocellulose membranes electrophoretically.
2. Near the end of the SDS–PAGE run as described above, prepare two 3 mm filter papers and one nitrocellulose membrane having the same dimensions as the polyacrylamide gel. Also re-wet the sponges and membrane in the transfer buffer which is prepared by mixing 100 mL 10× transfer buffer, 200 mL methanol, and 700 mL water.
3. When the SDS–PAGE run is completed, disassemble the apparatus and remove the stacking gel carefully with a sharp scalpel or similar instrument. Take care not to tear the separating gel, which is then washed in pure water and briefly placed in the transfer buffer until the next step.
4. Set up the transfer unit "sandwich" by putting the clamp into a glass tray filled with transfer buffer. From the negative to positive charged side, place the components in the following order: sponge–filter paper–gel–membrane–filter paper–sponge white side of the clamps. Make sure that there are no air bubbles between the gel and the nitrocellulose membrane. Put the transfer unit into the transfer cell unit in the correct orientation following manufacturer's instructions (see Note 6).
5. Pour the transfer buffer in the transfer cell unit, put the lid on in the correct orientation and connect the wires from the transfer cell chamber to the power supply. The electrophoretic transfer is performed at 20–25 V overnight at 4°C.
6. When the transfer is complete, disassemble the transfer unit. If efficient transfer of proteins from the gel to the nitrocellulose membrane has been achieved, the pre-stained protein ladders should be visible on the membrane. Mark the loading orientation and the surface closest to the gel for future reference. To save reagents, unrelated portions of the membrane are often trimmed off with a razor blade or scissors prior to further processing.
7. For the blocking step, the nitrocellulose membrane is incubated in a blocking solution of 5% (w/v) nonfat dry milk in PBS-T for 1 h on a rocking platform.
8. Discard the blocking buffer and rinse the membrane in PBS-T for 5 min. In the meantime, prepare primary antibodies by performing 1:1,000 (may vary) dilutions of anti-GST, anti-FLAG, or anti-GFP antibody into blocking solution and apply onto the surface of the membrane. Incubate for 1 h and then wash in three changes of PBS-T for 10 min.

9. During the last wash, prepare secondary antibodies dilutions of horse radish peroxidase-conjugated anti-mouse Ig or anti-rabbit Ig into blocking solution (1:5,000 to 1:10,000) as appropriate for the species of primary antibodies and apply onto the surface of the membrane. Incubate for 1 h and then wash 3 times with PBS-T.
10. During the last wash, prepare the ECL solution according to the manufacturer's directions. Make sure to use separate pipettes for each solution. Apply the ECL solution to the membrane and incubate at room temperature for 5 min in the dark.
11. Decant the ECL solution completely to minimize background. Put the membrane between two pieces of parafilm or acetate sheets, tape it on a board or filter paper, and place into the exposure cassette (see Note 7).
12. Obtain several exposures using appropriate X-ray films for the optimal exposure. The images are scanned in 300dpi resolution and converted to digital format for presentation (Fig. 2).

3.6. Immuno-fluorescence to Verify Ago2–GW182 Colocalization

1. HeLa cells are cultured and seeded into 6-well plates with a sterile coverslip in each well as described above (see Subheading 3.2). Transfection has also been described (see Subheading 3.2) except using FLAG-tagged Ago2 or its PAZ/PIWI domain with or without GFP-tagged middle region of GW182 construct GW1Δ10 (see Note 8).
2. Twenty-four hours after the transfection, cells are rinsed briefly with PBS once and fixed in 3% paraformaldehyde at room temperature for 10 min.
3. Remove the paraformaldehyde and rinse twice with PBS. To permeablize the cells, flood the cover slips with 0.5% Triton X-100 and incubate at room temperature for 5 min.
4. Remove the permeablization solution and wash three times with PBS. Coverslips containing transfected cells are then incubated with anti-FLAG diluted 1:500 in PBS for 1 h at room temperature.
5. Remove the primary antibody and wash the coverslips three times with PBS for 5 min each and then incubate with anti-mouse Alexa Fluor568 diluted 1:400 in PBS for 1 h at room temperature in a light tight chamber.
6. Remove the secondary antibody and then wash the coverslips in three changes of PBS for 5 min each. The coverslips are then transferred to a microscope slide and overlaid with one drop of mounting medium containing DAPI while using absorbent tissue (i.e., Kimwipes) to remove the excess fluid from the edges of the coverslip.

7. Seal the coverslip with nail polish and allow to air dry for 10 min in the dark.
8. Fluorescence images are obtained using a Zeiss Axiovert 200M microscope (or suitable instrument). Figure 3 shows the interaction of GW182 and Ago2 and transfection of the GW182 fragment can recruit Ago2 to form the GW body foci.

3.7. Tethering Assay and Translational Repression Reporter

1. HEK 293 cells are obtained from ATCC and cultured in DMEM containing 10% fetal bovine serum and penicillin/streptomycin at 37°C under 5% CO_2. One day before transfection, cells are trypsinized and seeded at 1×10^5 cells/well into 24-well plates with 0.5 mL medium (see Note 9).
2. On the day of transfection, mix 1 ng Renilla 5BoxB plasmid, 0.1 μg Firefly Luciferase internal control, and 0.8 μg plasmid-encoding NHA-tagged fusion protein with 50 μL Opti-MEM transfection medium in tube A. Firefly and Renilla plasmids were adopted from Dr. Wiltold Filipowicz, Friedrich Miescher Institute, Basel, Switzerland. In the meantime, mix 2 μL Lipofectamine 2000 with 50 μL Opti-MEM transfection medium in tube B. Mix well by repeated pipetting and then incubate at room temperature for 5 min (see Note 10). Transfections are generally performed in triplicate.
3. Add 50 μL of the Lipofactamine-Opti-MEM tube B mixture to the plasmid-Opti-MEM tube A mixture. Mix well by repeated pipetting and incubate at room temperature for 20 min.
4. Add 100 μL each transfection mixture to wells and incubate for 48 h. There is no need to change culture medium at this stage (see Note 9).
5. After 48 h, remove the medium carefully. Dilute 5× passive lysis buffer stock from the Dual Luciferase Reporter Assay kit with water into 1× concentration and add 100 μL into each well.
6. Vigorously shake the 24-well plate on an orbital shaker for 15 min and then collect each lysate into 1.5 tubes.
7. Pipette 20 μL lysate into a 96-well Opaque cell culture plate and luciferase activities are read on a FLUOstar Omega microplate reader. The Renilla/Firefly relative reporter activities are normalized to NHA vector negative control as shown in Fig. 4.

3.8. SiRNA Knockdown GW182 in Tethering Assay

1. HeLa cells are seeded at a concentration of 0.5×10^5 cells/well into 24-well plates in 0.5 mL medium. The final concentration of siGENOME SMARTpool siRNAs for TNRC6A (GW182) and Ago2 used for transfection is 100 nM (see Subheading 3.7).

2. Twenty-four hours after siRNA transfection, the tethering assay triplicate transfections are performed as described above under Subheading 3.7 (see Note 11).
3. Harvest the cell lysate and perform the luciferase assay as described above under Subheading 3.7.
4. To monitor the efficiency of the siRNA knock down, parallel experimental groups are set up. The total RNA samples are harvested using Qiagen kit 48 h after siRNA transfection following the manufacturer's instructions.
5. Perform reverse transcription using High Capacity cDNA Reverse Transcription Kit following the manufacturer's instructions. The reaction is carried out by mixing 2 μL 10× RT buffer, 0.8 μL 25× dNTP mix (100 mM), 2 μL 10× RT random primers, 1 μL Multiscribe Reverse Transcriptase, 1 μL RNAase Inhibitor, 3.2 μL molecular grade water, and 10 μL diluted total RNA containing 1–2 μg RNA. Thermocycler conditions are set as 25°C for 10 min, 37°C for 60 min, 85°C or 5 s and then 4°C for short-term storage. The RT cDNA samples can be stored at –20°C for a few weeks or at –80°C for extended times.
6. Set up quantitative real-time PCR by mixing 5 μL FAST Universal PCR Master Mix (2×), 0.5 μL Target PCR primer with Taqman probe, 0.5 μL Control Primer (Human 18s rRNA, 20×) (see Note 12), 1 μL RT product, and 3 μL molecular grade water. The thermocycler conditions are set as 95°C for 20 s, 95°C for 1 s, and 60°C for 20 s, repeat steps 2 and 3 for 40 cycles.
7. The relative mRNA level of GW182 or Ago2 was measured using ΔΔCt method and normalized to untransfected controls (Fig. 5).

4. Notes

1. We use the two-step PCR procedure described in the Gateway cloning manual (Invitrogen) to generate truncated constructs that are subcloned into the Gateway cloning system. This avoids employing very long primers that can increase the risk of nonspecific amplification.
2. It is important to determine the cell density to optimize transfection efficiency. For plasmid DNA transfection using Lipofectamine 2000, cells are seeded to achieve 85–95% confluence on the day of transfection. By comparison, a ~40% cell confluence is desired for siRNA transfection.
3. Since serum is believed to interfere with the lipid–nucleotide complex formation, the Opti-MEM is a necessary component

of the transfection mix because it contains minimum amounts of serum. To minimize cell toxicity for many cell lines including HeLa, after incubating the transfection mixture with cells for 4–6 h, the Opti-MEM is removed and replaced with regular culture medium. Note that transfected cells are often vulnerable to antibiotics and therefore antibiotics should not be added to the medium. For the transfection of HEK 293 cells, which have a tendency to wash off the glass support, exposure to the transfection reagent is left in place until the cells are harvested.

4. Glutathione Sepharose 4B beads from GE Healthcare is a chromatography medium with affinity for GST and other glutathione binding proteins. Glutathione Sepharose 4B beads are supplied as a 60% slurry in a 20% ethanol preservative. It is important to wash the beads extensively to remove the ethanol before adding to the lysate for binding. Also note that the salt concentration is important for the binding stringency. Protein–protein interactions are more readily disrupted under high salt buffer conditions.
5. When making the two layers of SDS–PAGE gels, it is crucial to ensure that the pH of Tris–HCl is appropriate for the various gel layers. After pouring the stacking gel, insert the comb gently to avoid generating bubbles. If bubbles appear, remove the comb, add more gel and then use a pipette to remove the bubbles immediately and insert the comb again. Once the APS and TEMED are added, pour the gel as soon as possible since it will solidify rapidly.
6. Make sure the orientation of the transfer "sandwich" is correct otherwise the proteins will transfer to the transfer buffer rather than to the nitrocellulose membrane. It is also important to make sure that the transfer unit is assembled in the correct orientation and that the anode and cathode of the power supply are attached with the wires to the respective poles of the transfer apparatus. The transfer can be performed at room temperature, but it is preferable to run at 4°C with a pre-chilled ice block to avoid overheating of the buffer. We use a constant voltage of 25 V for overnight transfers, but a higher voltage can reduce the transfer time. Semi-dry transfers have also been used with some success.
7. Remove the ECL from membrane completely by absorbing the surface absorbent tissues. After putting the membrane between two pieces of parafilm or acetate sheet, wipe several times to squeeze out any excess ECL fluid. Residual ECL reagent can produce membranes with poor resolution and high backgrounds.
8. When seeding cells for immunofluorescence, either 6-well plates with a sterile coverslip in each well or slides with 8-well

chambers can be used. Eight-well chamber slides require less transfection and staining reagents though. One advantage of using coverslips in 6-well plates is that the surface area of coverslips is often larger and thus hold more cells then 8-well chambers. Note that 1, 2, 4, or 8-well configurations are available from Nunc Lab-Tek II Chamber Slide system (Thermo Fisher Scientific, Pittsburgh, PA). Avoid generating air bubbles when mounting slides by applying coverslips on at a slightly tilted angle to the slide.

9. Both HeLa and HEK 293 cells can be used to perform tethering assays. For many protocols, HEK 293 cells have higher transfection efficiency but tend to be less adherent and can be easily flushed away during fixation and washing procedures. Thus, when using 293 cells in the tethering transfection assays, the transfection reagent is left in the medium and cells are harvested after 48 h. Since two transfection steps are performed in siRNA knockdown tethering experiments, HeLa cells are used instead of 293 cells.
10. The ratio of the two luciferase reporters and the tethered construct plasmid needs to be determined by initial optimization as described in the Promega Dual – Glo Luciferase Assay system technical manual TM085 (http://www.promega.com/tbs/tm058/tm058.pdf).
11. To achieve high knockdown efficiency, cells need to be maintained in a rapidly proliferating phase. In general, cells that have been passaged continuously for more than 2 months should not be used.
12. The Applied Biosystems TaqMan Real-Time PCR assay includes human 18S rRNA as an internal control to monitor the variability in the sample loading process. The 18S rRNA assay is designed to detect the VIC fluorescent dye (Applied Biosystems, Foster City, CA) as an internal control, whereas targets such as GW182 or Ago2 assay uses the FAM dye (6-carboxyfluorescein).

Acknowledgments

This work was supported in part by the Canadian Institutes for Health Research Grant MOP-38034, National Institute of Health grant AI47859, and grants from the Lupus Research Institute and the Andrew J. Semesco Foundation. We thank Dr. Witold Filipowicz (Friedrich Miescher Institute for Biomedical Research, Basel, Switzerland) for providing tethering assay plasmids and Dr. Tom Hobman (University of Alberta, Edmonton, Canada) for providing hAgo2 cDNA. M.J.F. holds the Arthritis Research Society Chair.

References

1. Grosshans, H., and Filipowicz, W. (2008) Molecular biology: the expanding world of small RNAs. *Nature* **451**, 414–416.
2. Filipowicz, W., Bhattacharyya, S. N., and Sonenberg, N. (2008) Mechanisms of post-transcriptional regulation by microRNAs: are the answers in sight?. *Nat Rev Genet* **9**, 102–114.
3. Hutvagner, G., and Simard, M. J. (2008) Argonaute proteins: key players in RNA silencing. *Nat Rev Mol Cell Biol* **9**, 22–32.
4. Liu, J., Carmell, M. A., Rivas, F. V., Marsden, C. G., Thomson, J. M., Song, J. J., Hammond, S. M., Joshua-Tor, L., and Hannon, G. J. (2004) Argonaute2 is the catalytic engine of mammalian RNAi. *Science* **305**, 1437–1441.
5. Meister, G., Landthaler, M., Patkaniowska, A., Dorsett, Y., Teng, G., and Tuschl, T. (2004) Human Argonaute2 mediates RNA cleavage targeted by miRNAs and siRNAs. *Mol Cell* **15**, 185–197.
6. Eystathioy, T., Chan, E. K. L., Tenenbaum, S. A., Keene, J. D., Griffith, K., and Fritzler, M. J. (2002) A phosphorylated cytoplasmic autoantigen, GW182, associates with a unique population of human mRNAs within novel cytoplasmic speckles. *Mol Biol Cell* **13**, 1338–1351.
7. Jakymiw, A., Lian, S., Eystathioy, T., Li, S., Satoh, M., Hamel, J. C., Fritzler, M. J., and Chan, E. K. L. (2005) Disruption of GW bodies impairs mammalian RNA interference. *Nat Cell Biol* **7**, 1267–1274.
8. Jakymiw, A., Pauley, K. M., Li, S., Ikeda, K., Lian, S., Eystathioy, T., Satoh, M., Fritzler, M. J., and Chan, E. K. L. (2007) the role of GW/P-bodies in RNA processing and silencing. *J Cell Sci* **120**, 1317–1323.
9. Li, S., Lian, S. L., Moser, J. J., Fritzler, M. L., Fritzler, M. J., Satoh, M., and Chan, E. K. L. (2008) Identification of GW182 and its novel isoform TNGW1 as translational repressors in Ago2-mediated silencing. *J Cell Sci* **121**, 4134–4144.
10. Zipprich, J. T., Bhattacharyya, S., Mathys, H., and Filipowicz, W. (2009) Importance of the C-terminal domain of the human GW182 protein TNRC6C for translational repression. *RNA* **15**, 781–793.
11. Takimoto, K., Wakiyama, M., and Yokoyama, S. (2009) Mammalian GW182 contains multiple Argonaute-binding sites and functions in microRNA-mediated translational repression. *RNA* **15**, 1078–1089.
12. Lian, S. L., Li, S., Abadal, G. X., Pauley, B. A., Fritzler, M. J., and Chan, E. K. L. (2009) The C-terminal half of human Ago2 binds to multiple GW-rich regions of GW182 and requires GW182 to mediate silencing. *RNA* **15**, 804–813.
13. Till, S., Lejeune, E., Thermann, R., Bortfeld, M., Hothorn, M., Enderle, D., Heinrich, C., Hentze, M. W., and Ladurner, A. G. (2007) A conserved motif in Argonaute-interacting proteins mediates functional interactions through the Argonaute PIWI domain. *Nat Struct Mol Biol* **14**, 897–903.
14. El-Shami, M., Pontier, D., Lahmy, S., Braun, L., Picart, C., Vega, D., Hakimi, M. A., Jacobsen, S. E., Cooke, R., and Lagrange, T. (2007) Reiterated WG/GW motifs form functionally and evolutionarily conserved ARGONAUTE-binding platforms in RNAi-related components. *Genes Dev* **21**, 2539–2544.
15. Pillai, R. S., Artus, C. G., and Filipowicz, W. (2004) Tethering of human Ago proteins to mRNA mimics the miRNA-mediated repression of protein synthesis. *RNA* **10**, 1518–1525.
16. Liu, J., Rivas, F. V., Wohlschlegel, J., Yates, J. R., 3rd, Parker, R., and Hannon, G. J. (2005) A role for the P-body component GW182 in microRNA function. *Nat Cell Biol* **7**, 1261–1266.

Chapter 5

Continuous Density Gradients to Study Argonaute and GW182 Complexes Associated with the Endocytic Pathway

Derrick Gibbings

Abstract

Most complexes involved in RNA silencing were thought to be concentrated in cytoplasmic sites called P-bodies in the absence of stress. Accumulating evidence suggests that distinct cellular organelles or sites may be involved in the maturation of RNA-induced silencing complexes (RISC), decapping and deadenylation of miRNA-repressed mRNA, transport of translationally repressed mRNA, and disassembly of RISC complexes. Significant fractions of proteins essential for RNA silencing associate with membranes in general (GW182, AGO, and DICER), or more specifically with endoplasmic reticulum and Golgi (AGO), or endosomes and multivesicular bodies (AGO, GW182). In contrast, mRNA decapping and decay occur mainly in the cytoplasm. Continuous density gradients capable of partitioning these cellular compartments are valuable tools in efforts to decipher the complexes, trafficking and regulation of RISC throughout its biogenesis, action and turnover.

Key words: Endosome, Multivesicular Body, Lysosome, Density Gradient, GW182, Argonaute, MicroRNA, mRNA localization, Endoplasmic reticulum, Silencing

1. Introduction

Proteins involved in mediating RNA silencing, such as argonaute (AGO) proteins, are concentrated in several subtypes of foci visible by confocal microscopy including mRNA transport granules, stress granules, P-bodies, and GW-bodies. The quintessential foci regrouping RNA-silencing proteins are P-bodies, whose defining markers are proteins involved in removing the 7-methyl-guanosine cap (decapping) or the 3′ poly-A tail (deadenylation) from mRNA (1). Proteins within P-bodies mediate RNA-decapping and deadenylation instigated by several RNA quality control mechanisms,

Tom C. Hobman and Thomas F. Duchaine (eds.), *Argonaute Proteins: Methods and Protocols*,
Methods in Molecular Biology, vol. 725, DOI 10.1007/978-1-61779-046-1_5, © Springer Science+Business Media, LLC 2011

including microRNA (miRNA)-mediated silencing (2) and nonsense-mediated decay (3).

A second subset of cellular AGO-containing foci, called GW-bodies, is similar to P-bodies (4–6). However, GW-bodies are distinguished by their enrichment in GW182, their relative paucity of DCP1A, and their association with endosomes and multivesicular bodies (5). GW-bodies, containing all essential components of RNA-induced silencing complexes (RISC), can be separated from P-bodies using continuous density gradients (5). Since P-bodies do not contain PABP or most ribosomal proteins (7), RISC complexes may be trafficked to P-bodies after translational inhibition of target mRNA. Translational inactivation of target mRNA may be effected in P-bodies by decapping and shortening of the 3′ poly(A)-tail. It has been proposed that RISC complex turnover, catalyzed by the removal of GW182 into multivesicular bodies, occurs at GW-bodies (5).

Since the first discovery of AGO and DICER proteins, cytoplasmic and membrane-associated fractions of these proteins have been described in biochemical detail (8, 9). Pools of AGO may associate with endoplasmic reticulum, Golgi (8), and several parts of the endolysosomal system (5, 10, 11). Proteins of the Biogenesis of Lysosome-related Organelles Complex-3 (BLOC-3), which is involved in trafficking in the endolysosomal system, are required for the formation of active RISC complexes (10). Specifically, hps1 and hsp4 are required for the disassociation of miRNA passenger strand from a complex containing DICER, AGO, and miRNA (10). Moreover, a significant fraction of DICER, the RNAse III enzyme required for the generation of mature ~22 nucleotide miRNAs from ~70 nucleotide pre-miRNAs, is membrane associated (9). This suggests that maturation (10) of active RISC may occur in membranous compartments including the endolysosomal system.

Granules involved in localizing mRNA for site-specific control of gene expression are known to transiently associate with AGO and other components of P-bodies (12). Unique RISC complexes may allow transient (13) translational repression of mRNA during transport (14). Some mRNA transport granules associate with cortical endoplasmic reticulum (15) or endosomes (16) during transport to their final destination.

Since mRNA transport granules, GW-bodies, and RISC maturation complexes may associate with distinct membranous compartments, while P-bodies are believed to be largely free of membrane attachments, continuous density gradients may be useful to distinguish complexes involved in various stages of the maturation, action, and disassembly or degradation of RISC.

RNA silencing can occur in the absence of visible P-bodies (17), suggesting that cellular aggregates too small for visualization by confocal microscopy are sufficient to mediate RNA silencing. For this reason, methods other than confocal microscopy are needed to separate and study molecular complexes involved in the formation, action, and disassembly of miRNA-mediated silencing complexes.

The method described in this chapter allows separation of endosomes, multivesicular bodies, endoplasmic reticulum, and cytoplasm on an iodixanol continuous density gradient (Fig. 1). As such, the method can be used to separate GW-bodies from P-bodies (Fig. 1) (5) and potentially mRNA transport granules associated with endoplasmic reticulum (15). The protocol is adapted to study the miRNA-dependent targeting of mRNA to

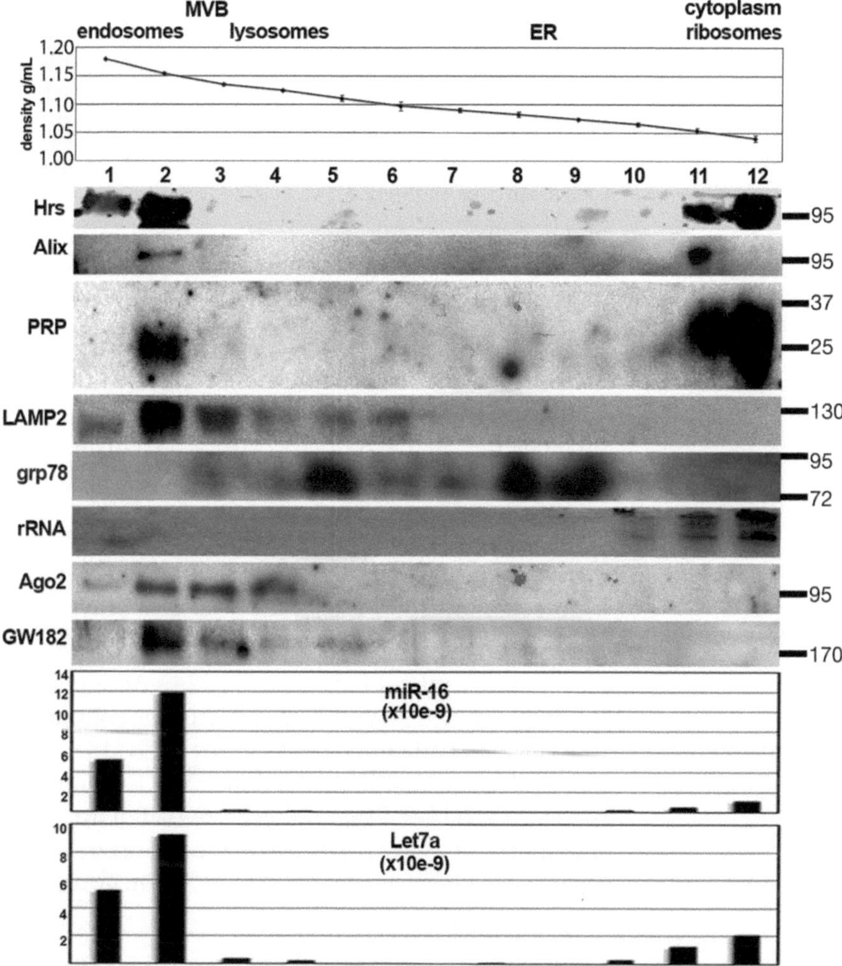

Fig. 1. GW182, AGO2, and miRNA partition in continuous density gradients with endosomes and multivesicular bodies. A 5–30% OptiPrep™ gradient was collected in 12 fractions and analyzed by western blot, ethidium bromide stained 1% agarose gel (rRNA), or qRT-PCR (miR-16, let-7a). Refractive index was used to calculate density (*top*) among three gradients (*error bars* denote standard error of the mean). Typical results for one such gradient are depicted. Reproduced with permission from ref. 5.

cytoplasmic and endolysosomal fractions of the gradients (Fig. 1). On the first day of the method, iodixanol continuous density gradients are prepared and cells are transfected with plasmids expressing miRNA-targeted mRNA reporters. On the second day, cells are mechanically lysed, loaded on the gradient and centrifugation is commenced. On the third day, at the termination of centrifugation, fractions are collected for further analysis by western blot, qRT-PCR, or other methods.

2. Materials

Double-distilled H_2O or MilliQ-filtered demineralized water that is RNAse-free (see Note 1) should be used at all steps. All reagents can be stored at room temperature unless otherwise indicated.

2.1. Cell Culture and Transfection

1. HeLa cells (American Type Culture Collection, CCL2) or Mono-Mac 6 (German Collection of Microorganisms and Cell Cultures (DSMZ), ACC 124).
2. HeLa cells are cultured in Dulbeco's modified essential medium (DMEM, Invitrogen) with GlutaMAX™, 4.5 g/L glucose supplemented with 10% fetal bovine serum (FBS) and 1% penicillin–streptomycin (media stored at 4°C).

   Mono-Mac-6 cells are cultured in RPMI 1640 (Invitrogen) with GlutaMAX™, 4.5 g/L glucose and supplemented with OPI supplement (Sigma–Aldrich), 10% FBS (Invitrogen) and 1% penicillin–streptomycin (media stored at 4°C).
3. Small cell scrapers.
4. Sterile phosphate-buffered saline solution (DPBS, Invitrogen).
5. Lipofectamine 2000 transfection reagent (Invitrogen).
6. Let-7a miRNA reporter system (18), consisting of three plasmids (a) pRL-TK-Let7A (11324, Addgene) expressing Renilla luciferase with two target sites for endogenous Let-7a miRNA in the 3′UTR, (b) pRL-TK-Let7B (11325, Addgene), identical to pRL-TK-Let7A but with two nucleotide mismatches in the Let-7a target seed site that dramatically reduces the ability of Let-7a to silence the reporter, and (c) pGL3 expressing Firefly luciferase for the normalization of Renilla luciferase levels.
7. Opti-MEM reduced serum media (Invitrogen).

2.2. Preparation of Continuous Iodixanol Density Gradient

1. 60% w/v Iodixanol solution (undiluted OptiPrep™, Sigma–Aldrich). Store at 4°C. Manipulate in sterile conditions (prone to biotic contamination).

2. Peristaltic pump.
3. Glass micropipettes (50 or 100 μL).
4. Magnetic stirring bar capable of spinning at bottom of gradient mixer.
5. Gradient mixer (e.g., Amersham Hoefer SG-50).
6. Rack for ultracentrifuge tubes that securely holds tubes in the vertical position.
7. Caps for ultracentrifuge tubes capable of firmly holding micropipettes at the bottom of tubes during gradient formation and fraction collection.
8. Ultracentrifuge tubes (preferentially 4–6 mL (e.g., Beckman Ultra-Clear™ 344062) to avoid concentration of fractions after gradient separation).
9. One hundred milliliters of the following stock solutions (stored at 4°C): (a) 1 M HEPES, (b) 1 M KCl, (c) 1 M $MgCl_2$, (d) 1 M $CaCl_2$, (e) 100 mM ethylene glycol tetraacetic acid, free acid (EGTA).
10. Working Solutions (store at 4°C):
    (a) OptiPrep™ diluent: Mix together the following stock solutions: 15 mL HEPES, 23.5 mL KCl, 1.2 mL $MgCl_2$, 2.5 mL $CaCl_2$, and 30 mL EGTA. Adjust to pH 7.0 with 1 M KOH and make up to 100 mL (Final concentration: 235 mM KCl, 12 mM $MgCl_2$, 25 mM $CaCl_2$, 30 mM EGTA, and 150 mM HEPES–NaOH pH 7.0).
    (b) 40% OptiPrep™: mix two parts supplied OptiPrep™ (60%) with one part OptiPrep™ diluent (above).
    (c) Working Solution diluent: Mix together the following stock solutions: 5 mL HEPES, 7.8 mL KCl, 0.4 mL $MgCl_2$, 0.84 mL $CaCl_2$, and 10 mL EGTA. Adjust to pH 7.0 with 1 M KOH and make up to 100 mL (Final concentration: 78 mM KCl, 4 mM $MgCl_2$, 8.4 mM $CaCl_2$, 10 mM EGTA, and 50 mM HEPES–NaOH pH 7.0).

2.3. Cell Homogenization

1. Cell lysis buffer: 0.25 M sucrose, 78 mM KCl, 4 mM $MgCl_2$, 8.4 mM $CaCl_2$, 10 mM EGTA, and 50 mM HEPES–NaOH pH 7.0. Add 8.5 g of sucrose to 75 mL of working solution diluent (see Subheading 2.2, item 10c), adjust pH to 7.0 with 1 M KOH and adjust volume to 100 mL. Store at 4°C. To 10 mL cell lysis buffer, add one tablet protease inhibitor, proteasome and ribonuclease inhibitors (see below) immediately before use. Dissolution of protease inhibitor requires vigorous, smooth vortexing.

(a) Protease inhibitor: Roche Complete Mini Protease Inhibitor without EDTA. Store at 4°C.

(b) Proteasome inhibitor MG132 (Calbiochem), add at 50 μM/mL gradient volume. Store at –20°C.

(c) Ribonuclease inhibitors: vanadyl ribonucleoside use at 1–2 mM/mL of gradient volume (Sigma–Aldrich), RNAseOut, 2 μL/mL gradient volume (Invitrogen). Store at –20°C. EGTA may cause disassociation of vanadyl ribonucleoside complexes and its concentration should, therefore, be reduced to submolar quantities (e.g., 1 mM) of vanadyl ribonucleoside when the latter is used.

2. Dounce homogenizer with an appropriately small interstitial space to rupture cells (e.g., Kimble Kontes 2 mL, clearance B).

3. Syringes (1 mL) and needles (25–27 Gauge).

2.4. Loading of Cell Lysates, Centrifugation, and Collection of Gradient Fractions

1. Ultracentrifuge and swinging bucket rotor (e.g., Beckman SW60 Ti for Beckman Ultra-Clear Tubes 344062).

2. BioLogic DuoFlow Chromatography System and Biofrac cell fraction collector (Bio-Rad) or similar apparatus. Alternative methods of fraction collection are possible that require only a peristaltic pump or syringe. Details for alternative methods of fraction collection can be found on the Axis-Shield website (address mentioned below).

2.5. Precipitation of Proteins from Gradient Fractions (Optional)

1. Acetone (pre-cool acetone to –20°C).

2. 100% (w/v) Trichloroacetic acid (TCA) solution. Manipulate in fumehood.

2.6. Western Blot of Gradient Fractions

3. Antibodies: Anti-GW182 anti-serum 18033 (human autoimmune serum, obtained from Marvin Fritzler, University of Calgary), anti-GW182 (clone 4B6, mouse IgG1, ABCAM), anti-AGO2 (rat mAb 11A9 Sigma–Aldrich), anti-DCP1A (rabbit polyclonal, obtained from Jens Lykke-Andersen, University of Colorado), anti-HRS (mouse IgG1, ab56468, ABCAM), and anti-ALIX (mouse IgG1, clone 2H12, Santa Cruz).

2.7. Analysis of RNA from Density Gradient Fractions

1. Trizol LS reagent (Invitrogen). This is a concentrated version of standard Trizol reagents designed for isolating RNA from solution rather than cell pellets. It contains phenol and other toxic compounds and therefore should be used in a fumehood with complete protection. Store at 4°C.

3. Methods

Timeline:

Day 1: Seed cells into culture flasks (40 min).

Day 2: Transfect cells and prepare continuous density gradients (2.5 h).

Day 3: Homogenize cells and start centrifugation (2.5 h).

Day 4: Collect fractions (2 h), isolate RNA (2 h), and precipitate proteins if necessary (4 h).

Day 5: SDS–PAGE and transfer to PVDF membrane (6 h).

Day 6: Immunoblot (5 h).

Cell lysis methods and/or buffers may need to be adapted for each cell type. Variations of iodixanol density gradients can be used to optimally separate other organelles (as well as the ones described here). For further detail on many aspects and variations of iodixanol continuous density gradients (gradient generation, centrifugation, fraction collection, cell type-specific lysis protocols, and protocols adapted for other organelles), the instructive website maintained by Axis-Shield, the maker of OptiPrep™ is a valuable resource (http://www.axis-shield-density-gradient-media.com/organelleindexes.htm). The following protocol was adapted from Axis-Shield Application Sheet S44 and the work of Sheff et al. (19) for the separation of GW-bodies and P-bodies.

3.1. Cell Culture and Transfection

For immunoblot detection of proteins in a 4 mL gradient with high-quality antibodies such as those described here, or qRT-PCR detection of abundant miRNA or mRNA, abundant miRNA or mRNA, it is sufficient to use one 90% confluent 25 cm^2 flask of Hela cells.

1. Cells are seeded 1-day before transfection and preparation of iodixanol gradients. HeLa cells are seeded (400–600 $\times$ 10^3 cells) in 10 mL complete DMEM media in enough 25 cm^2 flasks to satisfy experimental conditions (e.g., one for miRNA reporter and one for control reporter). Mono-Mac-6 cells are seeded at 500 $\times$ 10^3/mL in 20 mL in 75 cm^2 flasks in complete RPMI 1640 media.
2. HeLa cells can be transfected with Lipofectamine 2000 following the supplier's instructions. Mono-Mac-6 cells can be transfected with approximately 50% efficiency using a Lonza Nucleofector II electroporator (1 μg of DNA to transfect 5 $\times$ 10^6 cells per cuvette) in Solution V (Lonza) using program V-001. Transfect one batch of HeLa or MonoMac-6 cells with eight parts pRL-TK-Let-7A and one part pGL3.

Transfect a second batch of cells with eight parts pRL-TK-Let-7B and one part pGL3.

3. Remove cell culture media and replace with fresh, pre-warmed media 4 h after transfection.

3.2. Preparation of Continuous Density Gradients

Gradients can be prepared the day prior to the preparation of cell lysate and centrifugation. In this case, they are stored, away from sources of vibration, at 4°C to avoid the sedimentation of iodixanol. Two solutions are prepared: one each at the most and least dense iodixanol concentration in the final gradient. A gradient mixer is used to gradually and linearly mix the two solutions as a peristaltic pump draws off the mixed solution of gradually increasing density and delivers it to the base of the ultracentrifuge tube.

1. Set-up tubing connections from the gradient-mixer to the peristaltic pump and from the peristaltic pump to the ultracentrifuge tube using standard plastic tubing tipped with a glass micropipette inserted through the cap of the ultracentrifuge tube and terminating within millimeters of the bottom of the tube. Gradients can be formed in two tubes simultaneously if a Y-shaped branching adaptor is added to the tubing before the peristaltic pump. In this case, gradients may be more reproducible; however, it is important to ensure that the flow rates of the two branches are very similar. Flow rates of each branch can be independently adjusted by changing the pressure on each tube track of the peristaltic pump (often done with a screw mechanism).
2. Rinse system by filling gradient mixer with H_2O and allowing it to flow through system into a clean beaker. During this time, adjust the flow rate of the peristaltic pump and examine the system for leaks before attempting to form iodixanol gradients. Flow rates of ~1 mL/min work well.
3. To improve accurate filling of tubes, tubes can be filled with 3.75 mL of H_2O to mark the liquid level at which gradient loading should be stopped (to allow loading of 250 µL of cell homogenate in a 4 mL tube).
4. For two 4 mL ultracentrifuge tubes, first prepare a 40% iodixanol solution by mixing commercially prepared 60% iodixanol (undiluted OptiPrep™) with iodixanol diluent. Prepare 5 mL of 30% iodixanol and 5 mL of 5% iodixanol solutions. To make 5 mL of 30% iodixanol, mix 3.75 mL of 40% iodixanol with 1.25 mL working solution diluent. Mix 0.625 mL 40% iodixanol and 4.375 mL working solution diluent to make 5% iodixanol solution.
5. Close the valve linking the chambers of the gradient mixer, and the valve exiting the gradient mixer.

6. Remove any remaining water left from rinsing with a Pasteur pipette.
7. Add identical RNAse-free stir bars (wash well, autoclave, or use other methods) to each chamber of gradient mixer.
8. For each ultracentrifuge tube, add 2.0 mL 30% iodixanol solution to the gradient mixer chamber distant from the peristaltic pump and ultracentrifuge tube.
9. Quickly open and close the valve connecting the distant and proximal chambers of the gradient mixer to allow the valve between the two to fill with solution. This eliminates air bubbles in the valve that may block liquid flow between the chambers during gradient mixing.
10. Remove excess dense solution from the proximal chamber and return it to the distal chamber.
11. Fill the proximal chamber with 1.8 mL 5% iodixanol per ultracentrifuge tube.
12. Start stir bars mixing.
13. In quick sequence, start the peristaltic pump, open the exit valve from gradient mixer and open the valve between the distal and proximal chambers.
14. Stop pump when liquid in tube reaches 3.75 mL mark on ultracentrifuge tubes. Remember that you will add 250 μL of cell homogenate on top of the gradient tomorrow.
15. Gently remove the micropipette in a smooth but slow vertical motion.
16. Seal ultracentrifuge tube with parafilm and store at 4°C overnight.

3.3. Cell Homogenization

To minimize protein and RNA degradation, all steps should be performed with pre-cooled buffers on ice whenever possible until the loading of SDS–PAGE gels.

1. Wash cells three times in PBS (4°C) with gentle rocking (HeLa) or by centrifugation (Mono-Mac-6 cells) (see Note 2).
2. Prepare cell homogenization buffer (see Note 3).
3. Remove all supernatant and add cold 250 μL cell homogenization buffer. Scrape HeLa cells from flask surface. Two to three passages of scraper over surface is usually sufficient. Gently resuspend Mono-Mac-6 cells in homogenization buffer.
4. Transfer cells into Dounce homogenizer. Retain a small aliquot of cell suspension to evaluate the efficiency of subsequent cell disruption.
5. At regular intervals, monitor cell lysis using a light microscope and continue homogenization until approximately 80–90% of cells are lysed. Subject cells to approximately 30

(HeLa) to 70 (Mono-Mac-6) up and down strokes of the Dounce homogenizer (see Note 4).

6. Once cell disruption is deemed adequate, centrifuge cell homogenate once at 400 × *g* (5 min) and once at 1,000 × *g* (5 min) to remove remaining intact cells, nuclei, and other large debris (see Note 5).
7. Cell homogenate is now ready for loading on iodixanol gradients.

3.4. Loading of Cell Homogenate on Gradients and Centrifugation

1. Precool rotor, buckets, and ultracentrifuge (with vacuum on).
2. Balance ultracentrifuge tubes by placing them upright in a narrow beaker or graduated cylinder on the scale. From heavy tubes, remove liquid very slowly using a 200 μL pipette tip held at an angle against the side of the tube at the upper edge of the meniscus. Place pipetting elbow on bench and brace pipetting arm with other hand to ensure minimal disruption of the gradient.
3. Add equal volumes of cell homogenate to each ultracentrifuge tube by letting fluid gently run a minimal distance down the side of the tube onto the top of the gradient. You should be able to see a layer form on top of the gradient.
4. Gently slide tubes into rotor buckets and hand tighten the lid.
5. Avoid unnecessary movement in transporting gradients.
6. Centrifuge at 90,000–100,000 × *g* for 18–20 h (see Note 6).

3.5. Fraction Collection

The described protocol employs a BioRad BioLogic DuoFlow chromatography system in-line with a BioRad BioFrac fraction collector. If similar systems are not available to automate fraction collection, it can be done manually by inverting the peristaltic pump system used for gradient formation or by piercing the bottom of the ultracentrifuge tube with a syringe tip.

1. Lids of labeled microcentrifuge tubes to be used for fraction collection are removed with scissors. At least 12 fractions (and as many as 30) can be collected to allow maximal resolution of organelles.
2. Flow rate of the peristaltic pump of the fraction collector is adjusted to a rate of approximately 0.2–0.5 mL/min using H_2O, then the system is emptied of H_2O and turned off.
3. It is desirable to monitor the continuity of the density gradient using automated collection of conductivity profiles. In theory, any disruption of continuous gradients before centrifugation is corrected during centrifugation. However, in practice, this is not strictly the case. Vanadyl ribonucleoside is believed to interfere with the collection of absorbance measures at 254, 260, and 280 nm.
4. Ultracentrifuge tubes are placed in rack. Caps are placed on tubes as for the preparation of gradients. A glass micropipette

is connected in-line to the peristaltic pump and fraction collector. The glass micropipette is inserted to the bottom of the gradient in a smooth slow motion (through the center of the gradient) to minimize gradient disruption.

5. The peristaltic pump, then the fraction collector is started. Stop the peristaltic pump after no more drops are observed emitting into tubes.
6. Verify protein concentration using microplate scale BioRad DC protein assay that is compatible with components of the described density gradients.
7. Twenty or 30 μL of each fraction can be used to evaluate the refraction index (see Note 7) and confirm the linearity of the gradient along with conductivity traces.

3.6. Precipitation of Proteins from Gradient Fractions (Optional)

If the protein concentration is less than 300 μg/mL in most fractions, it may be necessary to concentrate proteins by precipitation prior to gel loading.

1. To 1 volume of gradient fraction, add 8 volumes of acetone (–20°C) and 1 volume of TCA solution.
2. Mix well and incubate for 1 h at –20°C.
3. Centrifuge 15 min, 16,000 × *g* at 4°C. A substantial white pellet should be visible.
4. Remove supernatant, add 1 mL acetone (pre-cooled to –20°C), pipette or vortex lightly and centrifuge 15 min, 16,000 × *g* at 4°C.
5. Repeat step 4. A second acetone wash is highly recommended to further eliminate salts precipitated by TCA solution.
6. Remove supernatant and let acetone evaporate from opened tubes in the fumehood (10 min).
7. Resuspend pellets in 30–40 μL 4× Laemmli buffer with heating (up to 60°C) and shaking or intermittent vortexing for up to 1 h. Fractions may be viscous and are best loaded when still warm using micropipette tips precut with scissors to widen their opening.

3.7. SDS–PAGE and Western Blot

Optimized procedures for western blots with commonly used antibodies to detect AGO, GW182, and DCP1A proteins are described.

1. SDS–PAGE analysis of fractions on 10 or 12% Tris–glycine buffered gels can be performed using standard protocols.
2. Proteins are transferred to PVDF membranes (Immobilon-P, Millipore) using a wet transfer system (1.5 h, 100 V, 4°C) in a 1× Tris–glycine solution containing 10% EtOH.
3. Membranes can be cut into horizontal strips to allow blotting of two to three antibodies per membrane. For example,

anti-GW182 antiserum (18033), anti-AGO2 antibodies (11A9), and anti-DCP1A antibodies can be blotted by cutting a single membrane into three strips (cut at 130 and 75 kDa). Membranes can be accurately cut by aligning with molecular weight markers on both ends of gels and/or with the vertical traces of lanes observable after transfer.

4. After cutting, membranes are rinsed three times in Tris-buffered saline and blocked using 5% milk in TBS.
5. Membranes are incubated with antibodies overnight (4°C) at the following concentrations: anti-GW182 [18033, 1/1,000, bands expected at 130 (potentially Ge-1 (20)) and 180 kDa [GW182]], anti-AGO2 (clone 11A9, 1/1000, band expected at 95 kDa), anti-DCP1A (rabbit polyclonal 1/5,000, bands expected at 60–72 kDa), anti-HRS (1/1,000, band expected at 90 kDa), and anti-ALIX (1/1,000, band expected at 95 kDa). Representative results using MonoMac-6 cells are shown in Fig. 1.

3.8. Analysis of RNA from Density Gradient Fractions

1. Trizol LS is allowed to warm to room temperature. Two hundred and fifty microliter of each fraction is mixed by pipetting with 750 μL of Trizol LS.
2. RNA isolation is performed according to manufacturer's instructions. Purified RNA containing miRNA and mRNA is treated with DNAse before qRT-PCR. (Representative results are shown in Fig. 1.)

3.9. Notes

1. We routinely use MilliQ water, hand-packed and subsequently autoclaved pipette tips and glassware. In our experience, more extreme measures to avoid or eliminate RNAse contamination such as the use of filtered pipette tips guaranteed to be RNAse-free, DEPC-treatment, or autoclaving of water and buffers (which are strategies of debatable efficacy) are unnecessary.
2. FBS contains substantial amounts of bovine exosomes (21) that may contain significant quantities of GW182 (our unpublished observations). Moreover, exosomes released by cultured cells may be loosely attached to the cell surface. Accordingly, washing cells is necessary to reduce these sources of GW182 that may otherwise complicate interpretations of GW182 and miRNA distribution in subcellular fractions.
3. Inorganic salts reduce ionic interactions and membrane aggregation. Divalent cations stabilize membranes and reduce nuclear breakage. Cell homogenization buffer is roughly iso-osmotic with the gradient. The cell homogenization buffer used here does not contain EDTA (either in the protease inhibitor or the buffer), and the distinct chelation properties of the low dose of EGTA is not believed to remove ribosomes

from the endoplasmic reticulum. It may be preferable to eliminate all divalent cation chelators from homogenization buffers since AGO and DICER require Mg^{+2} to cleave RNA (22). Removing divalent cations from AGO and DICER may disrupt not only their function but also key structural domains, thereby modulating their interactions with RNA or proteins.

4. Maintain equal depth and speed of strokes between samples. Differences in the severity of cell disruption can cause differences in integrity of organelles and vesicle size of tubular structures, leading to distinct fractionation profiles. Homogenization of each cell type may require significant optimization. Examples of homogenization protocols for many cell types can be found in the literature and among the curated methods references available on the website of Axis-Shield density gradients, the maker of OptiPrep™. Immune cells, such as Mono-Mac-6, are renowned for their resistance to similar homogenization methods. In most cases, it is necessary to pass the Mono-Mac-6 cell homogenate through a 1 mL syringe equipped with a 25–27 Guage needle 10–15 times to achieve homogenization.

5. Many organelles are depleted by centrifugation at 12,000–16,000 × *g* in standard preparations of cell lysates. Interestingly in this regard, "P-bodies" or a large fraction of GW182, AGO, and other RNA silencing proteins are eliminated by these centrifugation steps (5, 23, 24). Pelleting of these RNA silencing proteins may be due to their association with large cytoplasmic aggregates or with organelles. Accordingly, the inclusion of this centrifugation step may, in fact, bias the type of RNA silencing complexes that are isolated.

6. Centrifugation times and speeds can be adapted to achieve equivalent separation using rotor calculators (e.g., http://www.beckmancoulter.com/resourcecenter/labresources/centrifuges/rotorcalc.asp).Minimized acceleration and deceleration is necessary, in theory, only for fixed angle rotors but is probably good practice. Slow and long centrifugation, as suggested in this method, favors separation of organelles based on density, while short and fast centrifugation distorts this separation because organelle shape and size affects partitioning. Excessive speed may also cause sedimentation of iodixanol and distortion of gradients.

7. Refraction indexes are measured with a refractometer. Accuracy of the refractometer can be checked against known substances (e.g., refractive index of H_2O 1.3330, ethanol 1.362). Note that the molecular properties of iodixanol can cause organelles to fractionate at densities distinct from their fractionation on a sucrose gradient.

References

1. Chen CY, Zheng D, Xia Z, Shyu AB. (2009) Ago-TNRC6 triggers microRNA-mediated decay by promoting two deadenylation steps. Nat Struct Mol Biol; 16(11):1160–6.
2. Sen, GL, Blau HM. (2005) Argonaute 2/ RISC resides in sites of mammalian mRNA decay known as cytoplasmic bodies. Nat Cell Biol; 7(6):633–6.
3. Sheth U, Parker R. (2006) Targeting of aberrant mRNAs to cytoplasmic processing bodies. Cell; 125(6):1095–109.
4. Vasudevan S, Tong Y, Steitz JA. (2007) Switching from repression to activation: microRNAs can up-regulate translation. Science; 318(5858):1931–4.
5. Gibbings DJ, Ciaudo C, Erhardt M, Voinnet O. (2009) Multivesicular bodies associate with components of miRNA effector complexes and modulate miRNA activity. Nat Cell Biol; 11(9):1143–9.
6. Moser JJ, Eystathioy T, Chan EK, Fritzler MJ. (2007) Markers of mRNA stabilization and degradation, and RNAi within astrocytoma GW bodies. J Neurosci Res;85(16):3619–31.
7. Kedersha N, Stoecklin G, Ayodele M, et al. (2005) Stress granules and processing bodies are dynamically linked sites of mRNP remodeling. J Cell Biol; 169(6):871–84.
8. Cikaluk DE, Tahbaz N, Hendricks LC, et al. (1999) GERp95, a membrane-associated protein that belongs to a family of proteins involved in stem cell differentiation. Mol Biol Cell; 10(10):3357–72.
9. Tahbaz N, Kolb FA, Zhang H, Jaronczyk K, Filipowicz W, Hobman TC. (2004) Characterization of the interactions between mammalian PAZ PIWI domain proteins and Dicer. EMBO Rep; 5(2):189–94.
10. Lee YS, Pressman S, Andress AP, et al. (2009) Silencing by small RNAs is linked to endosomal trafficking. Nat Cell Biol; 11(9):1150–6.
11. Haraguchi CM, Mabuchi T, Hirata S, et al. (2005) Chromatoid bodies: aggresome-like characteristics and degradation sites for organelles of spermiogenic cells. J Histochem Cytochem; 53(4):455–65.
12. Barbee SA, Estes PS, Cziko AM, et al. (2006) Staufen- and FMRP-containing neuronal RNPs are structurally and functionally related to somatic P bodies. Neuron; 52(6):997–1009.
13. Bhattacharyya SN, Habermacher R, Martine U, Closs EI, Filipowicz W. (2006) Stress-induced reversal of microRNA repression and mRNA P-body localization in human cells. Cold Spring Harb Symp Quant Biol; 71:513–21.
14. Schratt G. (2009) microRNAs at the synapse. Nat Rev Neurosci.
15. Aronov S, Gelin-Licht R, Zipor G, Haim L, Safran E, Gerst JE. (2007) mRNAs encoding polarity and exocytosis factors are cotransported with the cortical endoplasmic reticulum to the incipient bud in Saccharomyces cerevisiae. Mol Cell Biol; 27(9):3441–55.
16. Tanaka T, Nakamura A. (2008) The endocytic pathway acts downstream of Oskar in Drosophila germ plasm assembly. Development; 135(6):1107–17.
17. Eulalio A, Behm-Ansmant I, Schweizer D, Izaurralde E. (2007) P-body formation is a consequence, not the cause, of RNA-mediated gene silencing. Mol Cell Biol; 27(11): 3970–81.
18. Doench JG, Sharp PA. (2004) Specificity of microRNA target selection in translational repression. Genes Dev; 18(5):504–11.
19. Sheff DR, Daro EA, Hull M, Mellman I. (1999) The receptor recycling pathway contains two distinct populations of early endosomes with different sorting functions. J Cell Biol; 145(1):123–39.
20. Bloch DB, Gulick T, Bloch KD, Yang WH. (2006) Processing body autoantibodies reconsidered. Rna; 12(5):707–9.
21. Thery C, Amigorena S, Raposo G, Clayton A. (2006) Isolation and characterization of exosomes from cell culture supernatants and biological fluids. Curr Protoc Cell Biol;Chapter 3:Unit 3 22.
22. Wang Y, Juranek S, Li H, et al. (2009) Nucleation, propagation and cleavage of target RNAs in Ago silencing complexes. Nature; 461(7265):754–61.
23. Lian SL, Li S, Abadal GX, Pauley BA, Fritzler MJ, Chan EK. (2009) The C-terminal half of human Ago2 binds to multiple GW-rich regions of GW182 and requires GW182 to mediate silencing. Rna; 15(5):804–13.
24. Sheth U, Parker R. (2003) Decapping and decay of messenger RNA occur in cytoplasmic processing bodies. Science; 300(5620):805–8.

Chapter 6

In Vitro RISC Cleavage Assay

Julia Stoehr and Gunter Meister

Abstract

RNA interference (RNAi) is a widely used tool for the analysis of gene expression. In this process, short-interfering RNAs (siRNAs) guide the RNA-induced silencing complex (RISC) to complementary target RNA molecules, which are sequence-specifically cleaved by the RISC. In vitro cleavage assays have proved to be powerful tools for the characterization of the RNAi pathway in many different organisms. Therefore, this chapter provides a detailed protocol for in vitro RISC assays.

Key words: RNA interference, RNAi, Argonaute proteins, Ago2, Slicer, microRNA, Short interfering RNAs, siRNAs

1. Introduction

First described in 1998 by Andrew Fire and Craig Mello, RNAi has now developed into a widely used lab tool allowing the analysis of gene function without labor-intensive genetic manipulations (1). Uses for RNAi have recently expanded beyond the lab, as the first RNAi-based drugs are being developed and have already reached clinical phase III levels (2). Depending on the organism, the trigger molecule for RNAi is either long or short double-stranded (ds) RNA. Long dsRNA is processed by the RNase III enzyme Dicer into short, approximately 21 nucleotides (nt) long dsRNAs, which are referred to as short-interfering RNAs (siRNAs) (3, 4). Such duplexes are subsequently processed and unwound, and one of the two strands, the guide strand, is incorporated into the RNA-induced silencing complex (RISC). The other strand, often referred to as the passenger strand, is degraded by cellular nucleases. Strand selection depends, among other parameters, on the relative thermodynamic stability of the duplex siRNA ends. The strand with the least stable 5′ end in

Tom C. Hobman and Thomas F. Duchaine (eds.), *Argonaute Proteins: Methods and Protocols*,
Methods in Molecular Biology, vol. 725, DOI 10.1007/978-1-61779-046-1_6, © Springer Science+Business Media, LLC 2011

the duplex is preferentially incorporated into the RISC and gives rise to the guide strand (5, 6). However, there are exceptions to this general rule.

The members of the Argonaute family of proteins constitute the central components of the RISC. The Argonaute protein family is divided into two subfamilies. The Ago subfamily is broadly expressed and mainly binds to siRNAs or endogenous microRNAs (miRNAs). The PIWI subfamily, named after the *P*-element-*i*nduced *wi*mpy testes locus in *Drosophila*, is expressed in the germ line of multicellular organisms and represses potentially harmful mobile genetic elements. PIWI proteins associate with the PIWI-interacting RNAs (piRNAs), which are specifically expressed in germ cells (7, 8).

In human, four members of the Ago subfamily, Ago1, Ago2, Ago3, and Ago4, have been identified and all four proteins interact with siRNAs or miRNAs irrespective of their sequence. Argonaute proteins are characterized by the presence of PAZ (PIWI–Argonaute–Zwille) and Piwi domains. Structural studies on bacterial and archaeal Argonaute proteins revealed that the PAZ domain interacts with the 3′ ends of the small RNAs (9). Based on these structures, a third functionally important domain was identified. The MID domain is located between the PAZ and the Piwi domain and specifically anchors the 5′ end of the small RNA (10, 11). The Piwi domain is structurally similar to RNase H and biochemical experiments unraveled that indeed Ago proteins themselves can function as endonucleases and cleave target RNAs that are complementary to their guide strand of small RNA, an activity referred to as "Slicer" (12, 13). Surprisingly, although the critical residues for such an activity are conserved among the human Ago proteins, only Ago2 is an active endonuclease (14, 15). The cellular function of the other human Ago proteins is currently unknown.

Biochemical studies were very helpful for the identification of human Slicer (14, 16), and in vitro RNase cleavage assays proved to be very powerful for the characterization of small RNA-guided gene silencing pathways in general. This book chapter provides a detailed method for in vitro RISC cleavage assays, providing an invaluable resource for the analysis of many different aspects of small RNA-guided gene silencing.

Specifically, this chapter describes the methods for:

- The preparation of the small RNA duplexes, the labeled RNA target, as well as a suitable RNA ladder (Subheadings 3.1–3.3).
- The preparation of a cell extract, and the immunoprecipitation of active Ago2 (Subheading 3.4 and 3.5).
- The RISC cleavage assay (Subheading 3.6).

2. Materials

2.1. Target RNA Preparation

1. Partially complementary DNA oligonucleotides containing the miR-19b cleavage site sequence: 5′-GAACAATTGCTTTTACAGATGCACATATCGAGGTGAACATCACGTACGTCAGTTTTGCATGGATTTGCACATCGGTTGGCAGAAGCTAT-3′ and 5′-GGCATAAAGAATTGAAGAGAGTTTTCACTGCATACGACGATTCTGTGATTTGTATTCAGCCCATATCGTTTCATAGCTTCTGCCAACCGA-3′.
2. T7 cleavage primer 5′-TAATACGACTCACTATAGAACAATTGCTTTTACAG-3′ and SP6 cleavage primer 5′-ATTTAGGTGACACTATAGGCATAAAGAATTGAAGA-3′ (concentration: 10 μM each).
3. Phusion Hot Start High-Fidelity DNA Polymerase and corresponding 5× HF buffer (Finnzymes, Espoo, Finland).
4. dNTP Mix: 2 mM each dATP, dCTP, dGTP, and dTTP (Fermentas, Burlington, Canada), in water. Store as aliquots at −20°C.
5. Gel Extraction Kit, e.g., NucleoSpin Extract II (Macherey-Nagel, Düren, Germany).
6. TOPO TA Cloning Kit (Invitrogen).
7. Sequencing primers M13 forward 5′-GTAAAACGACGGCCAGT-3′ and M13 reverse 5′-AGCGGATAACAATTTCACACAGG-3′.
8. 5× NTP Mix consisting of 5 mM ATP, 5 mM CTP, 8 mM GTP, 2 mM UTP (Fermentas) in water; store at −20°C, thaw on ice.
9. 1,4-Dithiothreitol (DTT), dissolve in water to yield a concentration of 1 M, store at −20°C.
10. T7 RNA Polymerase and the corresponding buffer (Fermentas).
11. 1× TBE: 89 mM Tris–HCl pH 8.3; 89 mM Boric acid in water.
12. UltraPure Sequagel Sequencing System (National diagnostics, Atlanta, GA), *N,N,N,N′*-tetramethyl-ethylenediamine (TEMED), and ammonium persulfate (APS): 10% (w/v) in water, store aliquots at −20°C.
13. Denaturing RNA sample buffer: 90% (v/v) formamide, 0.025% (w/v) xylene cyanol, and 0.025% bromophenol blue dissolved in 1× TBE buffer; store aliquots at −20°C.
14. Elution Buffer: 300 mM NaCl and 2 mM EDTA dissolved in water.

15. Sterile surgical blades, thin layer chromatography plate with fluorescence indicator (Roth), portable UV lamp ($\lambda = 254$ nm).
16. Ethanol, ≥ 99.9% (GC).

2.2. Cap Labeling and T1 Marker Preparation

1. Guanylyltransferase and the corresponding buffer (Gibco/BRL, Bethesda, MD).
2. RNasin (Promega), 500 μM *S*-adenosylmethionine (SAM, Sigma).
3. [α-^{32}P]-Guanosyl-triphosphate, 3,000 Ci/mmol (Perkin Elmer, Waltham, MA).
4. Biomax MR film (Kodak).
5. RNase T1 (1,000 U/μL, Fermentas).
6. Yeast tRNA (Invitrogen) dissolved in water to a concentration of 5 μg/μL. Store aliquots at –80°C.
7. Buffer A: 0.25 M Sodium citrate pH 5.0 in water, stored as aliquots at –20°C.
8. Urea buffer: 10 M Urea, 1.5 mM ethylenediaminetetraacetic acid (EDTA), 0.05% bromophenol blue, 0.05% xylene cyanol dissolved in water. Aliquot and store at –20°C.

2.3. Cell Culture and Lysis

1. Dulbecco's modified Eagle's medium (DMEM) supplemented with 10% fetal bovine serum (FBS), 100 U/mL penicillin, and 100 μg/mL streptomycin (Gibco).
2. Solution of trypsin (0.05%) and EDTA (0.02%).
3. Phosphate-buffered saline (PBS): 130 mM NaCl, 77.4 mL of 1 M Na_2HPO_4, 22.6 mL of 1 M NaH_2PO_4, complete with water to a volume of 1 L.
4. Cell lysis buffer: 0.5% Nonidet P-40 (NP-40), 150 mM KCl, 25 mM Tris–HCl pH 7.5, 2 mM EDTA, 1 mM NaF (handle with care, this is a toxic reagent), 0.5 mM DTT in water. Store at 4°C. Add 4-(2-aminoethyl)-benzenesulfonyl fluoride (AEBSF) to a final concentration of 0.5 mM immediately before use.
5. 2 M $CaCl_2$.
6. 2× HEPES buffer: 0.274 M NaCl, 1.5 mM Na_2HPO_4, 54.6 mM *N*-(2-hydroxyethyl)piperazine-*N'*-(2-ethanesulfonic acid) (HEPES).
7. Cell scrapers.

2.4. Immuno-precipitation of Argonaute Complexes

1. 300 mM IP wash buffer: 300 mM NaCl, 50 mM Tris–HCl pH 7.5, 5 mM $MgCl_2$, 0.05% NP-40 in water.
2. Phosphate-buffered saline (PBS) (see Subheading 2.3, item 3).

3. Anti-FLAG M2 agarose (Sigma, St Louis, MO), or Protein G sepharose (GE Healthcare) in combination with an anti-Ago2 monoclonal antibody (17).

2.5. RISC Assay

1. TM buffer: 3.33 mM ATP, 0.67 mM GTP, 33 U/mL RNAsin (Promega), 333 mM KCl, 5 mM $MgCl_2$, 1.67 mM DTT in water. Aliquot and store at –20°C.
2. Proteinase K storage buffer: 20 mg/mL Proteinase K, 50 mM Tris–HCl pH 8.0, 1 mM $CaCl_2$, 50% (v/v) glycerol in water. Aliquot and store at –20°C.
3. Proteinase K reaction buffer: 300 mM NaCl, 25 mM EDTA, 2% SDS (w/v), 200 mM Tris–HCl pH 7.5. Store at room temperature. Before use, add 1/100 volume of the Proteinase K storage buffer to yield a final concentration of 0.2 mg/mL of Proteinase K.
4. Roti-phenol/chloroform/isoamylalcohol (25:24:1) for RNA isolation (Roth).
5. Sequencing gel apparatus Model S2 (Gibco/BRL) with a corresponding set of glass plates and spacers.
6. Sigmacote (Sigma).
7. Whatman paper.
8. Gel drying apparatus.
9. X-Ray cassette.
10. TranScreen HE screen (Kodak).
11. Biomax MS film (Kodak).

3. Methods

The RISC cleavage assay method consists of several independent steps. One starts by designing and cloning a DNA fragment that encodes an RNA molecule bearing a target site for an endogenous miRNA, under the control of a T7 promoter. Following in vitro transcription and purification, the target RNA is cap-labeled radioactively. When such a substrate is cleaved, only its 5′-end remains visible by autoradiography. For the actual RISC assay, Ago2 is immunoprecipitated from cell lysates and incubated with the labeled target RNA. RNA is then extracted and separated on a sequencing gel, and the cleavage product is detected by autoradiography. The position of the cleavage site is inferred by comparing the electrophoretic migration of the cleavage product with a marker ladder obtained by the partial digestion of a fraction of the labeled target RNA with RNase T1.

The siRNA-guided cleavage is known to occur precisely ten nucleotides upstream of the nucleotide paired with the 5′ end of the siRNA (18).

3.1. Target RNA Preparation

1. Two partially complementary oligonucleotides encoding the miR-19b cleavage site sequence are subjected to a PCR fill-in reaction (see Note 1). Set up a reaction mix consisting of 100 ng of each oligonucleotide, 5 μL dNTP mix, 10 μL 5× HF buffer, and 0.5 μL Phusion DNA Polymerase in a 0.2-mL PCR tube. Add sterile H_2O to an end volume of 49.4 μL. Set up a PCR program as follows: 30 s at 98°C, 10 min at 72°C, followed by cooling to 4°C.
2. In a subsequent second PCR reaction, the T7 and SP6 promoter sequences are introduced on both ends of the DNA sequence obtained in the previous step. For this, directly add 0.3 μL each of T7 and SP6 cleavage primers from step 2 of Subheading 2.1 to the PCR fill-in mix. Use the following PCR program: initial denaturation (98°C for 30 s), 30 cycles (98°C for 30 s, 55°C for 20 s, and 72°C for 15 s), final elongation (72°C for 2 min) followed by cooling to 4°C. This PCR will yield a product of 195 bp, which should be purified by separation on a 2% agarose gel and subsequent gel extraction according to manufacturer's instructions.
3. Ligate and clone the PCR product from step 2 into the pCR2.1-TOPO vector using the TOPO TA Cloning Kit.
4. To verify the positive clones, sequence using the M13 forward and reverse sequencing primers.
5. Another PCR is performed using DNA from a sequence-verified clone as template. Set up a PCR reaction consisting of 1 μL template DNA, 10 μL 5× HF buffer, 5 μL dNTP mix, 1 μL each of T7 and SP6 cleavage primers, and 0.3 μL Phusion DNA polymerase. Fill up to a final volume of 50 μL with H_2O and use the PCR program from step 2 of Subheading 3.1. The DNA fragment resulting from this step will serve as template in the in vitro transcription.
6. Assemble a T7 in vitro transcription mix: 53.5 μL of water, 20 μL of 5× NTP mix, 20 μL of 5× T7 Polymerase Buffer, 0.5 μL of DTT (1 M), 5 μL of PCR template, and 1 μL of T7 polymerase (see Note 2). Be sure to use 5× NTP mix (not the dNTP mix used for PCR). Pipette on ice and incubate the reaction mix at 37°C for 2 h. Add 100 μL of the RNA sample buffer and mix carefully.
7. Meanwhile, prepare a 2-mm thick 8% acrylamide gel according to the instructions of the UltraPure Sequagel Sequencing System. Pour the gel and insert a comb with wide wells (5 mm). The gel should be allowed to polymerize for about

1 h. Pre-run at 300 V for approximately 20 min using 1× TBE running buffer. Before loading the RNA samples, rinse the wells carefully with running buffer using a syringe (see Note 3). Load the samples using long pipet tips and run the gel at 300 V for about 1 h. Ideally, dispense each sample into two wells to minimize the loading volumes.

8. Remove the glass plates and wrap the gel in a polyethylene membrane (Saran Wrap). Place it on a fluorescent thin layer chromatography plate and expose to UV light ($\lambda = 254$ nm). Usually, a weak band should be visible above the dark blue loading dye. Mark the position of the band on the Saran Wrap with a pen.
9. Cut the gel along the markings using a sterile scalpel and remove the Saran Wrap from the gel piece. Cut the gel fragment into small pieces and transfer to a fresh microcentrifuge (1.5 mL) tube. Add 400 μL of elution buffer and incubate over night at 4°C while shaking vigorously.
10. Transfer the supernatant to a fresh microcentrifuge tube and add 1 mL of ethanol. Mix and centrifuge for 30 min at 4°C and 17,000 × *g* in a tabletop centrifuge. Remove the supernatant and wash the RNA pellet with 1 mL of ethanol (70%). Centrifuge again at 4°C for 10 min.
11. Carefully remove the supernatant and let the RNA pellet dry (see Note 4). Dissolve the pellet in 20 μL of water and measure the RNA concentration at $\lambda = 260$ nm. Calculate molar values from the measured concentration, e.g., by using the following web site: http://www.molbiol.ru/eng/scripts/01_07.html assuming a RNA size of 195 nucleotides.

3.2. Cap Labeling of the Target RNA

Labeling and the consecutive assay steps must be carried out in an isotope lab. Be careful while handling the sample in all subsequent steps due to its high-energy radiation.

1. The minimum amount of in vitro-transcribed (ivt) RNA used in a cap-labeling reaction should be 40 pmol. Set up the labeling reaction by pipetting 2 μL of guanylyltransferase buffer, 0.25 μL RNAsin, 1 μL SAM (500 μM), 1 μL DTT (100 mM), 2 μL guanylyltransferase, 2 μL [α-^{32}P]-GTP, and the calculated volume of ivt RNA. Add water to a total reaction volume of 20 μL. Incubate at 37°C for 3 h, and then add 20 μL of RNA sample buffer.
2. Heat to 95°C for 5 min.
3. Prepare an 8% acrylamide gel as described above and let it pre-run for about 15 min. After rinsing the wells, load the labeling reaction into one well and run the gel at 300 V for about 1 h.

4. Wrap the gel in Saran Wrap and fix it to a support, e.g., an old film. Place it in the corner of an X-Ray cassette and expose to an MR film for 5 min. Develop the film, and place it below the wrapped gel in the same position as for the exposition. Mark the position of the radioactively labeled RNA with a pen. Usually a strong signal is visible below the expected RNA band at the bottom of the gel; this signal corresponds to unincorporated nucleotides and/or degraded RNA and should be discarded.
5. Cut the labeled target RNA according to the markings, elute, and precipitate as described in steps 9 and 10 of Subheading 3.1. After removing the gel fragment containing the target RNA, another MR film is exposed to the remaining gel for 5 min to verify the excision of the labeled target RNA.

3.3. RNase T1 Ladder Preparation

1. To check for the correct position of the cleavage signal, a small aliquot of the ^{32}P-cap-labeled RNA is partially digested by RNase T1 and used as a marker. RNase T1 cuts the RNA after each guanine residue, if digestion is complete. Partial digestion will yield a ladder that is to a certain degree characteristic to the RNA sequence. For the RNase T1 digestion, first prepare two separate solutions.
2. Assemble the RNase T1 solution: 28 μL of buffer A, 195 μL of urea buffer, 53.6 μL of water, and 1.4 μL of RNase T1. Mix carefully.
3. Assemble the carrier solution: 66 μL of buffer A, 463 μL of urea buffer, 14 μL of yeast tRNA, and 47 μL of water. Mix carefully.
4. Dilute an aliquot of RNase T1 solution 1:10 with water. 230 μL of the carrier solution is mixed with 11 μL of the diluted RNase T1 solution. This is the ready-to-use T1 mix.
5. Directly before loading the sequencing gel, 5 μL T1 mix is mixed with 1 μL of the labeled RNA on ice. The sample is incubated for 10 min at 50°C to yield the RNA ladder that is used as a marker. Keep the sample on ice afterwards and load as soon as possible (see Notes 5 and 6).

3.4. Cell Culture and Lysis

1. HEK293 cells are cultivated in DMEM medium supplemented with FBS, penicillin, and streptomycin.
2. Immunoprecipitation can be carried out on endogenous Ago2 when anti-Ago2 antibodies are available, or using a tagged Ago2 (e.g., FLAG/HA-Ago2) from transiently transfected HEK 293 cells. HEK 293 cells can easily and rather inexpensively be transfected using the calcium-phosphate method. For this, seed HEK293 cells to a confluency of 25% and culture for approximately 3–4 h prior to transfection. For one 10-cm plate, mix 435 μL sterile water, 4 μL vector DNA (1 μg/μL), and 61 μL 2 M $CaCl_2$ in a falcon tube.

Add 500 μL 2× HEPES buffer drop wise while shaking the falcon. Gently and carefully disperse the solution on the cells. Following transfection, let the cells in culture for 48 h before lysis. If you immunoprecipitate endogenous Ago2, expand HEK 293 cells to approximately 70% of confluency on 10-cm cell culture plates. Usually, one or two 10-cm plates will provide sufficient material for immunoprecipitation.

3. Prior to lysis, wash the cell cultures twice with cold PBS. Remove all of the supernatant from washing and add 200 μL of cell lysis buffer per 10-cm plate. Place the plates on ice and harvest the cells using a cell scraper. Resuspend cell clumps by pipetting to ensure that cells are properly lysed.
4. Transfer the cell lysates to 1.5-mL microcentrifuge tubes and incubate on ice for 5 min.
5. Centrifuge at 17,000 × *g* and 4°C for 10 min to remove cell debris.
6. Carefully transfer the supernatants to fresh microcentrifuge tubes. You now can directly proceed to immunoprecipitation, or you can snap-freeze the lysates in liquid nitrogen and store them at −80°C.

3.5. Immuno-precipitation of Argonaute Complexes

1. All immunoprecipitation steps are performed on ice or at 4°C (see Note 7).
2. For immunoprecipitation of FLAG/HA-Ago2, use 20 μL of Anti-FLAG agarose beads per sample. For immunoprecipitation of endogenous Ago2, the anti-Ago2 antibody first has to be coupled to the protein G sepharose beads. For this, we used 20 μL of beads per sample and washed the beads once with PBS before adding the appropriate volume of antibody. Fill the tube to 1.5 mL with PBS and incubate for 2–3 h at 4°C on a rotating wheel. Wash the beads twice with PBS and distribute the beads to fresh microcentrifuge tubes. As a negative control, a sample containing only beads is suitable.
3. Add cell lysates to the beads and fill the vial to 1.5 mL with PBS. Incubate for 2 h at 4°C on a rotating wheel.
4. Wash the samples three times with 300 mM IP wash buffer, and once with PBS. Transfer the beads to a fresh 1.5-mL safe seal (or screw cap) microcentrifuge tube.
5. At this point, you may want to take an aliquot of 5 μL beads/sample for future western blot analysis. Add 5 μL of protein sample buffer and store at −20°C (see Note 8).

3.6. RISC Assay

1. Carefully remove all the supernatant without disturbing the beads.
2. Set up a 25 μL reaction containing 15 μL of beads, 7.5 μL of TM buffer, 1 μL of labeled RNA, and sterile water.

Incubate at 30°C for 1.5 h while occasionally and gently stirring the beads.

3. Add 200 μL of Proteinase K solution per fraction, and incubate at 65°C for 10–20 min while shaking vigorously.
4. Add 200 μL of phenol/chloroform/isoamylalcohol solution and incubate at room temperature for 10–20 min while shaking. Be sure to use a safe seal microcentrifuge tube at this step, as the phenol and chloroform might cause the solution to leak out.
5. Centrifuge at 17,000 × *g* for 5 min in a tabletop centrifuge to separate organic from aqueous phase. Transfer aqueous phase to fresh microcentrifuge tubes (see Note 9).
6. Add 200 μL ethanol, mix, and keep samples at −20°C over night for RNA precipitation.
7. Centrifuge at 17,000 × *g* and 4°C for 30 min in a tabletop centrifuge (see Note 10). Carefully remove the supernatant without disturbing the tiny RNA pellet.
8. Let the pellet air-dry (see Note 11), then add 10 μL RNA sample buffer and incubate for 10–15 min at 37°C while shaking to dissolve the RNA pellet.
9. For sequencing gel electrophoresis, we used a sequencing gel apparatus Model S2 from Gibco/BRL. To prepare for the gel pouring, clean the glass plates of the gel apparatus with ethanol. Disperse 500 μL Sigmacote on each plate using a paper towel (see Note 12). To assemble the gel set up, place the larger glass plate with the Sigmacote-treated side facing upwards on some pipet boxes, so that it rests in an elevated position (this is important for the actual pouring, as you can easily tilt the glass plate to ensure equal distribution of the gel mixture). Place the spacers next to the side edges of the glass plate and put the smaller glass plate with the Sigmacote-treated side facing downwards on top. The spacers are very thin and have a foam rubber square at one end. This foam rubber part has to face upwards and must be sitting directly adjacent to the top edge of the smaller glass plate. When you fix the gel in the electrophoresis apparatus later, the foam rubber will work as a seal that prevents leaking of the buffer from the buffer reservoir (also see step 12 and Note 14). Fix the glass plates in their position with some clamps.
10. Prepare the gel mixture for 100 mL of an 8% sequencing gel according to UltraPure Sequagel Sequencing System manufacturer's instructions.
11. Slightly tilt the glass plates and carefully and steadily pour the gel mixture into the gap between the plates until the space is completely filled, and excess gel mixture leaks out at the

bottom. Be careful not to let any air bubbles form while pouring, as they can hardly be removed afterwards. Insert the comb (see Note 13), fix the glass plates around the comb with clamps and let the gel polymerize for about 1 h.

12. Remove the clamps and fix the gel in the gel apparatus. Fill the top and bottom tanks with 1× TBE. Remove the comb and let the empty gel pre-run for about 20 min at 65 W (see Note 14).
13. Before loading, incubate the samples in RNA sample buffer at 95°C for 1 min and spin down briefly.
14. Immediately before loading, carefully rinse the gel wells with 1× TBE buffer to remove accumulated urea (see Note 15). Load a sample volume of 5 and 1 μL of marker and run the gel for about 2 h at 65 W. The light blue loading dye should be in the bottom third of the gel.
15. Remove the gel from the apparatus, and drain the buffer from the top tank. Remove the spacers, and lift one glass plate from the gel by inserting a flat object (e.g., a surgical blade) and tilting the plates carefully. Be sure not to pull parts of the gel on the different plates as they separate, as this could tear the gel apart. Place a sheet of Whatman paper of adequate size on top of the gel, run your hand over it, and lift the Whatman paper slowly and carefully. The gel should remain on the Whatman paper.
16. Cover the gel with Saran Wrap, and dry it for 45 min at 80°C in a gel drying apparatus.
17. Place the dried gel in an X-Ray cassette with an amplifying screen, and expose to an MS film over night at –80°C (see Note 16).
18. After developing the film, you should be able to define the position of the expected cleavage signal in your samples (Fig. 1) by the comparing with the signal pattern of your marker.

4. Notes

1. Depending on the levels of miRNA expression in a given cell line, you may need to assay for an miRNA other than miR-19b. Therefore, you will have to replace miR-19b sequence in the first, partially complementary set of oligonucleotides with that of another miRNA and proceed as described.
2. RNA is very sensitive to degradation by RNases. Minimize the risk of RNase contamination by always using a fresh aliquot of sterile water and new pipet tip boxes.

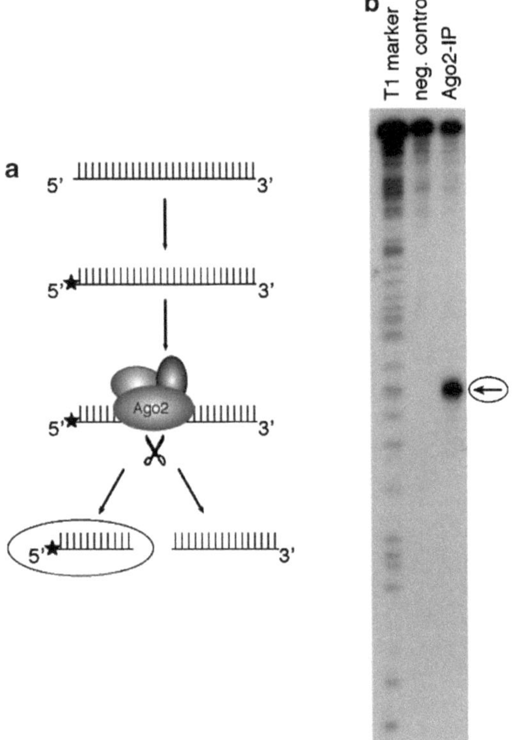

Fig. 1. RISC cleavage assay. (**a**) Schematic depiction of the RISC cleavage assay procedure. The *star* denotes the ^{32}P guanosine cap labeling of the target RNA. Ago2 complexes recovered by immunoprecipitation cleaves the labeled RNA substrate at a defined position. The cleaved fragment marked with a *circle* is detected by autoradiography. (**b**) Ago2 was immunoprecipitated from FLAG/HA-transfected HEK 293 cells, and incubated with a ^{32}P-cap-labeled RNA containing a sequence that is perfectly complementary to miR-19b. The lane labeled T1 marker shows an RNase T1 digestion of the RNA substrate. The signal corresponding to the cleaved miR-19b substrate is indicated by an *arrow*.

3. Rinsing the gel wells immediately before loading is very important for an optimal resolution. Also the RNA sample buffer does not contain glycerol, therefore the sample will not rapidly sink to the bottom of the wells. Use long pipet tips and make sure not to fill the wells completely.
4. Do not let the RNA pellets dry for too long, as over-drying will result in a difficult resuspension of the RNA.
5. Keep the RNase T1-digested RNA at –80°C and only thaw it on ice just before use. RNase T1 will become active as soon as the solution is thawed and digestion will continue.
6. When preparing the RNA marker for the first time or when using RNase T1 from a different manufacturer, it may be useful to optimize the digestion times. If the T1 digestion of your RNA is inadequate, only the full-length and large-digestion

fragments will be detected. Conversely, you will only see small fragments, if the digestion reaction has proceeded too far. Because an RNA marker ladder is crucial for defining the position of the cleavage fragment in your samples, we suggest doing a test run after your first T1 digestion (which is also advisable to get used to pouring the gels).

7. RISC assay is also possible with whole lysates (i.e., without immunoprecipitation). However, this risks the nonspecific RNA degradation by endogenous RNases, and unspecific signal may make the results difficult to interpret.
8. To control for equal Ago2 levels, it is advisable to perform a western blot with an aliquot from the immunoprecipitated samples prior to the RISC assay.
9. Be sure to monitor the transfer of your labeled RNA when changing to fresh tubes after phenol/chloroform extraction. For this, simply monitor the tube with a hand held radiation monitor before removing the aqueous phase. Afterwards, measure the fresh tube containing the aqueous phase again. Radioactivity should be associated only with the fresh tube.
10. Check for the radiation signal when removing ethanol after RNA precipitation. The RNA pellet will be very small or even invisible to the eye, so this will prevent an accidental loss of the RNA pellet by pipetting.
11. Only let the RNA air-dry until there is no visible liquid left. Excessive drying will prevent the subsequent redissolving of the RNA.
12. The sequencing gel is very thin and breaks easily. Coating the glass plates will facilitate the transfer of the gel from the glass plate to the Whatman paper.
13. Be sure to insert the comb only as far as necessary to get adequate wells. Inserting it fully will negatively affect sample loading.
14. Make sure the buffer does not leak from the top to the bottom tank. If leaking occurs, keep an eye on the buffer level in the top tank during the run and, if necessary, interrupt the run to refill the top tank with running buffer.
15. Rinsing the wells is important for the uniform shape of the bands on the autoradiograms. RNA sample buffer does not contain glycerol and therefore samples will not sink to the bottom of the wells. For sample loading, we use 10-μL pipet tips on a 2- to 20-μL pipet. Pipet slowly and carefully to prevent the diffusion of the sample into the buffer tank.
16. The screen will amplify the signal from the gel, if positioned correctly and according to the manufacturer's instructions. It will be most effective when used for exposure at –80°C.

However, make sure to develop the film quickly after removing the cassette from the freezer. Otherwise the film will become moist due to condensation, and water will appear on the developed film as black spots.

Acknowledgments

Our research is supported by the Max-Planck Society, Regensburg University and by grants from the Deutsche Forschungsgemeinschaft (DFG), the European Union (FP6, ERC), the German ministry for education and science (BMBF), and the Bavarian Genome research network (BayGene).

References

1. Fire, A., Xu, S., Montgomery, M. K., Kostas, S. A., Driver, S. E., and Mello, C. C. (1998) Potent and specific genetic interference by double-stranded RNA in *Caenorhabditis elegans*, *Nature* ***391***, 806–811.
2. Kim, D. H., and Rossi, J. J. (2007) Strategies for silencing human disease using RNA interference, *Nat Rev Genet* ***8***, 173–184.
3. Meister, G., and Tuschl, T. (2004) Mechanisms of gene silencing by double-stranded RNA, *Nature* ***431***, 343–349.
4. Siomi, H., and Siomi, M. C. (2009) On the road to reading the RNA-interference code, *Nature* ***457***, 396–404.
5. Khvorova, A., Reynolds, A., and Jayasena, S. D. (2003) Functional siRNAs and miRNAs exhibit strand bias, *Cell* ***115***, 209–216.
6. Schwarz, D. S., Hutvágner, G., Du, T., Xu, Z., Aronin, N., and Zamore, P. D. (2003) Asymmetry in the assembly of the RNAi enzyme complex, *Cell* ***115***, 199–208.
7. Peters, L., and Meister, G. (2007) Argonaute proteins: mediators of RNA silencing, *Mol Cell* ***26***, 611–623.
8. Hutvagner, G., and Simard, M. J. (2007) Argonaute proteins: key players in RNA silencing, *Nat Rev Mol Cell Biol* ***9***, 22–32.
9. Jinek, M., and Doudna, J. A. (2009) A three-dimensional view of the molecular machinery of RNA interference, *Nature* ***457***, 405–412.
10. Ma, J. B., Yuan, Y. R., Meister, G., Pei, Y., Tuschl, T., and Patel, D. J. (2005) Structural basis for 5'-end-specific recognition of guide RNA by the A. fulgidus Piwi protein, *Nature* ***434***, 666–670.
11. Parker, J. S., Roe, S. M., and Barford, D. (2005) Structural insights into mRNA recognition from a PIWI domain-siRNA guide complex, *Nature* ***434***, 663–666.
12. Wang, Y., Juranek, S., Li, H., Sheng, G., Wardle, G. S., Tuschl, T., and Patel, D. J. (2009) Nucleation, propagation and cleavage of target RNAs in Ago silencing complexes, *Nature* ***461***, 754–761.
13. Song, J. J., Smith, S. K., Hannon, G. J., and Joshua-Tor, L. (2004) Crystal structure of Argonaute and its implications for RISC slicer activity, *Science* ***305***, 1434–1437.
14. Meister, G., Landthaler, M., Patkaniowska, A., Dorsett, Y., Teng, G., and Tuschl, T. (2004) Human Argonaute2 Mediates RNA Cleavage Targeted by miRNAs and siRNAs, *Mol Cell* ***15***, 185–197.
15. Liu, J., Carmell, M. A., Rivas, F. V., Marsden, C. G., Thomson, J. M., Song, J. J., Hammond, S. M., Joshua-Tor, L., and Hannon, G. J. (2004) Argonaute2 is the catalytic engine of mammalian RNAi, *Science* ***305***, 1437–1441.
16. Martinez, J., Patkaniowska, A., Urlaub, H., Lührmann, R., and Tuschl, T. (2002) Single-stranded antisense siRNAs guide target RNA cleavage in RNAi, *Cell* ***110***, 563–574.
17. Rudel, S., Flatley, A., Weinmann, L., Kremmer, E., and Meister, G. (2008) A multifunctional human Argonaute2-specific monoclonal antibody, *Rna* ***14***, 1244–1253.
18. Elbashir, S. M., Lendeckel, W., and Tuschl, T. (2001) RNA interference is mediated by 21 and 22 nt RNAs, *Genes Dev* ***15***, 188–200.

Chapter 7

Native Gel Analysis for RISC Assembly

Tomoko Kawamata and Yukihide Tomari

Abstract

Small-interfering RNAs (siRNAs) and microRNAs (miRNAs) regulate expression of their target mRNAs via the RNA-induced silencing complex (RISC). A core component of RISC is the Argonaute (Ago) protein, which dictates the RISC function. In *Drosophila*, miRNAs and siRNAs are generally loaded into Ago1-containing RISC (Ago1-RISC) and Ago2-containing RISC (Ago2-RISC), respectively. We developed a native agarose gel system to directly detect Ago1-RISC, Ago2-RISC, and their precursor complexes. Methods presented here will provide powerful tools to biochemically dissect the RISC assembly pathways.

Key words: RNA interference, Argonaute, RISC, Small interfering RNA, MicroRNA, Native agarose gel, Ribonucleoprotein complex

1. Introduction

Small-interfering RNAs (siRNAs) and microRNAs (miRNAs) trigger gene silencing through the RNA-induced silencing complex (RISC), which contains an Argonaute (Ago) family protein as a core component (1, 2). Each small RNA species often binds to a specific Ago protein, which dictates the RISC function (3).

There are at least two steps in the RISC assembly pathway: the RISC-loading step, at which a small RNA duplex is incorporated into the Ago protein, and the RISC maturation step (or unwinding step), at which the two strands of the duplex are separated and only one of the strands is retained in the Ago protein (Fig. 1) (4, 5). The Ago complex containing the small RNA duplex is called pre-RISC, and the Ago complex containing the single-stranded small RNA is called mature RISC, holo-RISC, or simply RISC.

Drosophila melanogaster serves as one of the most powerful model organisms for dissecting the siRNA and miRNA pathways (Fig. 1).

Tom C. Hobman and Thomas F. Duchaine (eds.), *Argonaute Proteins: Methods and Protocols*, Methods in Molecular Biology, vol. 725, DOI 10.1007/978-1-61779-046-1_7, © Springer Science+Business Media, LLC 2011

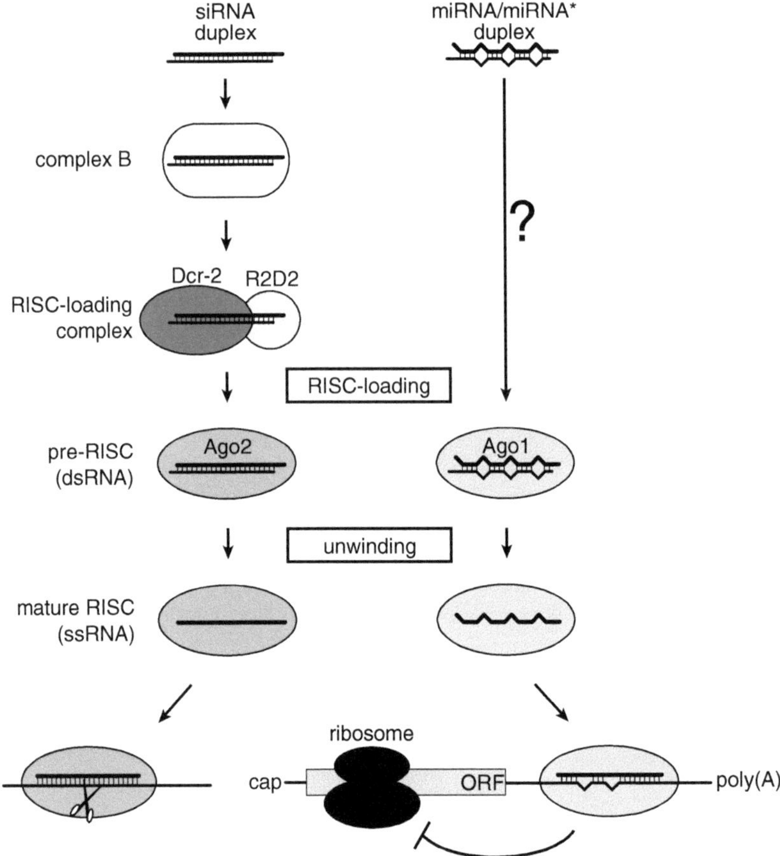

Fig. 1. siRNA and miRNA pathways in *Drosophila*.

In flies, miRNAs and siRNAs are actively sorted into Ago1-RISC and Ago2-RISC, respectively (6–8). In the Ago1-RISC assembly pathway, a long primary miRNA (pri-miRNA) transcript is first processed by the microprocessor complex (Drosha/Pasha) in the nucleus, and the resultant precursor miRNA (pre-miRNA) is then exported into the cytoplasm by Exportin-5. The pre-miRNA is further processed by Dicer-1 and Loquacious into miRNA/miRNA* duplex (9, 10). The miRNA/miRNA* duplex is then loaded into Ago1 to form the pre-Ago1-RISC. Although the protein factors required for Ago1-RISC loading remain unknown, mismatches in the central region (guide position 9–11) direct miRNA/miRNA* duplexes into pre-Ago1-RISC. miRNA/miRNA* duplexes are then passively unwound without the need for ATP or slicer activity of Ago1. Finally, the miRNA* strand is discarded and the miRNA strand is retained by Ago1. The unwinding requires mismatches in the seed region (position 2–8 in the guide RNA) and/or the middle of the 3′ region (position 12–15 of the guide RNA) (5). Mature Ago1-RISC generally induces translational repression of the partially complementary target mRNAs.

In the Ago2-RISC assembly pathway, long double-stranded RNAs (dsRNAs) are first processed by Dicer-2 into siRNA duplexes. siRNAs are actively loaded into Ago2 by the RISC-loading complex (RLC), which is a heterodimer of Dicer-2 and R2D2 (11–13). Dicer-2/R2D2 acts as a gatekeeper to selectively incorporate siRNA duplexes that are highly complementary into Ago2 (8) and to specifically exclude duplexes with central mismatches, contributing to the striking selectivity of small RNA sorting in flies (5, 8). Subsequently, one of the two strands (the passenger strand) is cleaved by Ago2 and discarded, while the other strand, the guide strand is retained in the RISC (4). Mature Ago2-RISC generally cleaves the perfectly complementary target mRNAs.

Here, we first describe an agarose native gel system, which can be used to directly detect pre-Ago1-RISC and mature Ago1-RISC (5). We then describe a similar agarose native gel system, which can be used to detect mature Ago2-RISC, Ago2-RLC, and complex B, a putative precursor of Ago2-RLC (13). We also show how to visualize mature Ago1-RISC or mature Ago2-RISC alone, without detecting the precursor complexes. These methods will provide powerful tools to biochemically dissect the RISC assembly pathways.

2. Materials

2.1. Lysate Preparation

1. Bleach solution.
2. Embryo collection cage (acryl, about 8.5 cm in diameter, with a height of 10 cm).
3. Apple juice agar plates (25 g sucrose, 250 ml of apple juice, 5 ml of 30% methylparaben in ethanol, 2% agar, complete to 1 L with water). Add yeast paste (ORIENTAL YEAST) onto the center of the plate before use.
4. Soft paint brush.
5. 1× lysis buffer: 30 mM HEPES–KOH (pH 7.4), 100 mM KOAc, 2 mM $Mg(OAc)_2$. Store at 4°C.
6. 1 M Dithiothreitol (DTT). Dissolve in water and store in aliquots at −20°C.
7. 25× protease inhibitors cocktail (25× PIC): Dissolve one tablet of Complete EDTA-free (Roche) in 2 ml water and store in aliquots at −20°C.
8. Dounce homogenizer (WHEATON; 7 ml, "TIGHT" pestle).

2.2. Preparation of the Vertical Agarose Gel for Gel-Shift Assay

1. Rain-X original glass treatment, and Rain-X anti-fog solutions.
2. Low Range Ultra Agarose (Bio-Rad Laboratories) (see Note 1).
3. 0.5× TBE: 45 mM Tris–HCl, 45 mM boric acid, and 1 mM EDTA (pH = 8.0). Store at room temperature or 4°C.

4. Tracking dye: 0.05% bromophenol blue, 0.05% xylene cyanol, and 30% glycerol. Store at room temperature or 4°C.
5. 500 mM $MgCl_2$: Dissolve in water and store at room temperature or 4°C (see Note 2).
6. A pair of 16 × 16 cm glass plates: One gel plate is "rabbit-ear" shaped with no spacers, and the back plate is flat with two spacers (15 cm × 5 mm × 1.5–2 mm) attached at the side (Fig. 2, see Note 3), and one spacer (12.5 × 5 × 0.5 mm) attached at the bottom (Fig. 2, see Note 4).
7. Silicone rubber gasket: 50-cm long and 2-mm diameter.
8. Electrophoresis apparatus and clips.

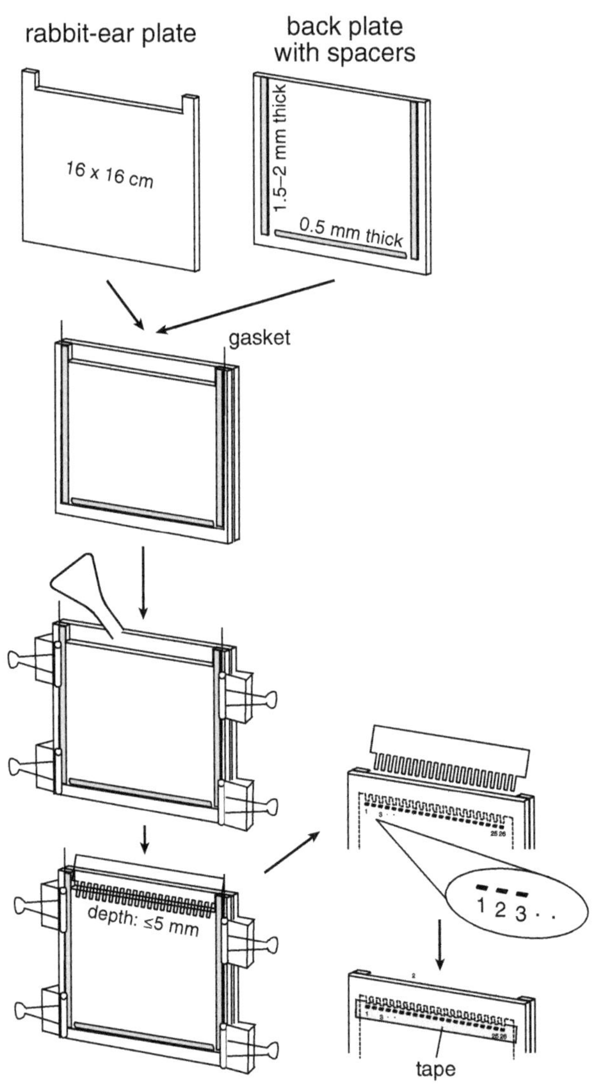

Fig. 2. Assembly of vertical native agarose gel system.

9. Positively charged nylon membrane: Hybond N+ (GE) or Immobilon-NY+ (Millipore): 16 × 16 cm.
10. Whatman 3MM paper (Whatman): 18 × 18 cm.
11. 25-μm plastic sheet (preferably polyethylene terephthalate): 20 × 20 cm.
12. Gel dryer (Bio-Rad).

2.3. Preparation of Radiolabeled Small RNA Duplexes and Target mRNAs

1. Synthetic small RNA duplexes: listed in Table 1.
2. *Renilla* Luciferase (RL) 1× target mRNA (100 nM): ~1,500 nt target mRNA harboring the *Renilla* Luciferase coding sequence and a target site complementary to the guide strand with a central bulge. The mRNA has an m^{7}G-cap and a 100–200 nt poly(A) tail (see Note 5). The RL 1× target may be prepared by PCR amplification using the primers: FW: 5′-GAATTCTAATACGACTCACTATAGG-3′ and RV: 5′-CGCCACTCCCTGAGGTAGTACGTTGTATAG TCCTCGCGCCTCCGGGTGAC-3′ on the psiCHECK2 (Promega) vector, followed by transcription, m^{7}G-capping, and polyadenylation using the mScript mRNA Production System (Epicenter). After NH_4OAc precipitation, adjust the mRNA to a final concentration of 100 nM.
3. [γ-^{32}P] ATP (7,000 Ci/mmol, >100 mCi/ml; MP Biomedicals).
4. 100 mM-unlabeled ATP: Dissolve in water, adjust pH to 7–8 with KOH, and store in aliquots at –20°C.
5. 20 mg/ml Glycogen (Roche): Store in aliquots at –20°C.
6. T4 Polynucleotide kinase (PNK) (Takara).
7. G-25 MicroSpin column (GE).

Table 1
Synthetic small RNA duplexes

Duplex A	5′-UGAGGUAGUUGGUUGUAUAGU-3′ | ||||| ||| |||| 3′-UCUCUCCAUCAUCCAACAUAU-5′
Duplex B	5′-UGAGGUAGUUGGUUGUAUAGU-3′ | ||||||||||||| 3′-UCUCUCGAUCAUC CAACAUAU-5′
Duplex C	5′-UGAGGUAGUAGGUUGUAUAGU-3′ |	||||||||||| 3′-UCUCUCCAUCAUCCAACAUAU-5′

The guide strand (*top*) of each small RNA duplex derived from *Drosophila let-7* miRNA

8. 2× lysis buffer: 60 mM HEPES–KOH (pH 7.4), 200 mM KOAc, and 4 mM $Mg(OAc)_2$. Store at 4°C.
9. Anti-*let-7* antisense oligonucleotide (ASO): 5′-mUmCmUmUmCmAmCmUmAmUmAmCmAmAmCmCmUmAmCmUmAmCmCmUmCmAmAmCmCmUmU-3′. This 2′-*O*-methylated ASO is complementary to the guide strand of duplex A and duplex B, except for one central mismatch (Table 1).
10. Anti-*let-7* seed mismatch ASO: 5′-mUmCmUmUmCmAmCmUmAmUmAmCmAmAmCmCmUmAmGmAmUmGmGmAmGmAmAmCmCmUmU-3′. The bases of anti-*let-7* ASO corresponding to the seed region of the guide strand (position 2–8) are substituted to introduce mismatches (Table 1).

2.4. In Vitro RISC Assembly

40× reaction mix: 120 μL of 40× reaction mix contains 50 μL of water, 20 μL of 500 mM creatine monophosphate (Fluka; prepared fresh from powder), 20 μL of 1 mM amino acid stock (Sigma; 1 mM each amino acid), 2 μL of 1 M DTT, 1 μL of 40 U/μL RNasin Plus (Promega), 4 μL of 100 mM ATP, 1 μL of 100 mM GTP, 16 μL of 1 M KOAc, and 6 μL of 2 U/μL creatine phosphokinase (Cal-biochem; freshly prepared by diluting 2 μL of a 10 U/μL stock in 8 μL of 1× lysis buffer) (see Note 6).

3. Methods

3.1. Preparation of a dcr-2 Null Embryo Lysate for In Vitro Ago1-RISC Assembly

In wild-type *Drosophila* embryo lysates, a strong signal from the Ago2-RISC assembly complexes can mask the Ago1 complexes on the native agarose gel. To detect complexes involved in Ago1-RISC assembly, we, therefore, use an embryo lysate prepared from *dcr-2* null mutant flies (*dcr-2*^{L811fsX} or *dcr-2*^{416X}) (14) (see Note 7) as Ago1-RISC assembly remains unaffected in such a lysate (8). The preparation of *Drosophila* embryo lysate presented here was previously described in detail in the literature (15).

To prepare a *dcr-2* null embryo lysate:

1. Set up the fly cages (~1 g of *dcr-2* null adult flies per cage) at 25°C. Change the apple juice-agar plate every 12 h (see Note 8).
2. Collect the embryos (0.2–1 g) on the apple juice-agar plate using a paint brush by washing them with running water and recovering the embryos into a mesh sieve.
3. Dechorionate the embryos with a 50% bleach solution for 2 min, and wash extensively with running tap water until the bleach smell dissipates.

4. Dry the embryos by blotting with paper towels from underneath of the mesh sieve.
5. Weigh the embryos and transfer them to a pre-chilled Dounce homogenizer.
6. Add 1 ml of ice-cold 1× lysis buffer per gram of embryo pellet, freshly supplemented with final concentrations of 5 mM DTT and 1× PIC.
7. Homogenize the embryos by 30 strokes on ice.
8. Clear the lysate by centrifugation at 17,000 × *g* for 20 min at 4°C.
9. Collect the supernatant and aliquot into new tubes.
10. Quickly freeze the lysates with liquid nitrogen and store at −80°C.

The lysates may be kept for up to ~6 months.

3.2. Preparation of Radiolabeled RNA for Ago1-RISC Assembly

Synthetic small RNA duplexes are listed in Table 1. These small RNA duplexes are derived from the natural *let-7/let-7** duplex. Duplex A and duplex B contain an identical guide strand sequence (5′-UGAGGUAGUUGGUUGUAUAGU-3′). A U–U mismatch is introduced at guide position 1 to ensure that the guide strand is always selectively retained in mature RISC, according to the "asymmetry rule" of the RISC assembly (5). Duplex A has a central mismatch at position 10. Duplex B bears an additional mismatch in the passenger strand across from guide position 5 in the seed region. The guide and passenger strands of duplex C are fully paired, except at guide position 1.

To label the 5′ end of the guide strand with ^{32}P:

1. Mix 1 μL of 10 μM single-stranded guide strand RNA, 0.7 μL of [γ-^{32}P]ATP, 1 μL of T4 PNK, 2 μL of PNK reaction buffer, and 15.3 μL of water.
2. Incubate the reaction mixture at 37°C for 1 h.
3. Adjust the volume to 40 μL with water, and run through a G-25 MicroSpin column to remove unincorporated [γ-^{32}P] ATP.
4. Precipitate the column flow through with 1/10× volume of 3 M NaOAc (pH 5.5), 1 μL of 20 mg/ml glycogen, and three volumes of absolute ethanol. Centrifuge at 20,400 × *g* for 15 min at 4°C.
5. Rinse the pellet with 70% ethanol. Let the pellet dry on bench.
6. Dissolve the precipitate in 20 μL of water. This will set the final concentration to 500 nM.
7. The 5′ end of the passenger strand should be phosphorylated by PNK using the same method, but with 1 mM of unlabeled ATP instead of [γ-^{32}P]ATP.

8. To make a 100 nM stock of radiolabeled duplex A or duplex B, add 4 μL of 500 nM radiolabeled guide strand, 6 μL of 500 nM nonradiolabeled passenger strand, and 10 μL of 2× lysis buffer and mix.
9. Heat the mixture at 90°C for 2 min and cool down to room temperature over a period of 30 min.
10. Store the annealed duplexes at −20°C.

3.3. Vertical Agarose Gel Preparation

The glass plates must be cleaned to avoid the formation of air bubbles while pouring the gel.

1. Coat the rabbit-ear plate with hydrophobic glass treatment (e.g. Rain-X original glass treatment), and coat the back plate with hydrophilic glass treatment (e.g. Rain-X anti-fog).
2. Assemble the glass plates, set the silicon rubber gasket and the clips, and position it in a standing (vertical) position (Fig. 2).
3. For a 16 cm × 16 cm × 1.5-mm plate, 40 ml of agarose-TBE solution is required. Add 0.56 g agarose to 40 ml 0.5× TBE buffer (final concentration of 1.4%) in a 500-ml conical flask.
4. Cover the top with Saran Wrap, melt the agarose in a microwave oven until the agarose dissolves completely.
5. Allow the melted agarose to cool down to 60–70°C (see Note 9).
6. Slowly pour the agarose into the glass plates in a continuous stream, and immediately insert a 1.5-mm thick, 26-well comb between the glass plates. Adjust the comb so that the teeth will enter the gel by only ~5 mm (see Note 10).
7. When the gel has solidified, mark the position of each well along the comb (Fig. 2), and protect the mark with a band of transparent tape (see Note 11).
8. Carefully remove the silicon-rubber gasket and comb so as not to break the wells.

The gel may be stored at 4°C for a few days (see Note 12).

3.4. Native Gel Analysis of Ago1-RISC Formation

To detect the complexes in Ago1-RISC assembly, the electrophoresis should be performed at 4°C (see Note 13).

1. Before starting the in vitro RISC assembly, set the gel in the electrophoresis apparatus and fill the upper and lower reservoirs of the electrophoresis tank with pre-chilled 0.5 × TBE.
2. Use a bent syringe needle to remove any air bubbles trapped beneath the bottom of the gel.
3. Perform the in vitro Ago1-RISC assembly as follows. For a 10 μL reaction, mix 3 μL of 40× reaction mix, 5 μL of *dcr-2*

embryo lysate, 1 μL of 100 nM ^{32}P-radiolabeled duplex A or duplex B, and 1 μL of 100 nM RL 1× target mRNA and incubate the reaction mixture at 15 or 25°C for the desired time (see Note 14).

4. Directly load 2 μL of the sample into the well (see Note 15). The density of the embryo lysate is usually high enough that it is not necessary to add ficoll or glycerol to the sample (see Note 16).
5. After loading all the samples, load 2 μL of the tracking dye in a few empty wells and perform electrophoresis at 300 V until the bromophenol blue reaches the bottom of the gel (see Note 17). In our hands, it usually takes 1.5 h to complete the electrophoresis.
6. After electrophoresis, slowly remove the rabbit-ear plate. The gel will stay attached to the back plate.
7. Overlay a Hybond N$^+$ membrane, the size of which is slightly larger than the gels, directly onto the native gel. Make sure that the gel is uniformly attached to the membrane (see Note 18).
8. Overlay a Whatman 3MM paper onto the membrane, and slowly and carefully peal off the layers of the gel, the membrane and the 3MM paper from the glass plate.
9. Place a thin plastic sheet (or Saran Wrap) at the top of the gel and let the gel dry for 30 min to 1 h under vacuum in a gel dryer at 80°C.
10. Perform autoradiography or phosphorimaging to detect the complexes.

Alternatively, and to monitor the kinetics of Ago1-RISC formation, start running the gel at 10 V just before loading the first sample. At each time point, directly load the sample in a sequential manner. Keep running the gel at 10 V during the entire course of the sampling, and when all of the samples are loaded, raise the voltage to 300 V.

At least five complexes can be detected using this method (Fig. 3). The top one corresponds to pre-Ago1-RISC, and the second one from the top corresponds to mature Ago1-RISC bound to the target mRNA. The other complexes (complexes III–V) have been shown to be irrelevant for Ago1-RISC assembly (5).

3.5. Alternative Native Gel Analysis to Exclusively Detect the Mature Ago1-RISC

A number of complexes can be detected using the above method, but these complexes can sometimes overlap with each other, which may interfere with their precise quantification. By radiolabeling the target RNA instead of the small RNA, it is possible to detect only the mature Ago1-RISC, i.e., the complex capable of binding to the target. This is especially useful to quantify the

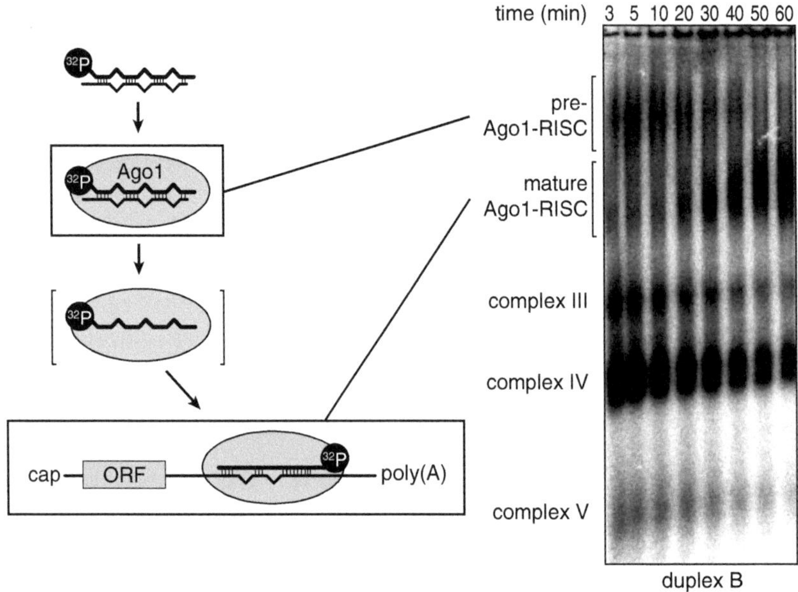

Fig. 3. Identification of pre-Ago1-RISC and mature Ago1-RISC. 5′ guide-radiolabeled duplex B were incubated with *dcr-2* embryo lysate at 25°C together with non-radiolabeled target mRNA (RL-1× mRNA). Complexes assembled at the indicated time points were then separated on a vertical agarose native gel.

amount of mature RISC, which quantitatively reflects the RISC activity. For this, we use a 2′-*O*-methylated antisense oligonucleotide (ASO) that is complementary to the guide strand, as an analog of the target RNA, because it is refractory to nonspecific endonucleases. To perform 5′ ^{32}P radiolabeling of 2′-*O*-methylated ASO, use the same protocol as for the guide strand RNA described above (Subheading 3.2), and phosphorylate each strand of the small RNA duplex with cold ATP.

Perform in vitro RISC assembly as follows:

1. For a 10 μL reaction, mix 3 μL of 40× reaction mix, 5 μL of *dcr-2* embryo lysate, 1 μL of 500 nM nonradiolabeled duplex A or duplex B. Incubate the reaction mixture at 25°C for 30 min.
2. Add 1 μL of 100 nM radiolabeled 2′-*O*-methylated ASO and incubate for 10 min.
3. Directly load 2 μL of the reaction into the 1.4% native agarose gel described above (see Note 19).
4. Load 2 μL of tracking dye in empty wells, and perform electrophoresis at 300 V in ice-cold 0.5× TBE buffer until the bromophenol blue reaches the bottom of the gel.
5. Proceed as in Subheading 3.4, with steps 6–10.

Only one complex can be detected on the gel, and a control experiment shows that the immunodepletion of Ago1 abolishes

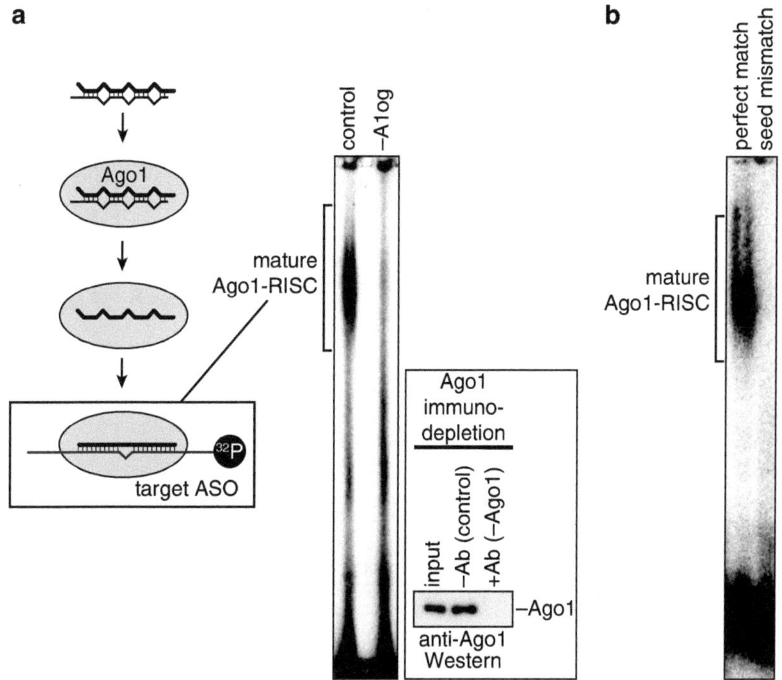

Fig. 4. Alternative native gel shift assay that detects only mature Ago1-RISC. (**a**) Non-radiolabeled duplex B was incubated in *dcr-2* lysate for 30 min, 5′[32]P-radiolabeled 2′-*O*-methylated ASO perfectly complementary to the guide strand was added and incubated for 10 min, and then the complexes were analyzed by native gel electrophoresis. Only one complex (control lane) was detected on the gel. This complex did not form when Ago1 was immunodepleted (−Ago1). Inset: Western blot analysis confirming efficient Ago1 immunodepletion. (**b**) Mature Ago1-RISC does form when the region of ASO complementary to the guide seed sequence was mutated. Native gel analysis was performed as in (**a**), using a 5′[32]P-radiolabeled perfect match ASO or seed mismatch ASO.

the formation of the complex (Fig. 4a). This complex does not form when mismatches are introduced between the seed of the guide and the target ASO (Fig. 4b), indicating that the complex is bona fide mature Ago1-RISC.

3.6. Native Gel Analysis of Ago2-RISC Assembly

The method to detect Ago2-RISC and its precursor complexes is slightly different from that for Ago1-RISC. Because genetic ablation of *ago1* is lethal, we cannot eliminate Ago1-related complexes. However, using a wild-type embryo lysate usually allows a satisfactory detection of the complexes involved in the Ago2-RISC assembly, as they are more abundant than Ago1-RISC complexes.

1. Before starting the in vitro Ago2-RISC assembly, supplement 0.5× TBE with 1.5 mM $MgCl_2$ and use it to prepare a 1.5% agarose gel and 0.5× TBE running buffer (see Note 20).
2. Perform the in vitro Ago2-RISC assembly as follows. For a 10 μL reaction, mix 3 μL of 40× reaction mix, 5 μL of wild-type embryo lysate, 1 μL of 100 nM 5′ ^{32}P-radiolabeled

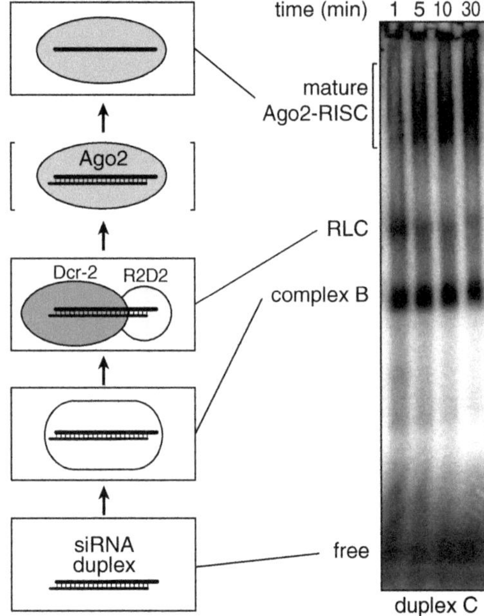

Fig. 5. Detection of complex B, Ago2-RLC, and mature Ago2-RISC. 5′ guide-radiolabeled duplex C was incubated with wild-type embryo lysate at 25°C. Complexes assembled at the indicated time points were then separated on a vertical agarose native gel.

duplex C, and 1 μL of water and incubate the reaction mixture at 25°C for 30 min.

3. Directly load 2 μL of reaction into the gel.
4. Perform electrophoresis at 300 V in 0.5× TBE buffer with 1.5 mM $MgCl_2$ until bromophenol blue reaches the bottom of the gel (see Note 17).
5. Proceed as in Subheading 3.4, with steps 6–10.
6. Three distinct complexes can be detected (Fig. 5): the top one is the mature Ago2-RISC; the middle one is the Ago2-RLC; and the bottom one is complex B, a putative precursor of Ago2-RLC. Visualizing the mature Ago2-RISC alone can also be achieved by using a radiolabeled 2′-*O*-methylated ASO complementary to the guide strand, as described above for the mature Ago1-RISC (13) (see Note 19). Sontheimer and colleagues have independently developed an acrylamide native gel system for similar purposes (12, 16).

4. Notes

1. Complex migration in agarose gels is affected by many factors. For example, the type and concentration of agarose influence the migration pattern of some complexes.

2. Concentration of Mg^{2+} in the gel and electrophoresis buffer also affects migration.
3. There are many types of electrophoresis apparatuses available commercially, and the arrangement of the glass plates and spacers should be adjusted accordingly. The spacing between the two plates should be 1.5–2 mm. Side spacers of ≤1 mm thickness are not suitable for vertical agarose gel, because it is difficult to pour melted agarose solution into a narrow space. In addition, such thin gels are very fragile and the wells may collapse easily when removing the comb.
4. A 0.5-mm spacer should be attached at the bottom, which prevents agarose gels from sliding off the plate. Instead of attaching the bottom spacer, using frosted or ground glass for either or both of the plates will also prevent gel slippage.
5. RL 1× target mRNA harbors *Renilla* Luciferase (RL) coding region with one *let-7*-binding site in the 3′ UTR. The target site is designed to have a central bulge to prevent endonucleolytic cleavage by Ago1-RISC. The FW primer contains T7 polymerase promoter sequence, and the RV primer contains the sequence corresponding to the *let-7* target site.
6. "40× Reaction mix" does not mean that it is concentrated 40-fold, but means that the mixture can afford 40 reactions of standard 10 μL RISC assembly. Amino acid stock and GTP are required for in vitro translation, but are dispensable for RISC assembly. Therefore, amino acid stock and GTP can be substituted with water.
7. Instead of a *dcr-2* mutant lysate, a lysate prepared from Schneider-2 (S2) cells overexpressing Ago1, and where *dcr-2* is knocked down can be also used.
8. Although we routinely use 0–12 h embryos, other stages of embryos (e.g., 0–2 h) may also be used.
9. If the agarose is too hot, the glass plate may break while pouring the agarose.
10. It is, particularly, important not to insert the comb too deeply, because vertical agarose gels are fragile. If the comb is inserted further than 5 mm, the wells can be corrupted when the comb is removed.
11. The marks will be very useful when applying the samples, because the samples without dyes are hardly visible.
12. For storage, the gels should be wrapped by water-soaked paper towels, and Saran Wrap.
13. It is ideal to perform the electrophoresis in the cold room. Alternatively, cool the electrophoresis buffer with coolant-filled plastic bars.

14. The internal structure of the duplex and the reaction temperature dramatically affect the formation of pre-Ago1-RISC and mature Ago1-RISC. Below 5°C, duplexes A and B do not form complexes with measurable efficiency. At 15°C, pre-Ago1-RISC is efficiently formed, but its conversion from pre-Ago1-RISC to mature Ago1-RISC is prevented. Therefore, in such conditions, both duplexes will only assemble within pre-Ago1-RISC. At 25°C, however, pre-Ago1-RISC is efficiently converted into mature Ago1-RISC. Hence, at this temperature, only duplex B will efficiently assemble into mature Ago1-RISC, while duplex A will predominantly assemble as pre-Ago1-RISC. At 37°C, both pre-Ago1-RISC and mature Ago1-RISC can be formed, but they are very unstable.
15. Although each well can hold up to 10 μL, it is better to load less than 5 μL of each sample for optimal resolution.
16. If necessary, Ficoll or Glycerol may be added at a final concentration of 3% to increase the density of the sample. Avoid adding bromophenol blue or other dyes to the samples, as they may interfere with complex formation.
17. Unincorporated, ^{32}P-radiolabeled duplex can be seen at the bottom of the gel, if the electrophoresis is stopped when the bromophenol blue has reached approximately two-thirds of the length of the gel.
18. In the absence of Hybond N^+ paper, the complexes will be dispersed on the 3MM paper.
19. Depending on the lysate preparations, nonspecific complexes, which form even in the absence of RISC programming, may be detected. In this case, addition of 1 mg/ml (final concentration) of Heparin after RISC assembly may be used to prevent their formation (13). However, note that the mobility of the mature RISC will be shifted downward when Heparin is used.
20. Mature Ago2-RISC and its precursors are better separated in the presence of 1.5 mM $MgCl_2$ (13). Note that 0.5× TBE contains 1 mM EDTA.

Acknowledgments

We thank M. Siomi and H. Siomi (Keio University) for the antibody to Ago1, R. Carthew (Northwestern University) for *dcr-2*^{L811fsX} and *dcr-2*^{416X} flies. We also thank members of the Tomari laboratory for helpful discussions. This work was supported in part by a Grant-in-Aid for Young Scientists (B) to T.K., and a

Grant-in-Aid for Young Scientists (A) and Grant-in-Aid for Scientific Research on Innovative Areas "Functional machinery for non-coding RNAs" to Y.T. from the Japan Ministry of Education, Culture, Sports, Science and Technology, and a Carrier Development Award from The International Human Frontier Science Program Organization to Y.T.

References

1. Bartel, D. P. (2004) MicroRNAs: genomics, biogenesis, mechanism, and function, *Cell 116*, 281–297.
2. Tomari, Y., and Zamore, P. D. (2005) Perspective: machines for RNAi, *Genes Dev 19*, 517–529.
3. Siomi, H., and Siomi, M. C. (2009) On the road to reading the RNA-interference code, *Nature 457*, 396–404.
4. Matranga, C., Tomari, Y., Shin, C., Bartel, D. P., and Zamore, P. D. (2005) Passenger-strand cleavage facilitates assembly of siRNA into Ago2-containing RNAi enzyme complexes, *Cell 123*, 607–620.
5. Kawamata, T., Seitz, H., and Tomari, Y. (2009) Structural determinants of miRNAs for RISC loading and slicer-independent unwinding, *Nat Struct Mol Biol 16*, 953–960.
6. Okamura, K., Ishizuka, A., Siomi, H., and Siomi, M. C. (2004) Distinct roles for Argonaute proteins in small RNA-directed RNA cleavage pathways, *Genes Dev 18*, 1655–1666.
7. Forstemann, K., Horwich, M. D., Wee, L., Tomari, Y., and Zamore, P. D. (2007) *Drosophila* microRNAs are sorted into functionally distinct argonaute complexes after production by Dicer-1, *Cell 130*, 287–297.
8. Tomari, Y., Du, T., and Zamore, P. D. (2007) Sorting of *Drosophila* small silencing RNAs, *Cell 130*, 299–308.
9. Forstemann, K., Tomari, Y., Du, T., Vagin, V. V., Denli, A. M., Bratu, D. P., Klattenhoff, C., Theurkauf, W. E., and Zamore, P. D. (2005) Normal microRNA maturation and germ-line stem cell maintenance requires Loquacious, a double-stranded RNA-binding domain protein, *PLoS Biol 3*, e236.
10. Saito, K., Ishizuka, A., Siomi, H., and Siomi, M. C. (2005) Processing of pre-microRNAs by the Dicer-1-Loquacious complex in *Drosophila* cells, *PLoS Biol 3*, e235.
11. Liu, Q., Rand, T. A., Kalidas, S., Du, F., Kim, H. E., Smith, D. P., and Wang, X. (2003) R2D2, a bridge between the initiation and effector steps of the *Drosophila* RNAi pathway, *Science 301*, 1921–1925.
12. Pham, J. W., Pellino, J. L., Lee, Y. S., Carthew, R. W., and Sontheimer, E. J. (2004) A Dicer-2-dependent 80s complex cleaves targeted mRNAs during RNAi in *Drosophila*, *Cell 117*, 83–94.
13. Tomari, Y., Du, T., Haley, B., Schwarz, D. S., Bennett, R., Cook, H. A., Koppetsch, B. S., Theurkauf, W. E., and Zamore, P. D. (2004) RISC assembly defects in the *Drosophila* RNAi mutant armitage, *Cell 116*, 831–841.
14. Lee, Y. S., Nakahara, K., Pham, J. W., Kim, K., He, Z., Sontheimer, E. J., and Carthew, R. W. (2004) Distinct roles for *Drosophila* Dicer-1 and Dicer-2 in the siRNA/miRNA silencing pathways, *Cell 117*, 69–81.
15. Haley, B., Tang, G., and Zamore, P. D. (2003) *In vitro* analysis of RNA interference in *Drosophila melanogaster*, *Methods 30*, 330–336.
16. Pham, J. W., and Sontheimer, E. J. (2005) Separation of *Drosophila* RNA silencing complexes by native gel electrophoresis, *Methods Mol Biol 309*, 11–16.

Chapter 8

Purification and Assembly of Human Argonaute, Dicer, and TRBP Complexes

Nabanita De and Ian J. MacRae

Abstract

The RNA-induced silencing complex (RISC) is a programmable gene-silencing machine involved in many aspects of eukaryotic biology. In humans, RISC is programmed or "loaded" with a small-guide RNA by the action of a tri-molecular assembly termed the RISC-loading complex (RLC). The human RLC is composed of the proteins Dicer, TRBP, and Argonaute2 (Ago2). To facilitate structural and biochemical dissection of the RISC-loading process, we have developed a system for the in vitro reconstitution of the human RLC. Here, we describe in detail methods for the expression and purification of recombinant Dicer, TRBP, and Ago2 and protocols for the assembly of RLCs and RLC subcomplexes. We also describe several simple assays to observe the biochemical activities of the assembled protein complexes.

Key words: RLC; RISC, Dicer; Argonaute, microRNA, RNAi

1. Introduction

RNA interference (RNAi) is a broad-spread eukaryotic mechanism of gene silencing that plays a fundamental role in many aspects of animal biology, including developmental timing, stem cell division, memory, and learning. On the molecular level, RNAi is mediated by a family of ribonucleoprotein complexes called RNA-induced silencing complexes (RISC), which silence genes by mediating translational repression and degradation of targeted message RNAs (mRNA) (1). The versatility and power of RNAi arises from the fact that RISC can be programmed to target any nucleic acid sequence for silencing. RISC programming is, therefore, a critical cellular function, requiring the action of a specialized macromolecular assembly called the RISC-loading complex (RLC) (2–4).

Tom C. Hobman and Thomas F. Duchaine (eds.), *Argonaute Proteins: Methods and Protocols*, Methods in Molecular Biology, vol. 725, DOI 10.1007/978-1-61779-046-1_8, © Springer Science+Business Media, LLC 2011

On the molecular level, the RLC programs RISC with target sequence information by mediating the non-covalent binding, or "loading", of a ~22 nucleotide RNA onto Argonaute proteins which are the core subunits of RISC. The small RNA functions as a guide for gene silencing through base pairing recognition of target mRNAs (5–8).

The mammalian RLC is a trimeric complex (350 kDa) composed of the proteins Dicer, TRBP, and Argonaute-2 (Ago2) (3, 9). The main function of the human RLC is to load Ago2 with microRNAs (miRNA), an abundant class of 22-nucleotide regulatory RNAs that arise from endogenous pre-miRNA hairpin structures. The RLC first recognizes pre-miRNA and cleaves, or "dices," it into a 22-nucleotide RNA duplex. Based on the stability properties of the duplex, one strand of RNA is selected to be the guide RNA for subsequent gene silencing and loaded into Ago2. The overall reaction is spontaneous and does not require any factors beyond the three proteins and a pre-miRNA (9).

The loading of Argonaute with an miRNA is perhaps the most important step in the mammalian RNAi pathway, because this is the point at which RISC is programmed with its target sequence information. To insure fidelity in the process, Ago2 loading is coupled to the pre-miRNA recognition and dicing steps. The RLC also has the ability to distinguish which strand in an miRNA duplex is to be loaded into Ago2 as the silencing guide and which RNA strand is to be discarded as the "passenger" (10, 11). This is an essential function because loading the incorrect RNA strand could lead to targeted silencing of an entirely different and unintended set of genes.

Here, we describe detailed methods for the expression and purification of each of the three protein components in the human RLC (9). We also describe methods for assembling the purified components into RLCs and assaying RLC activity. Methods described here should facilitate detailed biochemical and structural characterization of the RISC-loading mechanism and might be used to characterize various Ago-associated proteins.

2. Materials

2.1. RNA Oligonucleotides

1. The following RNAs (synthesized by Dharmacon) are used: *Drosophila* pre-let-7 (pre-let-7), 5′-AAUGAGGUAGUAGG UUGUAUAGUAGUAAUUACACAUCAUACUAUA CAAUGUGCUAGCUUUCU-3′; 37-nt A, 5′-UGAGGUA GUAGGUUGUAUAGUUUGAAAGUUCACGAUU-3′ and its complementary partner 37-nt B, 5′-UCGUGAACUUU CAAACUAUACAACCUACUACCUCAUU-3′; 21-nt guide RNA, 5′-UGAGGUAGUAGGUUGUAUAGU-3′; and 21-nt passenger RNA, 5′-UAUACAAUGUGCUAGCUUUCU-3′.

2. For radiolabelling, adenosine 5′-triphosphate, [γ-^{32}P] (3,000 Ci/mmol) (Perkin Elmer) and T4 Polynucleotide kinase (New England Biolabs) were used.
3. For preparing denaturing polyacrylamide gels, acrylamide 40% w/v solution (EMD); Urea (Fisher Chemicals); *N*, *N*, *N*′, *N*′-tetramethyl ethylenediamine (TEMED, Fisher Bioreagents); and ammonium persulfate (MP Biomedicals).
4. 2× denaturing loading buffer: 95% formamide, 18 mM EDTA, 0.025% SDS, 0.1% xylene cyanol, and 0.1% bromophenol blue.
5. RNA gel running buffer: 0.5× Buffer TBE (1× TBE: 89 mM Tris base, 89 mM boric acid, and 2.5 mM EDTA).
6. RNA gel apparatus (Dan-Kar Corp) and power supply (VWR).
7. For visualization, storage phosphor screen and Phosphorimager (Amersham, Healthcare Life Sciences).
8. For precipitating RNA, ethanol 200 proof (Sigma Aldrich) and 3 M sodium acetate, pH 5.2 (Fisher Bioreagents).

2.2. Baculovirus Production and Amplification

1. Human Dicer, Argonaute (Ago2) and TRBP cDNA clones (Open Biosystems).
2. pFastBac HTa vector (Invitrogen).
3. MAX Efficiency DH10BAC Competent cells (Invitrogen, Carlsbad, CA) and LB media.
4. LB agar plates containing 50 μg/ml kanamycin, 7 μg/ml gentamicin, 10 μg/ml tetracycline, 40 μg/ml isopropyl β-D-1-thiogalactopyranoside (IPTG), and 100 μg/ml bromo-chloro-indolyl-galactopyranoside (X-gal) (100 μg/ml).
5. DNA Miniprep solutions P1, P2, and N3 (Qiagen), isopropanol, 70% ethanol.
6. Fugene6 (Roche Applied Science).
7. ESF-921, Sf-9 cell media (Expression Systems, Woodland, CA).

2.3. Protein Expression and Purification

1. ESF-921, Sf-9 cell media (Expression Systems, Woodland, CA).
2. Lysis buffer: 300 mM NaCl, 50 mM sodium phosphate dibasic pH 8.0, 10 mM imidazole pH 8.0, 0.5% Triton X-100, 5% glycerol, 1 mM tris(2-carboxyethyl)phosphine (TCEP), and one tablet of EDTA-free protease inhibitor cocktail (Roche) per 25 ml buffer.
3. Lysis: Dounce tissue grinder (Kimble Chase Life Science and Research Products LLC).
4. Ni-NTA resin (Qiagen).

5. Wash Buffer: 300 mM NaCl, 50 mM Sodium phosphate dibasic heptahydrate pH 8.0, 20 mM imidazole pH 8.0, 5% glycerol, and 1 mM TCEP [tris(2-carboxyethyl)phosphine].
6. Elution Buffer: 300 mM NaCl, 50 mM sodium phosphate dibasic heptahydrate pH 8.0, 300 mM imidazole pH 8.0, 5% glycerol, and 1 mM TCEP [tris(2-carboxyethyl)phosphine].
7. TEV protease: purified in house by standard Nickel affinity purification, or alternatively may be purchased (Invitrogen).
8. Dialysis membrane (10,000 Da molecular weight cut-off) (Spectrum Labs).
9. HisTrap NiNTA column (Pharmacia) (GE Healthcare).
10. Superdex 200 16/60 column (Pharmacia) (GE Healthcare).
11. Gel filtration buffer: 300 mM NaCl, 50 mM HEPES pH 8.0, 5% glycerol, and 1 mM TCEP [tris(2-carboxyethyl) phosphine].
12. Bradford dye reagent (Bio-rad).

2.4. RLC Reconstitution

1. Gel filtration buffer: 300 mM NaCl, 50 mM HEPES pH 8.0, 5% glycerol, and 1 mM TCEP [tris(2-carboxyethyl) phosphine].
2. Superose 6 10/30 column (GE Healthcare).
3. SDS cassette gel (Expedeon).
4. Tris–Tricine–SDS running buffer with bisulfite (Expedeon).

2.5. RNA Filter-Binding Assay

1. BA-85 nitrocellulose filter (to retain protein–RNA complexes) (Whatman).
2. Hybond-N+ nylon membrane (to retain free RNA) (Amersham Biosciences).
3. MiniFold-1 Dot-Blot System (Whatman).
4. Membrane soaking buffer: 20 mM HEPES, pH 7.5.
5. RNA renaturing buffer: 10 mM Tris–HCl (pH 7.5), 1.5 mM Mg^{2+}, and 50 mM NaCl.
6. Reaction buffer: 20 mM HEPES (pH 7.5), 60 mM KCl, 5 mM EDTA, 1 mM DTT, 0.01% Igepal-680, and 0.1 mg/ml tRNA.

2.6. Dicing Assay

1. Reaction buffer: 100 mM NaCl, 40 mM HEPES, pH 7.5, 1 mM DTT, and 3 mM $MgCl_2$.
2. Pre-let-7 hairpin RNA, ^{32}P 5′ end-labeled 37 nt-A and cold 37 nt-B RNA.
3. RNA gel: 14% acrylamide gel (23 g Urea, 2.5 ml 10× TBE buffer, 17.5 ml 40% polyacrylamide, water to 50 ml, 71 μl TEMED, and 350 μl 10% ammonium persulfate).

4. 2× denaturing loading buffer: 95% formamide, 18 mM EDTA, 0.025% SDS, 0.1% xylene cyanol, and 0.1% bromophenol blue.

2.7. Slicing Assay

1. Reaction buffer: 0.1 mg/ml yeast tRNA, 20 mM Tris–HCl (pH 7), 50 mM KCl, 5% glycerol, and 1.5 mM $MgCl_2$.
2. 21-nt Guide RNA, 37 nt-A sense target RNA.

2.8. RISC-Loading Activity Assay

1. Reaction buffer: 0.1 mg/ml yeast tRNA, 20 mM Tris–HCl (pH 7), 50 mM KCl, 5% glycerol, and 1.5 mM $MgCl_2$.
2. Pre-let-7 hairpin RNA, 37 nt-A and -B RNAs, 21-nt guide, and passenger RNAs.

3. Methods

3.1. Radiolabeling RNA Oligos

1. 7 μl synthetic RNA substrates (5 μg/μl) are mixed with 1 μl adenosine 5 -triphosphate, [γ-^{32}P] (3,000 Ci/mmol), 1 μl T4 polynucleotide kinase, and 1 μl T4 polynucleotide kinase buffer and incubated for 1 h at 37°C.
2. Unreacted ATP is removed by passing the reaction mixture through an Illustra MicroSpin G-25 column. 2× denaturing loading buffer is then added to each sample. The volume of the flow through is estimated and an equal volume of 2× denaturing loading buffer is added prior to gel purification.
3. RNA samples are loaded into a denaturing 14% polyacrylamide gel poured by mixing 23 g urea, 2.5 ml 10× TBE buffer, 20 ml 40% polyacrylamide, and water to a final volume of 50 ml. Once the urea is dissolved 71 μl TEMED and 350 μl 10% ammonium persulfate are added to induce polymerization. The gel is run with 0.5× TBE buffer at a constant power of 20 W.
4. The RNA gel is wrapped in a layer of plastic wrap and exposed to a storage phosphor screen for 1–3 min and visualized by phosphorimaging. The imaged gel is then printed at full size (100%) so that the printed image has dimensions identical to those of the gel. The printed image is placed under the gel to help identify the physical position of the desired RNA. To align the printed image of the gel with the actual gel, "hot dots", or spots of ^{32}P placed on small pieces of filter paper which are then placed on top of the gel prior to exposure and imaging, can be employed. The desired band is cut out of the gel with a clean razorblade, crushed with a sterile needle, resuspended in approximately twice the volume of water, and incubated on a rocker overnight at 4°C.

5. The following day the aqueous solution is moved to a fresh Eppendorf tube and centrifuged briefly to pellet any residual gel fragments. The volume of the liquid is then estimated and added to a new Eppendorf tube containing 2.5 times (v/v) of 100% ethanol and 0.1 times (v/v) of 3 M sodium acetate, pH 5.2. The mixture is incubated at –80°C for 30 min to precipitate the RNA.
6. The precipitated RNA is centrifuged at maximum speed in a tabletop centrifuge for 10 min and the supernatant is carefully removed and discarded as radioactive waste. The pellet is subsequently washed with 1 ml of 70% ethanol, and centrifuged briefly again. The supernatant solution is removed and the pellet resuspended in 20 μl double-distilled water.

3.2. Baculovirus Production and Amplification

1. Baculoviruses separately expressing His_6-tagged Dicer, TRBP, and Ago2 are generated using a modified version of the Bac2Bac system (Invitrogen). This protocol begins after the insertion of cDNA clones of human Dicer, TRBP, and Ago2 individually into the plasmid pFastBac HTA, which appends a His_6-tag and recognition sequence for the Tobacco Etch Virus (TEV) protease appended to the N terminus of each protein. Plasmid DNAs are transformed into DH10BAC cells by mixing 250 ng of DNA with 10 μl of competent cells and incubating on ice for 20 min.
2. Cells are then heat shocked at 42°C for 45 s and immediately moved back onto ice. After 2 more minutes, 800 μl of LB media is added to each transformation and cells are allowed to recover at 37°C for 5 h with constant shaking.
3. After recovery, 10 μl of the transformation mixture is plated out on LB agar plates containing kanamycin, gentamicin, tetracycline, IPTG, and X-gal. Plates are incubated at 37°C for 48 h to allow large colonies to grow.
4. A large, well-isolated, white colony from each DH10BAC plate is identified and used to inoculate 2.5 ml of LB media containing 50 μg/ml kanamycin, 7 μg/ml gentamicin, and 10 μg/ml tetracycline. Cultures are grown overnight (12–18 h) at 37°C with vigorous shaking.
5. "Bacmid" DNA is isolated from each bacterial culture using the buffers from a QIAprep Spin Miniprep Kit; however, the spin columns are not employed. Cells from 1.5 ml of each bacterial culture are pelleted by brief centrifugation in 1.5 ml Eppendorf tubes using a tabletop microfuge. The supernatant solution is removed and the cell pellet resuspended in 250 μl buffer P1. Following resuspension of the cell pellet, 250 μl of buffer P2 is added to induce cell lysis. After a 3-min incubation at room temperature, 300 μl of buffer N3 is added

and the tube is gently mixed and then centrifuged for 15 min at maximum speed on a tabletop microfuge. The supernatant liquid is then carefully transferred to a new microfuge tube containing 800 μl of 100% isopropanol. After mixing, the tube is centrifuged for 5 min at maximum speed to pellet the precipitated DNA. The supernatant solution is discarded and the DNA pellet is washed once with 800 μl of 70% ethanol. The DNA pellet is air dried for 5 min (or until all the ethanol is evaporated) before it is resuspended in 100 μl buffer EB. DNA concentration is determined by spectrometry. A typical yield is 100 μl of a 1 mg/ml DNA solution.

6. Sf-9 cells grown in Excel 420 media are added to a 6-well plate so that they are 70% confluent ($\sim 1 \times 10^6$ cells per well). Plated cells are incubated at 27°C for 10 min to allow cells to attach. The media is removed and the cells are covered with 3.5 ml of fresh Excel 420.
7. 94 μl Excel 420 and 6 μl Fugene 6 are mixed in a tube by gentle tapping.
8. After 5 min, 1 μg bacmid DNA is added and mixed by tapping.
9. After 15 min, the transfection mixture is added to the cells in 1 well of a 6-well dish, swirling after every few drops. The plate is then incubated at 27°C for 4–5 days for the initial generation of the virus.
10. After 4–5 days, the liquid media, which contains the virus, is harvested and any cell debris are removed by centrifugation. The virus is then amplified by applying 200 μl of the virus-containing media to 1×10^6 of Sf-9 cells freshly plated in a 6-well dish. The remaining media, containing the initial virus is stored at 4°C.
11. After 2–4 days, step 11 is repeated two more times to obtain second and third amplifications of the virus. To obtain a large quantity of virus, the third amplification can be done in a sterile 500-ml flask containing 250×10^6 cells in 250 ml of Excel 420, infected with 5 ml of the second amplification of the virus and shaken at ~140 rpm.

3.3. Protein Expression and Purification

1. N-terminally His_6-tagged Human Dicer, Ago2, and TRBP can be purified separately from Sf-9 cells infected with baculovirus bearing the cDNA copy of the desired protein. For each protein, 1.5×10^9 cells in 750 ml of Excel 420 media are infected with 25 ml of virus (third amplification) at 27°C and then harvested 60–72 h after infection.
2. Cells are pelleted by centrifugation at $2{,}000 \times g$ for 10 min. Cell pellets are resuspended in 25 ml of ice-cold Lysis buffer. All subsequent steps are carried out on ice or at 4°C.

3. Cells are lysed by seven strokes with a B pestle of a 40-ml Dounce tissue grinder.
4. Insoluble material is pelleted by centrifugation (15 min, 20,000 × *g*) and the supernatant solution is applied to 2.5 ml (packed) of Ni-NTA resin in a 50-ml Falcon tube and gently rocked for 40 min (see Note 1).
5. The resin is pelleted by brief centrifugation and washed by resuspending in 45 ml of Wash buffer. The resuspended resin is pelleted again and subjected to four more rounds of washing.
6. Protein is eluted from the washed resin with 7.5 ml of Elution buffer.
7. To remove the N-terminal His_6-tag, 0.5 mg of TEV protease is added to the eluted protein. The protein solution is then dialyzed using a 10,000 Da molecular weight cut-off dialysis membrane against 1 L of wash buffer overnight.
8. Dialyzed protein is then passed through a 5-ml His-Trap column. The unbound material (protein without His_6) is collected and concentrated to 1–2 ml using a 15-ml Amicon Ultra centrifugal filter.
9. The concentrated protein sample is applied to a Superdex 200 16/60 column equilibrated in gel filtration buffer (100 mM KCl, 5% glycerol, 1 mM DTT, 20 mM HEPES, pH 7.5). The column is run with gel filtration buffer at a flow rate of 1 ml/min, collecting 2.5 ml fractions. Fractions containing non-aggregated protein are pooled, concentrated to 5–10 mg/ml, and used in subsequent reconstitution experiments.
10. Protein concentrations are determined by the Bradford Assay.

3.4. RLC Reconstitution

1. Dicer (700 μg or 3 nmol), Ago2 (600 μg or 6 nmol), and TRBP (550 μg or 11 nmol, assuming a dimer) are mixed in 250 μl (final volume) of gel filtration buffer and incubated on ice for 10 min (see Notes 2 and 3).
2. The protein solution is applied to a Superose 6 10/30 column equilibrated in gel filtration buffer. The column is run at a flow rate of 0.5 ml/min while collecting 0.5 ml fractions. Examples of the elution profiles of the various RLC components are shown in Fig 1.
3. Fractions are analyzed by SDS–PAGE and those containing the RLC are pooled and concentrated to ~1.5 mg/ml (~4 μM). Aliquots are stored frozen at −80°C.

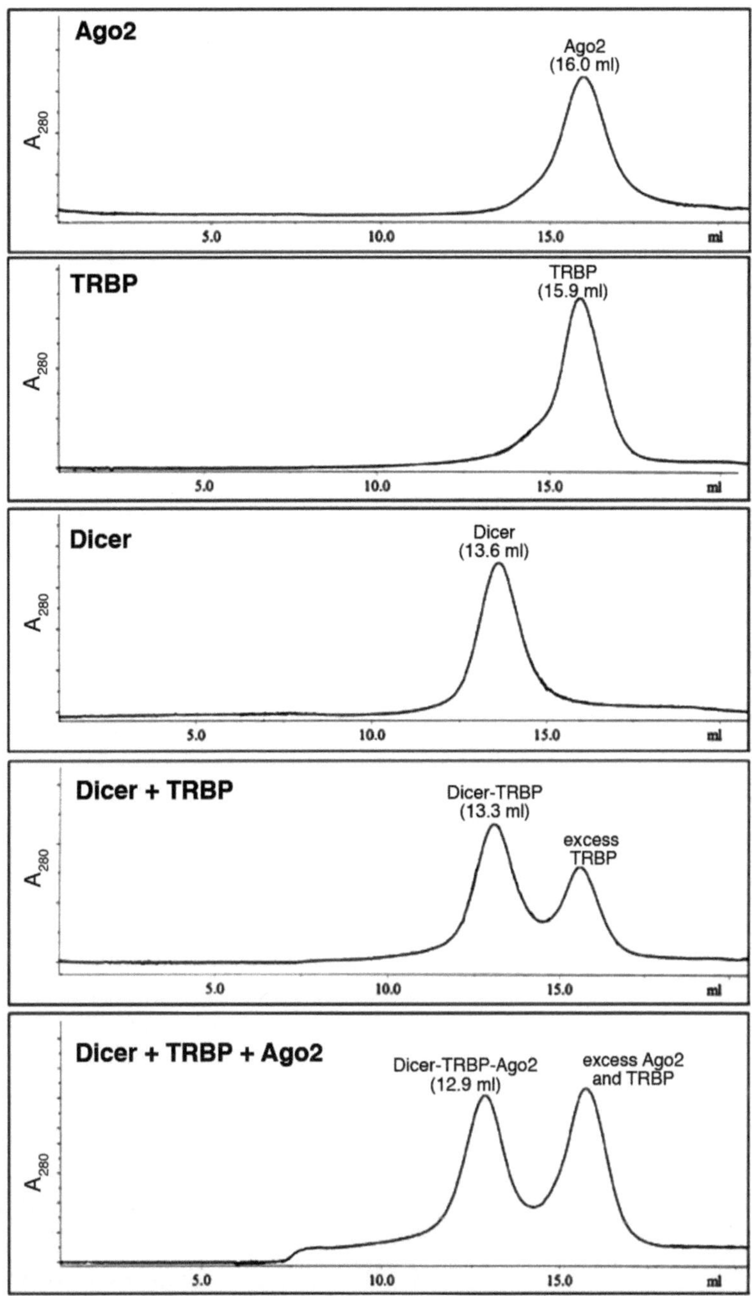

Fig. 1. Superose 6 elution profile of RLC and RLC components. The absorbance at 280 nm of the eluate is plotted against the elution volume for each protein sample. Protein components and elution volumes are indicated above each absorbance peak.

3.5. RNA Filter-Binding Assay

1. Prior to setting up the binding reaction, the RNA is annealed in a buffer containing 10 mM Tris–HCl (pH 7.5), 1.5 mM Mg^{2+}, and 50 mM NaCl by heating to 65°C for 7 min followed by snap cooling on ice.

2. 5′ end-labeled pre-let-7 hairpin RNA (<2 pM) is incubated in a 50-μl reaction volume containing 20 mM HEPES (pH 7.5), 60 mM KCl, 5 mM EDTA, 1 mM DTT, and 0.1 mg/ml tRNA.
3. The RNA is incubated with protein for 1 h prior to application to filters.
4. For each experiment, two filters are used: a BA-85 nitrocellulose filter to retain protein–RNA complexes and a Hybond-N+ nylon membrane to retain free RNA (12). These filters are soaked in a buffer containing 20 mM HEPES, pH 7.5, for 1 h prior to use in a 96-well dot blot apparatus.
5. A 40 μl aliquot from each reaction is applied to the top filter and then vacuum is applied to the apparatus to draw the samples through. The filters are *not* washed with additional buffer after the sample is drawn through.
6. After brief air-drying, the free and bound RNA is quantified by phosphorimaging of the filters.

3.6. Dicing Assay

1. Dicing substrates can be either an RNA hairpin (pre-let-7) or a dsRNA composed of two 37-nt RNA oligonucleotides (designated strand A and B) (see Note 4). For the hairpin RNA, ^{32}P 5 end-labeled RNA (~10 nM) is incubated in reaction buffer (100 mM NaCl, 40 mM HEPES, pH 7.5, 1 mM DTT, and 3 mM $MgCl_2$) in a metal heat block at 65°C for 7 min, then snap-cooled by immediately placing the sample on ice prior to every assay. To form the 37 nt duplex substrate, ^{32}P 5 end-labeled strand A (~10 nM) is mixed with an equal amount of unlabeled strand B RNA in reaction buffer. The RNA mixture is then denatured by placing in a 65°C metal heat block for 7 min and then slowly cooled to room temperature by moving the heat block (still containing the RNA sample) to the bench top.
2. Reactions are carried out in a total volume of 10 μl containing 100 mM NaCl, 40 mM HEPES, pH 7.5, 1 mM DTT, and 3 mM $MgCl_2$ with radiolabeled RNA at 37°C for 10–60 min in the presence of 100–200 nM RLC or Dicer alone.
3. RNA products are then resolved by denaturing PAGE (14% acrylamide–7 M urea in 0.5× TBE).
4. The gel is dried and visualized by phosphorimaging. An example of the results typically observed is shown in Fig. 2a.

3.7. Slicing Assay

1. RLC or Ago2 alone (100–200 nM) is loaded with a single-stranded guide 21-nt RNA by incubating with 100–200 nM guide RNA in a 10-μl reaction containing 50 mM KCl, 5% glycerol, 1.5 mM $MgCl_2$, and 20 mM Tris–HCl pH 7 for 30 min at 30°C.

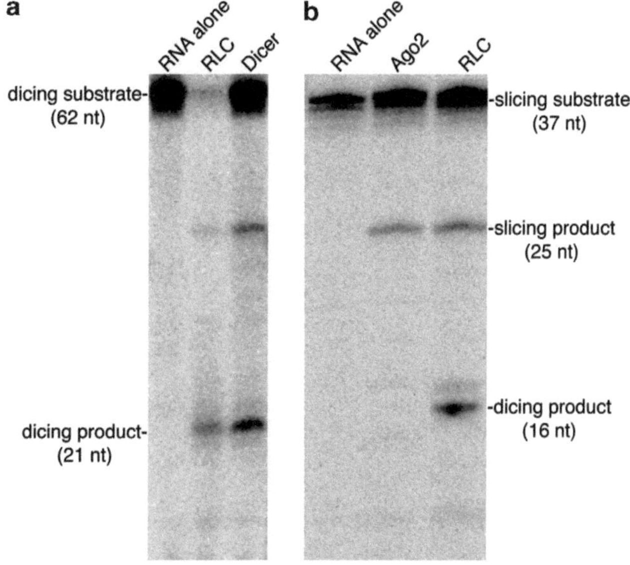

Fig. 2. Dicing and slicing assays. (**a**) Dicing assay – both the RLC or Dicer alone can process a pre-let-7 hairpin (63 nt) into a 21-nt guide RNA. (**b**) Slicing assay – once loaded with a single-stranded guide RNA, both Ago2 and the RLC will cleave the corresponding target RNA. The RLC can also produce a smaller 16 nt product. This is generated by the Dicer subunit of the RLC acting on target RNA that is annealed to excess guide RNA. Dicer removes 21 nt from the 37-nt target, which generates a 16-nt product.

2. The slicing reaction is started by the addition of 1–100 nM of 5′-end ^{32}P labeled single-stranded target 37 nt-A RNA and then carried out at 30°C for 10–60 min.
3. The reaction is stopped by adding 10 μl of formamide denaturing loading buffer and heating the samples to 75°C for 5 min.
4. Denatured samples are analyzed by electrophoresis on a 14% polyacrylamide–7 M urea gel.
5. The gel is dried and the labeled RNA substrates and products are detected by phosphorimager. An example of the results produced is shown in Fig. 2b.

3.8. RISC-Loading Activity Assay

1. The RISC-loading assay is very similar to the slicing assay except that dsRNA (Dicer substrate) is used as the source of the guide RNA. Reconstituted RLC (200 nM) is incubated with 200 nM pre-let-7 hairpin pre-miRNA in slicing assay buffer at 30°C for 30 min.
2. Radiolabeled target 37 nt-A or anti-target 37 nt-B RNA oligo (10 nM) is added and slicing activity is measured as described above. An example of the results produced from this assay is shown in Fig. 3.

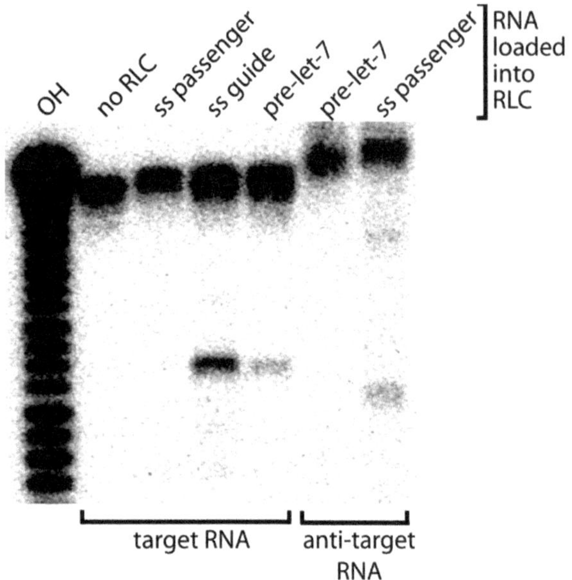

Fig. 3. RISC-loading assay. RLC can cleave target or anti-target (reverse complement of the target) RNA based on complementarity to the RNA loaded into the Ago2 subunit of the RLC. Lanes indicate the addition of unlabeled single-stranded 21-nt guide RNA (ss guide), single-stranded 21 nt passenger RNA (ss passenger) or pre-let7 RNA to RLC.

4. Notes

1. Proteins are purified from cell lysates using Ni-NTA resin in batch instead of using a Ni-NTA column. This is very important because crude Sf-9 cell lysates are often very viscous and will often clog conventional chromatography columns. Once the bound resin has been washed a few times it can be moved to a column format, if desired.
2. A key step in reconstitution of Dicer containing complexes, such as the RLC, is ensuring that Dicer is the limiting reagent in the assembly reaction. This is because Dicer runs anomalously large through size exclusion columns and it is, therefore, essentially impossible to separate the assembled RLC from free Dicer by this method. Fortunately, keeping Dicer-associated proteins (TRBP and Ago2) in twofold excess of Dicer eliminates any detectable free Dicer from the reaction.
3. The reconstitution protocols described here are specific for the human RLC. However, the methods are quite general and therefore (with some optimization) may be extended to other Ago2-containing molecular assemblies. Likewise, omitting either Ago2 or TRBP from the reconstitution reaction allows the generation of Dicer-TRPB or Dicer-Ago2 complexes, respectively. We have not been successful in producing a stable Ago2–TRBP complex.

4. The Dicing assays described here use either an RNA hairpin or a 35-bp RNA duplex. However, Dicer enzymes display flexibility in substrate specificity and can act on basically any dsRNA that posses (1) an open helical end, preferably with a 3 overhang; (2) a region of duplex structure that extends at least 20 bp from the helical end. When presented with long dsRNA, Dicer will cleave the substrate multiple times, each time ~21 bp from the open helical end, until all products are equal to or less than 23 bp in length.

Acknowledgments

We thank the members of the MacRae lab for helpful suggestions. Work on the Human RLC is funded by the NIH (R01 GM086701). I.J.M. is a Pew Scholar in the Biomedical Sciences.

References

1. Filipowicz, W. (2005) RNAi: the nuts and bolts of the RISC machine. *Cell* **122**, 17–20.
2. Gregory, R.I., Chendrimada, T.P., Cooch, N., & Shiekhattar, R. (2005) Human RISC couples microRNA biogenesis and posttranscriptional gene silencing. *Cell* **123**, 631–640.
3. Maniataki, E. & Mourelatos, Z. (2005) A human, ATP-independent, RISC assembly machine fueled by pre-miRNA. *Genes Dev.* **19**, 2979–2990.
4. Tomari, Y., Matranga, C., Haley, B., Martinez, N., & Zamore, P.D. (2004) A protein sensor for siRNA asymmetry. *Science* **306**, 1377–1380.
5. Meister, G. *et al.* (2004) Human Argonaute2 mediates RNA cleavage targeted by miRNAs and siRNAs. *Mol. Cell* **15**, 185–197.
6. Liu, J. *et al.* (2004) Argonaute2 is the catalytic engine of mammalian RNAi. *Science* **305**, 1437–1441.
7. Song, J.J., Smith, S.K., Hannon, G.J. & Joshua-Tor, L. (2004) Crystal structure of Argonaute and its implications for RISC slicer activity. *Science* **305**, 1434–1437.
8. Rivas, F.V., Tolia, N.H., Song, J.J., Aragon, J.P., Liu, J., Hannon, G.J., & Joshua-Tor, L. (2005) Purified Argonaute2 and an siRNA form recombinant human RISC. *Nature Struct. Mol. Biol.* **12**, 340–349.
9. MacRae, I.J., Ma, E., Zhou, M., Robinson, C.V., Doudna, J.A. (2008) In vitro reconstitution of the human RISC-loading complex. *Proc Natl Acad Sci U S A.* **105**, 512–7.
10. Schwarz DS, Hutvagner G, Du T, Xu Z, Aronin N, & Zamore PD (2003) Asymmetry in the assembly of the RNAi enzyme complex. *Cell* **115**, 199–208.
11. Khvorova A, Reynolds A, & Jayasena SD (2003) Functional siRNAs and miRNAs exhibit strand bias. *Cell* **115**, 209–216.
12. Batey RT, Sagar MB, & Doudna JA (2001) Structural and energetic analysis of RNA recognition by a universally conserved protein from the signal recognition particle. *J. Mol. Biol.* **307**, 229–246.

Chapter 9

Detection of Human Dicer and Argonaute 2 Catalytic Activity

Marjorie P. Perron, Patricia Landry, Isabelle Plante, and Patrick Provost

Abstract

The microRNA (miRNA)-guided RNA silencing pathway is a central and well-defined cellular process involved in messenger RNA (mRNA) translational control. This complex regulatory process is achieved by a well orchestrated machinery composed of a relatively few protein components, among which the ribonuclease III (RNase III) Dicer and Argonaute 2 (Ago2) play a central role. These two proteins are essential and it is of particular interest to measure and detect their catalytic activity under various situations and/or conditions. In this chapter, we describe different protocols that aim to study and determine the catalytic activity of Dicer and Ago2 in cell extracts, immune complexes, and size-fractionated cell extracts. Another protocol aimed at assessing miRNA binding to Ago2 is also described. These experimental approaches are likely to be useful to researchers investigating the main steps of miRNA biogenesis and function in human health and diseases.

Key words: Dicer, Argonaute 2, Enzyme activity, MicroRNA, MicroRNA precursor, Messenger RNA Target, Gene regulation, Method

1. Introduction

The microRNA (miRNA)-guided RNA silencing pathway is a recently discovered gene regulatory process present in almost all eukaryotic cells and based on miRNAs. These small RNA species approximately 21–23 nucleotides (nt) are encoded by the genome and are responsible for the recognition and translational control of specific messenger RNA (mRNAs). Involving relatively few protein components, this complex and well-integrated regulatory pathway plays a key role in recognizing a multitude of mRNA targets (1). Recent estimates suggest that up to 90% of the genes may be regulated by miRNAs in humans (2). Understanding the biological role and importance, as well as the possible defects of

Tom C. Hobman and Thomas F. Duchaine (eds.), *Argonaute Proteins: Methods and Protocols*, Methods in Molecular Biology, vol. 725, DOI 10.1007/978-1-61779-046-1_9, © Springer Science+Business Media, LLC 2011

the miRNA-guided RNA silencing pathway is of great interest, and some protein components have been already implicated in some human diseases (3).

MiRNA genes are transcribed by RNA polymerase II (RNA polII) into primary miRNAs (pri-miRNA) transcripts that adopt hairpin folds. The pri-miRNAs are recognized by the nuclear microprocessor complex, composed of the ribonuclease III (RNase III) Drosha and the DiGeorge syndrome Critical Region gene 8 (DGCR8) protein (4–7), and processed into a miRNA precursor (pre-miRNA). After being exported to the cytoplasm via Exportin-5 (8), the pre-miRNA is recognized by the pre-miRNA processing complex, composed of the RNAse III Dicer (9, 10), the TAR RNA binding protein (TRBP) (11, 12), and the PKR-activating protein (PACT) (13), to generate a miRNA:miRNA* duplex. The complex is then joined by the Argonaute 2 (Ago2) protein, and the miRNA guide strand is selected based on the relative stability of the duplex extremities, to form a miRNA-containing ribonucleoprotein (miRNP) complex (12). The associated miRNA confers to the miRNP complex the ability to recognize specific binding sites generally located in the 3 untranslated region (UTR) of different mRNAs. The mRNA will be cleaved if the complementarity between the miRNA and its binding site is perfect, or its translation regulated if the complementarity is imperfect (14). In this latter case, the repressed mRNA is translocated to the P-bodies, after which the mRNA can either be degraded or returned to the translational machinery for expression upon a specific cellular signal (15, 16).

Two of the major components of the miRNA-guided RNA silencing pathway are the RNase III Dicer and Ago2. These proteins are essential, and deregulation of their expression can have a major impact on normal cell functions (for a recent review, see Perron and Provost, (3)). Dicer recognizes its pre-miRNA substrates via its PAZ domain through the characteristic extremity harboring of pre-miRNA, formed by a 5′ phosphate and a 3′ hydroxylated end with 2-nt overhang, which represent the cleavage signature of members of the RNases III family of enzymes (17). The miRNA:miRNA* duplex is then excised by Dicer through the concerted intramolecular homodimerization of its two RNase III domains (18, 19).

As for Ago2, it is a member of the PAZ and PIWI domain (PPD) protein family expressed in metazoans and fungi, with the notable exception of the budding yeast *Saccharomyces cerevisiae* (20, 21). Ago2 harbors a binding pocket for miRNAs, in its PAZ domain, that mediate recognition of the characteristic 2-nt 3′ overhangs of miRNA duplexes (22–25). Acting in concert with the PAZ domain, the PIWI domain cleaves the mRNA strand between the nucleotides paired with the miRNA nt 10 and 11 if the complementarity is perfect. An active miRNP complex can then be regenerated and initiate a new round of mRNA cleavage, along a process known to amplify RNA silencing (26, 27).

In this chapter, we describe different protocols aimed to study and assess the specific catalytic activity of Dicer and Ago2 under various situations and/or conditions. The protocols can be easily transposed to different experimental contexts, i.e. cell types, cell lysates or immune complexes, and use various RNA substrates. Therefore, it is possible to compare wild-type and mutated proteins, as well as cells and tissues related to human diseases. We first propose a protocol for measuring Dicer and Ago2 catalytic activities in cell extracts and immune complexes, followed by the analytical methods (denaturing PAGE and an efficient Northern blot protocol to detect miRNAs) required to visualize the results. We also present a variation of these protocols to facilitate the study of fractionated cell extracts. Finally, we describe an efficient method to validate the presence of our miRNA of interest in Ago2 complexes.

2. Materials

It is important to use diethyl pyrocarbonate (DEPC)-treated water (see Note 1) in all preparation of the different solutions. Use RNase/DNase-free material in all conditions and always wear gloves for protection.

2.1. Cell Culture

1. Dulbecco's modified Eagle's medium (DMEM) or Roswell Park Memorial Institute 1,640 medium (RPMI) supplemented with 10% (v/v) fetal bovine serum (FBS), 1 mM sodium pyruvate, 100 units/mL penicillin, 100 μg/mL streptomycin, and 2 mM l-glutamine.
2. Phosphate-buffered saline (PBS).
3. 1× Trypsin- ethylenediaminetetraacetic acid (EDTA) solution.

2.2. In Vitro Transcription and Radiolabeling of RNA Transcript

1. MEGAshortscript T7 kit (Ambion).
2. DNA template (see Subheading 3.1 and Notes 2–4).
3. α ^{32}P UTP, (10 μCi/μL, ~3,000 Ci/mmol) (Perkin Life Science, 250 μCi).
4. γ ^{32}P ATP (10 μCi/μL, ~3,000 Ci/mmol) (Perkin Life Science, 250 μCi).
5. RNase/DNase-free screw-cap tubes (VWR).
6. 0.5 M EDTA at pH 8.0.
7. Calf intestine alkaline phosphatase (CIAP) with its 10× reaction buffer (GE Healthcare).
8. Opti-kinase with its 10× reaction buffer (USB Affimetrix).
9. RNase/DNase-free microcentrifuge tubes (VWR).

10. Sephadex-G25 column (GE Healthcare).
11. Glycogen (20 mg/m).
12. RNA annealing buffer: 10 mM Tris–HCl, pH 7.5, 1 mM EDTA, and 25 mM NaCl.
13. RNA gel extraction buffer: 0.5 M Ammonium Acetate, 1 mM EDTA and 0.2% SDS.

2.3. Dicer RNase Assay

2.3.1. S10 Cell Extracts

1. Dicer lysis buffer: 20 mM Tris–HCl pH 7.5, 150 mM NaCl, 1.5 mM $MgCl_2$, and 0.25% NP-40 (stored at 4°C). Add 1 mM phenylmethylsulfonyl fluoride (PMSF) and protease cocktail inhibitor without EDTA 1× (Roche), prior to use.
2. 2× Dicer assay buffer: 20 mM Tris–HCl pH 7.5, 2 mM $MgCl_2$, 75 mM NaCl, 10% glycerol. Add 1 mM PMSF and protease inhibitor cocktail mix without EDTA 1×, prior to use.

2.3.2. Immune Complexes

1. Dicer immunoprecipitation (IP) lysis buffer: 50 mM Tris–HCl pH 8.0, 137 mM NaCl, 1% Triton X-100 (stored at 4°C). Add 1 mM PMSF and protease cocktail inhibitor without EDTA 1×, prior to use.
2. Dicer IP washing buffer: 20 mM Tris–HCl pH 7.5, 2 mM $MgCl_2$ (stored at 4°C).
3. 2× Dicer IP assay buffer: 20 mM Tris–HCl pH 7.5, 10 mM $MgCl_2$, 2 mM dithiothreitol (DTT), and 2 mM adenosine triphosphate (28) (see Note 5), 10% Superase·In (Ambion). Prepare immediately prior to use.
4. Protein G agarose beads (Roche).
5. Control IgG and a suitable anti-Dicer antibody, such as our rabbit polyclonal anti-Dicer antibody (9).

2.4. Ago2 Cleavage Assay

2.4.1. S100 Cell Extracts

1. 2× Ago2 lysis buffer: 100 mM KOAc, 40 mM HEPES, 5 mM $MgCl_2$, 2 mM DTT, 0.35% Triton X-100, adjust pH to 7.6 (stored at 4°C). Add 1 mM PMSF and protease cocktail inhibitor without EDTA 1×, prior to use.
2. 10 mM ATP/2 mM guanosine triphosphate (GTP) solution (see Note 5).
3. Superase·In (Ambion).

2.4.2. Immune Complexes

1. Ago2 IP lysis buffer: 20 mM Tris–HCl pH 7.5, 150 mM NaCl, 1.5 mM $MgCl_2$, and 0.25% NP-40 (stored at 4°C). Add 1 mM PMSF and protease cocktail inhibitor without EDTA 1×, prior to use.
2. Ago2 IP washing buffer: 50 mM Tris–HCl pH 7.5, 300 mM NaCl, 5 mM $MgCl_2$, and 0.1% NP-40 (stored at 4°C).
3. 2× Ago2 lysis buffer (see step 1 in Subheading 3.4).

4. 10 mM ATP/2 mM GTP solution (see Note 5).
5. Superase·In (Ambion, cat. no. 2694).
6. Control IgG and anti-Ago2/EIF2C2 antibody. Our laboratory uses a mouse monoclonal antibody against human Ago2 (Abnova).

2.5. miRNA Detection in Ago2 Immune Complex

1. Ago2 IP lysis buffer (see Subheading 2.4.2).
2. Ago2 IP washing buffer (see Subheading 2.4.2).
3. Yeast tRNA (5 μg/mL) (Ambion).
4. Protein G agarose beads (Roche).
5. Control IgG and Ago2 antibody (Abnova).

2.6. Size-Fractionation of Cell Extracts Using a Fast Protein Liquid Chromatography System

1. FPLC lysis buffer: 50 mM Tris–HCl pH 8.0, 137 mM NaCl, and 1% Triton X-100 (stored at 4°C). Add 1 mM PMSF and protease cocktail inhibitor without EDTA 1×, prior to use.
2. 0.2 μm filter (Pall).
3. Tris elution buffer: 20 mM Tris–HCl pH 7.5 and 150 mM NaCl (stored at 4°C).
4. 2× Dicer FPLC assay buffer: 20 mM Tris–HCl pH 7.5, 4 mM $MgCl_2$, 75 mM NaCl, and 10% Glycerol. Add 2 mM PMSF and protease cocktail inhibitor without EDTA 2×, prior to use.
5. 2× Ago2 FPLC assay buffer: 100 mM KOAc, 40 mM HEPES, 5 mM $MgCl_2$, adjust pH to 7.6 (stored at 4°C) Add 4 mM DTT, 2 mM ATP and 0.2 mM GTP, prior to use.

2.7. RNA Extraction

1. RNase/DNase-free screw-cap tubes (VWR).
2. Glycogen (20 mg/mL) (Roche).
3. Yeast tRNA (5 μg/mL) (Ambion).
4. 0.5 M EDTA pH 8.0.
5. Proteinase K (20 mg/mL) (Ambion).
6. Proteinase K buffer: 200 mM NaCl, 20 mM Tris pH 8.0, 2 mM EDTA, 1% SDS.
7. Acid Phenol:$CHCl_3$ 5:1 solution pH 4.5 (Ambion).
8. 3 M sodium acetate, pH 5.5.
9. 5 M ammonium acetate.
10. 70 and 100% ethanol (ETOH).

2.8. Denaturing PAGE and Northern Blot Analysis of Small RNAs with EDC Cross-Linking

1. Acrylamide/Bis-Acrylamide 19:1 40% (Bio-Rad) (this is a neurotoxin when unpolymerized and care should be taken to avoid exposure).
2. Urea.
3. 10× Tris-borate-EDTA (TBE 10×): 89 mM Tris, 89 mM boric acid, and 2 mM EDTA.

4. 10× MOPS/NaOH solution: 200 mM MOPS, 0.05 M sodium acetate, and 0.1 M EDTA pH 8.0, adjust at pH 7.0 with NaOH. Keep at room temperature (RT) for 2 weeks. Protect from light.
5. *N,N,N,N′*-Tetramethyl-ethylenediamine (TEMED). TEMED is very corrosive and flammable, and should be handled with care. Wear gloves and work in a chemical hood.
6. Ammonium persulfate (APS) (prepare a 10% stock solution in water and stored at −20°C).
7. Hybond NX nylon membrane (GE Healthcare) and 3 MM chromatography paper.
8. 1-ethyl-3-(3-dimethylaminopropyl) carbodiimide (EDC) (Thermo Scientific).
9. 1-methylimidazole (Sigma).
10. 20× SSC: 3 M NaCl, 0.3 M sodium citrate, adjust at pH 7.0 with HCl.
11. Prehybridization solution: 2× SSC, 1% SDS, 100 μg/mL ssDNA.
12. Wash solution: 0.2× SSC, 2% SDS.
13. Stripping buffer: 10 mM Tris/HCl, pH 8.5, 5 mM EDTA, 0.1% SDS.
14. Decade marker (Ambion) (prepare as described in the company protocol with γ ^{32}P ATP).

2.9. Equipment

1. Fast protein liquid chromatography (FPLC) Akta system.
2. Refrigerated ultracentrifuge and microcentrifuge.
3. Spectrophotometer.
4. Vertical gel electrophoresis system with a small or large tank (GE Healthcare).
5. Semidry electroblotter (Fisher Biotech).
6. Roller for radioactivity. Hybridization oven.
7. End-over-end mixer.
8. Phosphorimager and storage phosphor screen, or X-ray film with intensifying screen, and an X-ray film processor.
9. Beta scintillation counter and Geiger counter.

3. Methods

3.1. In Vitro Transcription and Radiolabeling of RNA Transcript

These protocols are useful for different RNA substrates as well for RNA probes for Northern blot analysis. DNA templates will be different and prepared according to your conditions and specific requirements. T7-mediated in vitro transcription requires a DNA template with a T7 RNA polymerase promoter site (see the MEGAshort script T7 kit protocol). The template can be prepared by annealing of two complementary oligonucleotides, or by PCR filling with a T7 RNA polymerase promoter site oligonucleotide and a complementary oligonucleotide containing this sequence and the sequence to be transcribed. For Dicer RNA substrate, select your pre-miRNA sequence of interest (see Note 2) in miRBase http://microrna.sanger.ac.uk, which is a searchable database of published miRNA sequences and annotation (29), from which a complementary probe can be prepared (see Note 3). Finally, to detect Ago2 catalytic activity, the most efficient RNA substrate is the open–open RNA probe described by Ameres et al. (30).

3.1.1. *In Vitro Transcription for RNA Radiolabeling with [α ^{32}P] UTP*

1. Since working with radioactive material requires particular attention, be sure to have all the relevant information at hand (see Note 6).
2. Mix, according to the MEGAshort script T7 kit protocol, water, DNA template, ATP, CTP, GTP, transcription buffer, T7 RNA polymerase and add, at the end, the 20 μCi of [α ^{32}P] UTP. Incubate for 3 h at 37°C. Add DNase I, mix and incubate for 15 min at 37°C. Stop the reaction by adding 1 μL of EDTA 0.5 M, mix and incubate for 2 min at RT (see Note 7, RNA probe for Northern blot analysis). Add 20 μL of gel loading buffer (GLB) II (provided in the MEGAshort script T7 kit). Heat for 5 min at 95°C and quickly put on ice for 5 min.
3. Load all 42 μL on a denaturing PAGE (see the protocol below, Subheading 3.6).
4. Run at 275 V in TBE 1× as gel running buffer, until the bromophenol blue reaches 2 in. from bottom of the gel (duration: ~2 h 30 min).
5. When the migration is finished, keep the gel on one of the two glass plates and wrap using a plastic wrap. Put the gel in a plexiglass box (see Note 8).
6. In the dark room, expose the gel to an X-ray film for 30 s and 2 min. Back in the radioactivity room, cut out the band corresponding to your RNA probe.
7. Cut the gel slice in small pieces and put into a 1.5-mL screw-cap tube. Add 400 μL of gel extraction buffer and incubate overnight at 37°C.
8. Centrifuge at 600 × *g* for 1 min at 4°C. Transfer the supernatant into a new 1.5-mL screw-cap tube. Add 1 μL of glycogen and 1 mL of ice-cold ethanol 100%. Incubate for 30 min at −80°C, then centrifuge at 16,000 × *g* for 30 min at 4°C.

9. Remove the supernatant with a pipet and wash with 900 μL of ice-cold ethanol 70%, then centrifuge at 16,000×*g* for 5 min at 4°C.
10. Dry the RNA pellet and resuspend the Dicer substrate probe in 20 μL of annealing buffer. Anneal the RNA probe by heating for 5 min at 85°C and remove the block from the heating block and allow temperature to cool down to RT.
11. Remove 1 μL, add to 5 mL of scintillation liquid and count in the beta scintillation counter.
12. Aliquot in different tubes (make a tube ready to use at 40,000 cpm/μL) and store at −20°C.

3.1.2. In Vitro Transcription for RNA Radiolabeling with [γ-^{32}P] ATP

1. Mix, according to the MEGAshort script T7 kit protocol, water, DNA template, ATP, CTP, GTP, UTP, transcription buffer, and T7 RNA polymerase to obtain a final volume of 20 μL. Incubate for 3 h at 37°C. Add DNase I, mix and incubate for 15 min at 37°C. Stop the reaction by adding 1 μL of EDTA 0.5 M, mix and incubate for 2 min at RT. Keep the reaction on ice.
2. Prepare a Sephadex-G25 column according to the manufacturer's protocol.
3. Put the RNA on the column and centrifuge at 600×*g* for 2 min at 4°C.
4. Add 180 μL of water and 200 μL of Acid Phenol:$CHCl_3$ (5:1 solution, pH 4.5). Vortex for 20 s. Centrifuge at 16,000×*g* for 4 min at 4°C. Transfer the supernatant into a new 1.5-mL screw-cap tube.
5. Add 1 μL of glycogen and 1 mL of ice-cold ethanol 100%. Incubate for 30 min at −80°C, then centrifuge at 16,000×*g* for 30 min at 4°C.
6. Remove the supernatant with a pipet and wash with 900 μL of ice-cold ethanol 70%, then centrifuge at 16,000×*g* for 5 min at 4°C.
7. Dry the RNA pellet and resuspend in 40 μL of water. Add 5 μL of 10× CIAP buffer, 4 μL of water, and 1 μL of CIAP. Incubate for 1 h at 37°C.
8. Add 150 μL of water and 200 μL of Acid Phenol:$CHCl_3$ (5:1 solution, pH 4.5). Vortex for 20 s. Centrifuge at 16,000×*g* for 4 min at 4°C. Transfer the supernatant into a new 1.5-mL screw-cap tube.
9. Add 1 μL of glycogen, 50 μL of 5 M ammonium acetate, and 700 μL of ice-cold ethanol 100%. Incubate for 30 min at −80°C, then centrifuge at 16,000×*g* for 30 min at 4°C.

10. Remove the supernatant with a pipet and wash with 900 μL of ice-cold ethanol 70%, then centrifuge at 16,000 ×*g* for 5 min at 4°C.
11. Dry the RNA pellet and resuspend it in 20 μL of water. Evaluate the RNA concentration (see Note 9).
12. To 5 pmol of dephosphorylated RNA, add 2 μL of 10× kinase buffer, 20 μCi of [γ-^{32}P] ATP, 10 U of Opti-kinase and complete with water to have a 20-μL final volume. Incubate for 1 h at 37°C. Add 20 μL of GLB II. Heat for 5 min at 95°C and quickly put on ice for 5 min.
13. Purify the RNA on a denaturing PAGE as described in Subheading 3.1.1, steps 3–8.
14. If the RNA is an Ago2 RNA target, resuspend it in 20 μL of RNase free water. If the RNA is a Dicer substrate, resuspend it in 20 μL of annealing buffer and anneal the probe by heating for 5 min at 85°C and remove the block from the heating block and allow temperature to cool down to RT.
15. Remove 1 μL, add to 5 mL of scintillation liquid and count in the beta scintillation counter.
16. Aliquot in different tubes (make a tube ready to use at 40,000 cpm/μL for Dicer RNase assay or 10,000 cpm/μL for Ago2 cleavage assay) and store at −20°C.

3.2. Detection of Human Dicer Activity

3.2.1. Detection of Human Dicer Activity in S10 Cell Extracts

This protocol is adapted from Haase et al. (11).

1. Lyse mammalian cells (e.g. HEK293) from one 100-mm petri dish with 250 μL of *Dicer lysis buffer*, incubate for 15 min on ice, and centrifuge at 10,000 ×*g* for 10 min at 4°C. Determine the protein concentration by the method of Bradford (31) using the Bio-Rad dye reagent, with bovine serum albumin as standard. Adjust protein concentration to 5 mg/mL with *Dicer lysis buffer*.
2. Prepare the 50-μL reaction on ice in 1.5-mL screw-cap tubes as follows: 25 μL of S10 cell extract at 5 mg/mL (125 μg protein in total), 24 μL of *2× Dicer assay buffer*, and 1 μL of a radiolabeled dsRNA substrate probe (40,000 cpm/μL). Vortex.
3. Incubate for 10, 30, 60, and 120 min at 37°C.
4. To stop the reaction, quickly add 150 μL of water and 200 μL of Acid Phenol:$CHCl_3$ (5:1 solution, pH 4.5). Vortex for 20 s. Separate the aqueous and organic phases by centrifugation at 16,000 ×*g* for 4 min at 4°C.
5. Precipitate RNA by adding 50 μL of 5 M ammonium acetate, 700 μL of ice-cold ethanol 100%, 1 μL of yeast tRNA, and 1 μL of glycogen. Mix well and incubate for at least 1 h at −80°C.

6. Centrifuge at 16,000 × *g* for 30 min at 4°C. Wash with 400 μL of ice-cold ethanol 70%. Centrifuge at 16,000 × *g* for 5 min at 4°C.
7. Dry the RNA pellet, add 10 μL of GLB II, and heat for 5 min at 95°C.
8. Load RNA on a denaturing PAGE (see the detailed protocol in Subheading 3.6). A typical result is shown in Fig. 1a.

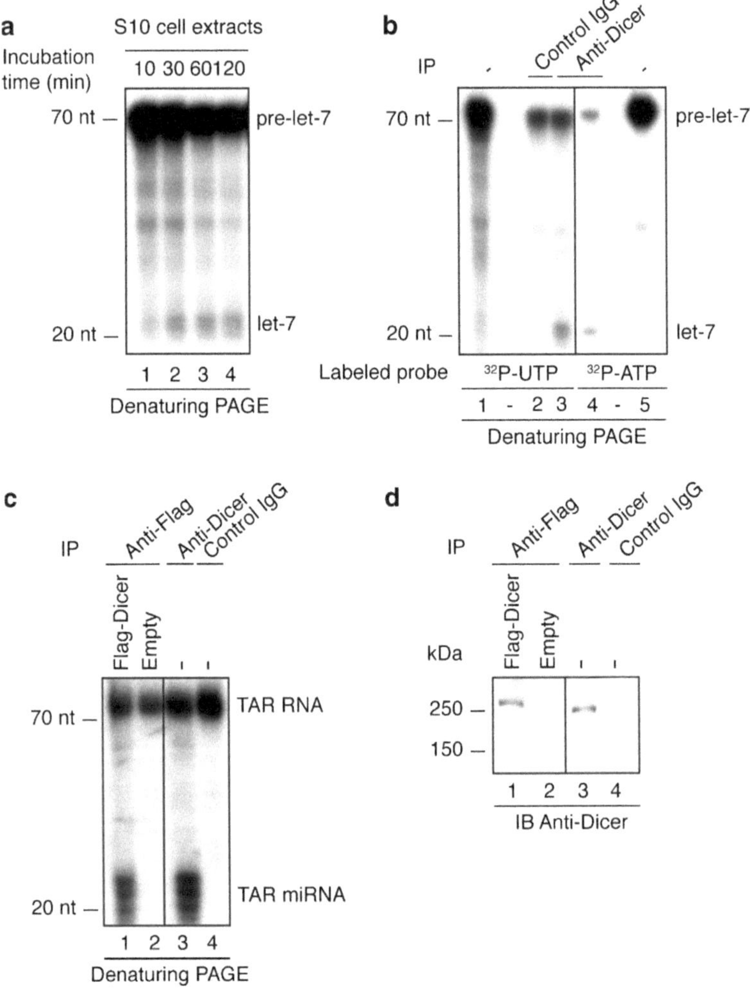

Fig. 1. Detection of human Dicer activity in HEK293 cells. (**a**) S10 protein extracts were incubated in the presence of a ^{32}P-UTP labeled human let-7a-3 pre-miRNA substrate for the indicated period of time. (**b**) Endogenous Dicer immune complexes were incubated with a ^{32}P-UTP or ^{32}P-ATP labeled human let-7a-3 pre-miRNA substrate for 60 min. Lanes 1 and 5 represent the untreated probe (–). (**c**) Anti-Flag immune complexes derived from cells overexpressing Flag-Dicer, or transfected with empty plasmid, and endogenous Dicer immune complexes derived from untransfected HEK293 cells (–) were incubated in the presence of ^{32}P-UTP-labeled TAR RNA substrate for 60 min. (**a**–**c**) The reactions were analyzed by denaturing PAGE and autoradiography. A 10-nt RNA ladder was used as a size marker. (**d**) The immune complexes from (**c**) were analyzed for the presence of Dicer protein by 7% SDS-PAGE and immunoblotting using anti-Dicer antibody (9).

3.2.2. Detection of Human Dicer Activity in Immune Complexes

This protocol is adapted from Provost et al. (9).

1. For IP of endogenous proteins, plate the cells and incubate in a CO_2 incubator overnight at 37°C. For IP of overexpressed proteins, plate the cells so they reach appropriate confluency for transfection by calcium phosphate, or another transfection procedure, on the following day and incubate the cells an additional 24–48 h.
2. Lyse cells from one 100-mm petri dish with 1 mL of *Dicer IP lysis buffer*, incubate for 15 min on ice, and centrifuge at 10,000 × *g* for 10 min at 4°C. Determine the protein concentration by the method of Bradford (31).
3. Prepare the IP by incubating 1 mg of protein extract with the appropriate antibody (see Note 10) for 1 h at 4°C under continuous rotation. During this time, pre-wash Protein-G agarose beads extensively (100 volumes, two times) with *Dicer IP lysis buffer* and resuspend the beads in 1 volume of buffer. Then, add 20 μL of beads (50% slurry) to the reaction and continue the incubation for an additional 3 h.
4. Wash the immune complexes three times with 1 mL of *Dicer IP lysis buffer*. Then, wash once with the *Dicer IP washing buffer* and transfer the beads into a clean 1.5-mL screw-cap tube. Be sure to leave ~10 μL of beads in the tube, by gently removing the supernatant with a pipette.
5. Prepare the 20-μL reaction on ice in a 1.5-mL screw-cap tube as follows: To the beads, add 9 μL of *2× Dicer IP assay buffer* and 1 μL of a radiolabeled dsRNA substrate probe (40,000 cpm/μL). Vortex.
6. Incubate for 60 min at 37°C.
7. Stop the reaction by quickly adding 180 μL of water and 200 μL of Acid Phenol:$CHCl_3$ (5:1 solution, pH 4.5). Vortex for 20 s. Separate the aqueous and organic phases by centrifugation at 16,000 × *g* for 4 min at 4°C.
8. Perform RNA precipitation and gel electrophoresis as described in Subheading 3.2.1, steps 5–8. Typical results are shown in Fig. 1b, c when using different RNA labeled probe.

3.3. Detection of Ago2 Cleavage Activity

3.3.1. Detection of Ago2 Cleavage Activity in S100 Cell Extracts

This protocol is adapted from Ameres et al. (30).

1. Lyse mammalian cells from one 100-mm petri dish with 150 μL *2× Ago2 lysis buffer*, incubate for 15 min on ice and perform an ultracentrifugation at 100,000 × *g* for 1 h at 4°C. Determine the protein concentration by the method of Bradford (31). Adjust the protein concentration of the extract at 5 mg/mL with *Ago2 cleavage assay lysis buffer*.

2. Prepare the 20-μL reaction on ice in a 1.5-mL screw-cap tube as follows: 10 μL of S100 cell extract at 5 mg/mL (50 μg protein in total), 2 μL of 10 mM ATP/2 mM GTP solution, 0.5 μL of Superase·In, 6.5 μL of water and, finally, 1 μL of a radiolabeled RNA target probe (10,000 cpm/μL). Vortex.
3. Incubate for 90 min at 30°C.
4. Add 20 μL of proteinase K buffer and 1 μL of proteinase K. Incubate for 20 min at 50°C.
5. Stop the reaction by quickly adding 160 μL of water and 200 μL of Acid Phenol:$CHCl_3$ (5:1 solution, pH 4.5). Vortex for 20 s. Separate the aqueous and organic phases by centrifugation at 16,000 ×*g* for 4 min at 4°C.
6. Precipitate RNA by adding 20 μL of 3 M sodium acetate pH 5.2, 500 μL of ice-cold ethanol 100%, 1 μL yeast tRNA, and 1 μL of glycogen. Mix well and incubate for at least 1 h at –80°C.
7. Centrifuge at 16,000 ×*g* for 30 min at 4°C. Wash with 500 μL of ice-cold ethanol 70%. Centrifuge at 16,000 ×*g* for 5 min at 4°C.
8. Dry the RNA pellet and add 10 μL of GLB II.
9. Load RNA on a denaturing PAGE (see the detailed protocol in Subheading 3.6). A typical result is shown in Fig. 2a.

3.3.2. Detection of Ago2 Cleavage Activity in Immune Complexes

This protocol is adapted from Ameres et al. (30), Ender et al. (32), Rudel et al. (33).

1. For IP of endogenous proteins, plate the cells and incubate in a CO_2 incubator overnight at 37°C. For IP of overexpressed proteins, plate the cells so they reach the appropriate

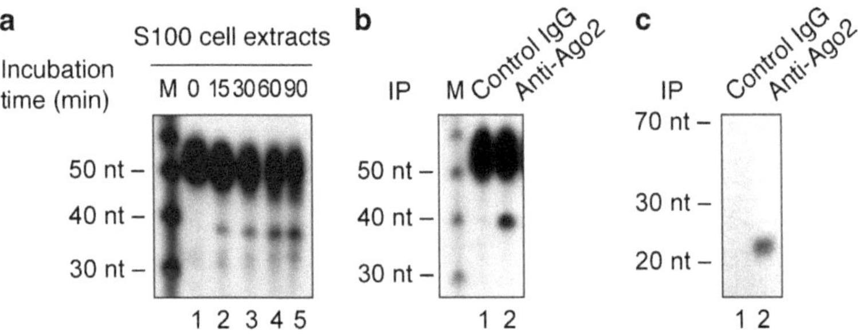

Fig. 2. Detection of human Ago2 activity in megakaryocytes. RISC activity assays were performed using S100 protein extracts (**a**) or Ago2 immune complex (**b**) from MEG-01 cell line and incubated in the presence of a ^{32}P-labeled sensor RNA bearing a binding site complementary to human miR-223. (**c**) Ago2 immune complexes were analyzed by Northern blot for the presence of miR-223. The reactions were analyzed by denaturing PAGE and autoradiography. A 10-nt RNA ladder was used as a size marker (M). Adapted from Landry et al. (35), with permission from The Nature Publishing Group.

confluency for transfection by calcium phosphate, or another transfection procedure, on the following day and incubate the cells for an additional 24–48 h.

2. Lyse cells from one 100-mm petri dish with 1 mL of *2× Ago2 IP lysis buffer*, incubate for 15 min on ice, and centrifuge at 10,000 × *g* for 10 min at 4°C. Determine the protein concentration by the method of Bradford (31).
3. Prepare the IP by incubating 1 mg of protein extract with the appropriate antibody (see Note 11) for 1 h at 4°C under continuous rotation. During this time, prewash Protein-G agarose beads extensively (100 volumes, two times) with *2× Ago2 lysis buffer* and resuspend the beads in 1 volume of buffer. Then, add 20 µL of beads (50% slurry) to the reaction and continue the incubation for an additional 3 h.
4. Wash the immune complexes three times with 1 mL of *Ago2 IP washing buffer*. Then, wash a last time with the *2× Ago2 lysis buffer* and transfer the beads into a new 1.5 mL-screw-cap tube. Be sure to leave ~10 µL of beads in the tube, by gently removing the supernatant with a pipette.
5. Prepare the reaction on ice in a 1.5-mL screw-cap tube as follows: To the beads, add 10 µL of *1× Ago2 lysis buffer* (a dilution of the *2× Ago2 lysis buffer*), 2 µL of 10 mM ATP/2 mM GTP solution, 0.5 µL of Superase•In, and finally 1 µL of a radiolabeled RNA target probe (10,000 cpm/µL). Vortex.
6. Incubate for 90 min at 30°C.
7. Add 20 µL of proteinase K buffer and 1 µL of proteinase K. Incubate for 20 min at 50°C.
8. Stop the reaction by quickly adding 180 µL of water and 200 µL of Acid Phenol:$CHCl_3$ (5:1 solution, pH 4.5). Vortex for 20 s. Separate the aqueous and organic phases by centrifugation at 16,000 × *g* for 4 min at 4°C.
9. Perform RNA precipitation as described in Subheading 3.3.1, steps 6–9. A typical result is shown in Fig. 2b.

3.4. Detection of miRNAs Bound to Ago2 Immune Complexes

This protocol is adapted from Ameres et al. (30), Ender et al. (32), and Rudel et al. (33).

1. Prepare cell extracts as described in Subheading 3.3.2, steps 1 and 2.
2. To 1 mg of proteins, add 6 µL of yeast tRNA, 1.5 µL of Superase·In, and adjust the final reaction volume to 300 µL with *2× Ago2 IP lysis buffer* (used in Subheading 3.3.2). Preclear the reaction mixture with 10 µL of Protein-G agarose beads prewashed with *2× Ago2 IP lysis buffer*. Incubate for 45 min at 4°C under continuous rotation.

3. Centrifuge at 600 × *g* for 1 min at 4°C, and transfer the supernatant to a new 1.5-mL screw-cap tube. Add 10 μL of fresh prewashed Protein-G agarose beads and the appropriate antibody (see Note 11) and incubate for 3 h at 4°C under continuous rotation.
4. Wash six times with 1 mL of *Ago2 IP washing buffer*.
5. Add 20 μL of proteinase K buffer and 1 μL of proteinase K. Incubate for 20 min at 50°C.
6. Stop the reaction by quickly adding 150 μL of water and 200 μL of Acid Phenol:$CHCl_3$ (5:1 solution, pH 4.5). Vortex for 20 s. Separate the aqueous and organic phases by centrifugation at 16,000 × *g* for 4 min at 4°C.
7. Precipitate RNA by adding 20 μL of 3 M sodium acetate pH 5.2, 500 μL of ice-cold ethanol 100% and 1 μL of glycogen. Mix well and incubate for at least 1 h at −80°C.
8. Centrifuge at 16,000 × *g* for 30 min at 4°C. Wash with 500 μL of ice-cold ethanol 70%. Centrifuge at 16,000 × *g* for 5 min at 4°C.
9. Dry the RNA pellet, add 10 μL of GLB II and heat for 5 min at 95°C.
10. Load RNA on a MOPS denaturing PAGE to perform Northern blot analysis (see the detailed protocol in Subheading 3.6). A typical result is shown in Fig. 2c.

3.5. Detection of Human Dicer and Ago2 Activity in Size-Fractionated Cell Extracts

1. Lyse cells from 3 to 4 100-mm petri dishes with *FPLC lysis buffer*, incubate for 15 min on ice and perform an ultracentrifugation at 100,000 × *g* for 45 min at 4°C. Filter the supernatant through a 0.2 μm filter and determine the protein concentration by the method of Bradford (31). Load 1 mg (100 μL of 10 mg/mL) of proteins derived from this S100 cell extract on a Superose 6 column (10/300 GL) using an ÄKTA FPLC system to give fractions of 400 μL in the *Tris elution buffer* (0.3 mL/min) (see Note 12).
2. Divide each fraction as follows; transfer 100 μL for Dicer RNase activity assay and 100 μL for Ago2 cleavage assay in separate 1.5-mL screw-cap tubes. For the input, take 5 μL of the S100 cell extract (10 mg/mL) and add 95 μL of elution buffer. Elution buffer (100 μL) will serve as a negative loading control (see Note 13).
3. For Dicer RNase activity assay: To prepare the 200-μL reaction mixture, add 99 μL of *2× Dicer FPLC assay buffer* and 1 μL of a radiolabeled dsRNA substrate, mix and incubate for 60 min at 37°C. To stop the reaction quickly add 200 μL of Acid Phenol:$CHCl_3$ (5:1 solution, pH 4.5). Vortex for 20 s. Separate the aqueous and organic phases by centrifugation at

16,000×g for 4 min at 4°C. Perform RNA precipitation, as described in Subheading 3.2.1, steps 5–8. A typical result is shown in Fig. 3b.

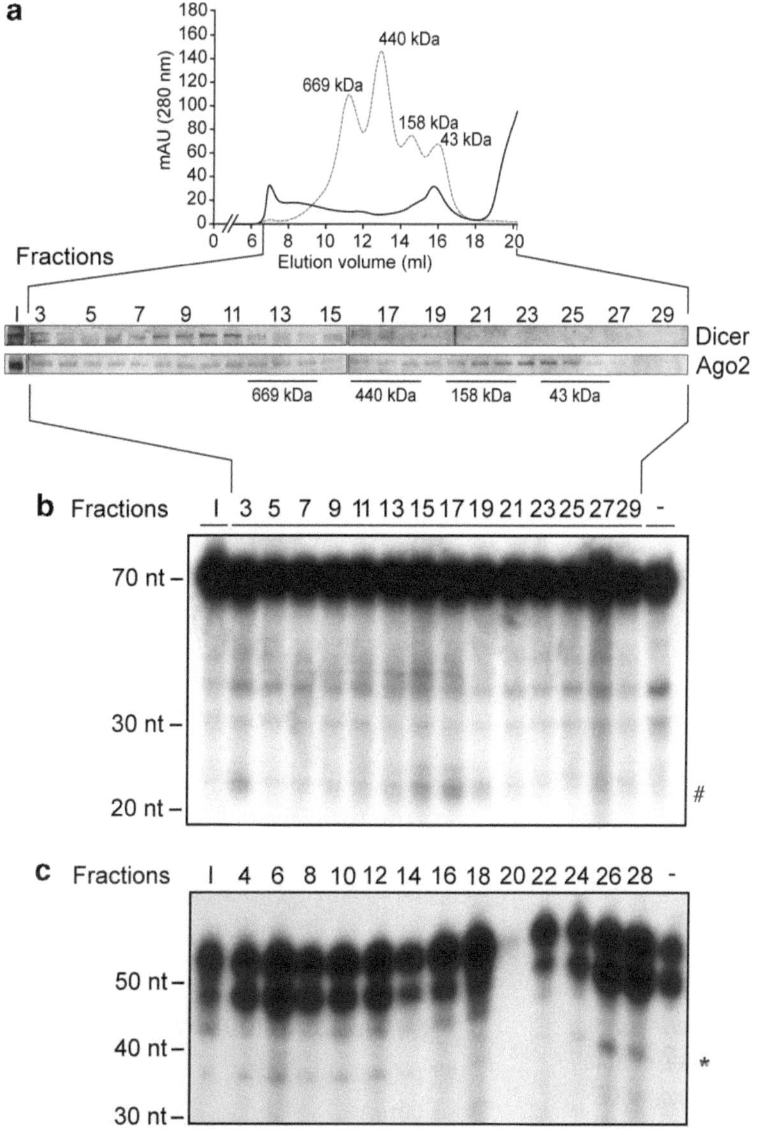

Fig. 3. Characterization of an Ago2-containing effector complex competent in RNA silencing in HeLa cells. (**a**) Extracts from HeLa cells were separated by gel filtration on a Superose six column and the fractions were analyzed by immunoblot analysis for the presence of Dicer and Ago2. (**b**) Selected (odd) fractions were tested for their intrinsic Dicer activity upon addition of a ^{32}P-labeled human let-7a-3 pre-miRNA substrate. # indicates the expected miRNA product. (**c**) RISC activity assays were performed by using a ^{32}P-labeled let-7c sensor RNA transcript. * Indicates the expected 38-nt cleavage products. The reactions were analyzed by denaturing PAGE and autoradiography. A 10-nt RNA ladder was used as a size marker. Adapted from Landry et al. (35), with permission from The Nature Publishing Group.

4. For Ago2 cleavage assay: To prepare the 200-μL reaction mixture, add 1 μL of Superase·In to each tube and incubate for 10 min at RT. Then, add 99 μL of *2× Ago2 FPLC assay buffer* and 1 μL of a radiolabeled RNA target, mix and incubate for 90 min at 30°C. To stop the reaction, add 200 μL of proteinase K buffer and 1 μL of proteinase K, and incubate for 15 min at 45°C. Then, add 400 μL of Acid Phenol:$CHCl_3$ (5:1 solution, pH 4.5). Vortex for 20 s. Separate the aqueous and organic phases by centrifugation at 16,000 × *g* for 4 min at 4°C. Perform RNA precipitation, as described in Subheading 3.3.1, steps 6–9. A typical result is shown in Fig. 3c.

3.6. Denaturing PAGE

1. Prepare a 0.75-mm thick 10% polyacrylamide gel (19:1) containing 7 M urea. For a 30-mL preparation, mix 7.5 mL of polyacrylamide (19:1) (stock 40%), 12.6 g of urea, 3 mL of 10× TBE solution and complete with DEPC water. Incubate at 37°C and mix until urea is completely dissolved. Add 150 μL of 10% APS, 30 μL of TEMED, mix gently, and pour the gel immediately. The gel should polymerize in about 30 min.
2. Pre-run the gel at 250 V for 30 min in 1× TBE at RT. Use of an electrophoresis apparatus with a big tank will contribute to a constant migration temperature.
3. Rinse two times the wells with the 1× TBE buffer before loading the RNA samples on the denaturating 10% polyacrylamide gel. Load a radiolabeled 10-nt Decade size marker in order to estimate the size of RNA species under study. Electrophorese at 275 V for 2–3 h.
4. Stop the electrophoresis when the bromophenol blue reaches 2 in. from bottom of the gel (duration: ~2 h 30 min).
5. Wrap the gel in a plastic wrap and expose it to an X-ray film with an intensifying screen at −80°C or analyze with the phosphorimager.

3.7. Northern Blot Analysis of Small RNAs with EDC Cross-Linking

Protocol adapted from Pall et al. (34).

1. Prepare a 0.75-mm thick 10% polyacrylamide gel (19:1) containing 7 M urea. For a 30-mL preparation, mix 7.5 mL of polyacrylamide (19:1) (stock 40%), 12.6 g of urea, 3 mL of 10× MOPS solution, and complete with DEPC water. Incubate at 37°C and mix until urea is completely dissolved. Add 150 μL of APS 10%, 30 μL of TEMED, mix gently, and pour the gel immediately. The gel should polymerize in about 30 min.
2. Pre-run the gel at 250 V for 30 min in 1× MOPS solution at RT. Use of an electrophoresis apparatus with a big tank will contribute to a constant migration temperature.

3. Rinse two times the wells with the 1× MOPS buffer before loading the RNA samples on the denaturating 10% polyacrylamide gel. Load a radiolabeled 10-nt Decade size marker in order to estimate the size of RNA species under study. Electrophorese at 275 V for 2–3 h.
4. Stop the electrophoresis when the bromophenol blue reaches 2 in. from bottom of the gel (duration: ~2 h 30 min). Wash the gel with DEPC water for 1 min.
5. Prepare six filter papers and a membrane Hybond-NX that fit the size of the gel. Hydrate the filter papers and the membrane in the water for at least 5 min.
6. Place three filter papers on the bottom half of the semidry electroblotter. Eliminate air bubbles by rolling over with a glass pipette. Place the gel on the filter papers, cover with the membrane and remove air bubbles by rolling over the membrane with a glass pipette. Complete the sandwich with other three filter papers. Remove air bubbles. Place the top of the electroblotter and screw the notches not too tightly.
7. Transfer at 500 mA for 1 h at 4°C.
8. After the transfer, prepare the cross-linking reaction. Cut a sheet of filter paper slightly larger than the membrane.
9. Immediately prior to use, prepare a fresh solution of EDC in 0.13 M 1-methylimidazole at pH 8.0. For a 20 × 16 cm Hybond NX nylon membrane, add 122.5 μL of 1-methylimidazole to 10 mL of water and adjust to pH 8.0 with 1 M HCl. Then add 0.373 g of EDC and complete to 12 mL.
10. Saturate the filter paper with the EDC solution. On a plastic wrap, place the nylon membrane face down and the EDC saturated paper over, and wrap it. Incubate for 2 h at 60°C.
11. Wash residual EDC with distilled water prior to pre-hybridization or dry and store the membrane at –20°C.
12. Prepare the probe, as described in Subheading 3.1.1.
13. Prehybridize the membrane in 10 mL of prehybridization solution with gentle agitation for at least 1 h at 50°C.
14. Hybridize the membrane in 10 mL of hybridization solution (same as pre-hybridization) containing at least 1×10^7 cpm of labeled antisense RNA probe with gentle agitation for 8–24 h at 50°C.
15. After the hybridization, wash the membrane five times with 25 mL of wash solution with gentle agitation for 10 min at 50°C.
16. With the Geiger counter, ensure that the membrane has been washed sufficiently. If not, wash once for 5 min at 60°C (or ~10°C lower than the estimated melting temperature of the probe) under agitation.

17. Wrap the blot in plastic wrap and expose to an X-ray film in the presence of an intensifying screen at −80°C or analyze with a phosphorimager.
18. If stripping is needed, place the membrane in stripping buffer and incubate for 1 min at 100°C.

4. Notes

1. Although DEPC water is commercially available, it can be prepared by adding 1 mL of DEPC to 1 L of milliQ water, mixing overnight, and autoclaving. DEPC is carcinogenic and should be handled with care. Wear gloves and work under a chemical hood.
2. Human pre-let-7a-3 DNA template for in vitro transcription was prepared by PCR amplification of the following oligonucleotides: The human pre-let-7a-3 sequence oligonucleotide 5′-GGAAAGACAGTAGATTGTATAGTTATCCCATA GCAGGGCAGAGCCCCAAACTATACAACCTAC TACCTCATATAGTGAGTCGTATTA-3′ and the T7 promoter oligonucleotide 5′-TAATACGACTCACTATA-3′. The PCR product was isolated on a 1% agarose gel and utilized as a DNA template (see Fig. 1a, b).
3. The DNA template used for in vitro transcription of the human miR-223 Northern blot probe was prepared by PCR amplification of the following oligonucleotides: The miR-223 detection probe oligonucleotide 5′-TGTCAGTTTGTCAAA TACCCCACCCTATAGTGAGTCGTATTA-3′ and the T7 promoter oligonucleotide: 5′-TAATACGACTCACTATA-3′. The PCR product was isolated on a 1% agarose gel and utilized as a DNA template (see Fig. 2c).
4. The DNA template used for in vitro transcription of the sensor bearing a binding site complementary to human miR-223 was prepared by PCR amplification of the following oligonucleotides: The miR-223 binding site oligonucleotide 5′-TGTTCTAGTTGTCTATGTTAATCTGATTGTCAGTT TGTCAAATACCCCAGTGTTGTTGTGTCTATGTT AATGTGCCTATAGTGAGTCGTATTAAATT-3′ and the T7 promoter oligonucleotide: 5′-TAATACGACTCACTATA-3′. The PCR product was isolated on a 1% agarose gel and utilized as a DNA template (see Fig. 2a, b).
5. Preparation of the 1 M stock of ATP (200 μL final): 110 mg of ATP (ATP-5′-triphosphate, disodium salt, Sigma, cat. no. A-2383) are dissolved in 150 μL of 1 M Tris, pH 8.0. The pH should be verified on a pH paper strip and adjusted to 7.0

by adding 10 N NaOH, 10 μL at a time. Complete by adding water to 200 μL. Preparation of the 200 mM stock of GTP (400 μL final): 45 mg of GTP (GTP-5′-triphosphate, disodium salt) are dissolved in 320 μL of water. The pH should be verified on a pH paper strip and adjusted to 7.0 by adding 10 N NaOH, 10 μL at a time. Complete by adding water to 400 μL. Freeze at −20°C in small aliquots.

6. All radioactive substances should be used in accordance with local regulation and safety recommendations. Work in a safe environment behind plexiglass protection. The level of radioactivity used in these protocols is important enough to monitor any possible contamination at every step with a Geiger counter.

7. For Northern blot RNA probes, prepare a Sephadex-G25 column according to the manufacturer's instructions. Place the labeled RNA on the column and centrifuge at 600 × *g* for 2 min at 4°C. Remove 1 μL and add 5 mL of scintillation liquid for counting in a beta scintillation counter. The probe is then ready to be used in Northern blot analyses (see Subheading 3.7).

8. Be sure that the plastic wrap is not contaminated by carefully applying it on the gel. If necessary, change your glove protection before each manipulation. Verify that a piece of paper rubbed softly on the plastic wrap covering the gel is not contaminated with a Geiger counter to avoid contaminating the X-ray film processor in the dark room.

9. To measure RNA concentration, dilute your RNA sample 1/100 (3 μL RNA + 297 μL water) and read the absorbance at 260 and 280 nm on a spectrophotometer. For short nucleic acids (<200 nt), the concentration is determined by the following formula: A_{260} × dilution factor × 33 μg/mL. The purity of the RNA can be estimated from the A_{260}/A_{280} ratio. A ratio of >1.8 and up to 2.1 is expected from highly pure RNA.

10. Determine the appropriate amount of antibody to be used for IP. The IP can be divided in half for concomitant Dicer RNase activity assay and Western blot analysis of the IP content in Dicer protein (see Fig. 1 c, d).

11. Determine the appropriate amount of antibody to be used for IP. When using the anti-Ago2 antibody from Abnova, 2 μL was optimal.

12. It is important to utilize a *Tris elution buffer* in order not to interfere with RNA precipitation.

13. Reserve an additional 75 μL for the detection of protein(s) of interest in each fraction by Western blot. Another alternative is to perform IP of Dicer or Ago2 protein on each fraction and document the activity concealed in these immune complexes in vitro (see Subheadings 3.2.2 and 3.3.2).

Acknowledgments

M.P.P. was supported by a doctoral studentship from Natural Sciences and Engineering Research Council of Canada (NSERC). P.P. is a Senior Scholar from the Fonds de la Recherche en Santé du Québec (FRSQ). This work was supported by a Discovery Grant from the Natural Sciences and Engineering Research Council of Canada (NSERC) (262938-2008), an Operating Grant from Health Canada/Canadian Institutes of Health Research (CIHR) (HOP-83069) and a CIHR/Rx&D Collaborative Research Grant (IRO-86239) to P.P.

References

1. Perron, M. P., and Provost, P. (2008) Protein interactions and complexes in human microRNA biogenesis and function, *Front Biosci 13*, 2537–2547.
2. Miranda, K. C., Huynh, T., Tay, Y., Ang, Y. S., Tam, W. L., Thomson, A. M., Lim, B., and Rigoutsos, I. (2006) A pattern-based method for the identification of MicroRNA binding sites and their corresponding heteroduplexes, *Cell 126*, 1203–1217.
3. Perron, M. P., and Provost, P. (2009) Protein components of the microRNA pathway and human diseases, *Methods Mol Biol 487*, 369–385.
4. Lee, Y., Jeon, K., Lee, J. T., Kim, S., and Kim, V. N. (2002) MicroRNA maturation: stepwise processing and subcellular localization, *Embo J 21*, 4663–4670.
5. Lee, Y., Kim, M., Han, J., Yeom, K. H., Lee, S., Baek, S. H., and Kim, V. N. (2004) MicroRNA genes are transcribed by RNA polymerase II, *Embo J 23*, 4051–4060.
6. Lee, Y., Ahn, C., Han, J., Choi, H., Kim, J., Yim, J., Lee, J., Provost, P., Radmark, O., Kim, S., and Kim, V. N. (2003) The nuclear RNase III Drosha initiates microRNA processing, *Nature 425*, 415–419.
7. Gregory, R. I., Yan, K. P., Amuthan, G., Chendrimada, T., Doratotaj, B., Cooch, N., and Shiekhattar, R. (2004) The Microprocessor complex mediates the genesis of microRNAs, *Nature 432*, 235–240.
8. Yi, R., Qin, Y., Macara, I. G., and Cullen, B. R. (2003) Exportin-5 mediates the nuclear export of pre-microRNAs and short hairpin RNAs, *Genes Dev 17*, 3011–3016.
9. Provost, P., Dishart, D., Doucet, J., Frendewey, D., Samuelsson, B., and Radmark, O. (2002) Ribonuclease activity and RNA binding of recombinant human Dicer, *Embo J 21*, 5864–5874.
10. Zhang, H., Kolb, F. A., Brondani, V., Billy, E., and Filipowicz, W. (2002) Human Dicer preferentially cleaves dsRNAs at their termini without a requirement for ATP, *Embo J 21*, 5875–5885.
11. Haase, A. D., Jaskiewicz, L., Zhang, H., Laine, S., Sack, R., Gatignol, A., and Filipowicz, W. (2005) TRBP, a regulator of cellular PKR and HIV-1 virus expression, interacts with Dicer and functions in RNA silencing, *EMBO Rep 6*, 961–967.
12. Chendrimada, T. P., Gregory, R. I., Kumaraswamy, E., Norman, J., Cooch, N., Nishikura, K., and Shiekhattar, R. (2005) TRBP recruits the Dicer complex to Ago2 for microRNA processing and gene silencing, *Nature 436*, 740–744.
13. Lee, Y., Hur, I., Park, S. Y., Kim, Y. K., Suh, M. R., and Kim, V. N. (2006) The role of PACT in the RNA silencing pathway, *EMBO J 25*, 522–532.
14. Bartel, D. P. (2004) MicroRNAs: genomics, biogenesis, mechanism, and function, *Cell 116*, 281–297.
15. Bhattacharyya, S. N., Habermacher, R., Martine, U., Closs, E. I., and Filipowicz, W. (2006) Relief of microRNA-mediated translational repression in human cells subjected to stress, *Cell 125*, 1111–1124.
16. Liu, J., Rivas, F. V., Wohlschlegel, J., Yates, J. R., 3rd, Parker, R., and Hannon, G. J. (2005) A role for the P-body component GW182 in microRNA function, *Nat Cell Biol 7*, 1261–1266.
17. Elbashir, S. M., Lendeckel, W., and Tuschl, T. (2001) RNA interference is mediated by 21- and 22-nucleotide RNAs, *Genes Dev 15*, 188–200.

18. Macrae, I. J., Zhou, K., Li, F., Repic, A., Brooks, A. N., Cande, W. Z., Adams, P. D., and Doudna, J. A. (2006) Structural basis for double-stranded RNA processing by Dicer, *Science 311*, 195–198.
19. Zhang, H., Kolb, F. A., Jaskiewicz, L., Westhof, E., and Filipowicz, W. (2004) Single processing center models for human Dicer and bacterial RNase III, *Cell 118*, 57–68.
20. Kataoka, Y., Takeichi, M., and Uemura, T. (2001) Developmental roles and molecular characterization of a Drosophila homologue of Arabidopsis Argonaute1, the founder of a novel gene superfamily, *Genes Cells 6*, 313–325.
21. Cerutti, L., Mian, N., and Bateman, A. (2000) Domains in gene silencing and cell differentiation proteins: the novel PAZ domain and redefinition of the Piwi domain, *Trends Biochem Sci 25*, 481–482.
22. Lingel, A., Simon, B., Izaurralde, E., and Sattler, M. (2004) Nucleic acid 3'-end recognition by the Argonaute2 PAZ domain, *Nat Struct Mol Biol 11*, 576–577.
23. Yan, K. S., Yan, S., Farooq, A., Han, A., Zeng, L., and Zhou, M. M. (2003) Structure and conserved RNA binding of the PAZ domain, *Nature 426*, 468–474.
24. Song, J. J., Smith, S. K., Hannon, G. J., and Joshua-Tor, L. (2004) Crystal structure of Argonaute and its implications for RISC slicer activity, *Science 305*, 1434–1437.
25. Lingel, A., Simon, B., Izaurralde, E., and Sattler, M. (2003) Structure and nucleic-acid binding of the Drosophila Argonaute 2 PAZ domain, *Nature 426*, 465–469.
26. Yuan, Y. R., Pei, Y., Ma, J. B., Kuryavyi, V., Zhadina, M., Meister, G., Chen, H. Y., Dauter, Z., Tuschl, T., and Patel, D. J. (2005) Crystal structure of A. aeolicus argonaute, a site-specific DNA-guided endoribonuclease, provides insights into RISC-mediated mRNA cleavage, *Mol Cell 19*, 405–419.
27. Hutvagner, G., and Zamore, P. D. (2002) A microRNA in a multiple-turnover RNAi enzyme complex, *Science 297*, 2056–2060.
28. Leu, Y. W., Rahmatpanah, F., Shi, H., Wei, S. H., Liu, J. C., Yan, P. S., and Huang, T. H. (2003) Double RNA interference of DNMT3b and DNMT1 enhances DNA demethylation and gene reactivation, *Cancer Res 63*, 6110–6115.
29. Griffiths-Jones, S., Saini, H. K., van Dongen, S., and Enright, A. J. (2008) miRBase: tools for microRNA genomics, *Nucleic Acids Res 36*, D154–158.
30. Ameres, S. L., Martinez, J., and Schroeder, R. (2007) Molecular basis for target RNA recognition and cleavage by human RISC, *Cell 130*, 101–112.
31. Bradford, M. M. (1976) A rapid and sensitive method for the quantitation of microgram quantities of protein utilizing the principle of protein-dye binding, *Anal Biochem 72*, 248–254.
32. Ender, C., Krek, A., Friedlander, M. R., Beitzinger, M., Weinmann, L., Chen, W., Pfeffer, S., Rajewsky, N., and Meister, G. (2008) A human snoRNA with microRNA-like functions, *Mol Cell 32*, 519–528.
33. Rudel, S., Flatley, A., Weinmann, L., Kremmer, E., and Meister, G. (2008) A multifunctional human Argonaute2-specific monoclonal antibody, *Rna 14*, 1244–1253.
34. Pall, G. S., Codony-Servat, C., Byrne, J., Ritchie, L., and Hamilton, A. (2007) Carbodiimide-mediated cross-linking of RNA to nylon membranes improves the detection of siRNA, miRNA and piRNA by northern blot, *Nucleic Acids Res 35*, e60.
35. Landry, P., Plante, I., Ouellet, D. L., Perron, M. P., Rousseau, G., and Provost, P. (2009) Existence of a microRNA pathway in anucleate platelets, *Nat Struct Mol Biol 16*, 961–966.

Chapter 10

Imaging the Cellular Dynamics of *Drosophila* Argonaute Proteins

Jing Li, Nima Najand, Wendy Long, and Andrew Simmonds

Abstract

Drosophila melanogaster is used extensively as a model system to uncover genetic and molecular pathways that regulate various cellular activities. There are five members of the Argonaute protein family in *Drosophila*. Argonautes have been found to be localized to cytoplasmic ribonucleoprotein containing structures in both cultured *Drosophila* cells and developing embryos. However, in fixed cell preparations some *Drosophila* Argonaute family proteins co-localize with structures containing known as RNA processing (P) body components while others do not. The ability to image Argonaute family proteins in live *Drosophila* cells, (both cultured and within developing embryos) allows for accurate genetic dissection of the pathways involved in the assembly, mobility, disassembly, and other dynamic processes of Argonaute-containing bodies. Here we describe a method of rapidly creating vectors for, and assay the activity of, fluorescently tagged Argonaute proteins in cultured *Drosophila* cells and embryos.

Key words: Argonaute, Confocal imaging, Live cell imaging, Fluorescent proteins

1. Introduction

Drosophila has several advantages for imaging dynamic events in both cultured cells and live embryos. Schneider 2 (S2) cells and embryos grow and develop quite well at room temperature (25°C), and there is no requirement for CO_2 or extensive humidity control. This obviates the need for expensive and elaborate incubation systems. Creation of vectors suitable for expressing fluorescently tagged proteins in cultured *Drosophila* cells or in whole embryos has become routine with the advent of a large set of freely available Gateway vectors, (the *Drosophila* Gateway Vector Collection, Terrence Murphy), that allow for rapid production of plasmids that express protein fusions with one of many

Tom C. Hobman and Thomas F. Duchaine (eds.), *Argonaute Proteins: Methods and Protocols*, Methods in Molecular Biology, vol. 725, DOI 10.1007/978-1-61779-046-1_10, © Springer Science+Business Media, LLC 2011

Table 1
The *Drosophila* Argonaute gene family

Gene	Chromosome location	Phenotype(s)
Argonaute 1	50C9-50C17	Most mutations are lethal with an array of defects manifested in different organs and tissues
Argonaute 2	71C3-71E5	Lethal, semi-lethal, and viable mutations have been isolated. Multiple phenotypes are observed in different organs tissues
Argonaute 3	80F9	Viable, female sterile, male partially sterile, manifested in male dorsal appendage and maternal effect
PIWI	32C1-32C1	Maternal lethal. Sterile and fertile alleles have been found
Aubergine	32C1-32C1	Recessive lethal, sterile

different fluorescent proteins, which can then be imaged via high-speed confocal imaging.

There are five Argonaute family proteins encoded by the *Drosophila* genome (Table 1) (1). Argonaute 1 (AGO1) is thought to be primarily involved in micro-RNA (miRNA)-mediated gene silencing, associating with Dicer-1. Argonaute 2 (AGO2) is a component of the RNA-induced silencing complex (RISC), and is the catalytic core of RISC associated with RNA interference (RNAi). Argonaute 3 (AGO3) is involved in the piwi RNA (piRNA) pathway which is important for transposon silencing, through binding to piRNAs that are derived from retrotransposons. PIWI also functions in piRNA-mediated transposon silencing, germline stem-cell maintenance, and RNAi. PIWI is thought to bind to heterochromatin protein-1α (HP-1α), an interaction that is important for PIWI epigenetic function (2). Finally, Aubergine is known to function in piRNA, transposon silencing, and RNAi. It binds to piRNAs derived from the antisense strands of retrotransposons and can induce additional production of piRNAs (3, 4).

Imaging cellular bodies that accumulate Argonaute proteins, such as P-bodies, stress granules, GW-bodies, and other cytoplasmic RNA-protein bodies embodies several challenges. Many of these structures appear to be exquisitely sensitive to changes in the cellular environment (e.g. stress or cell differentiation) thereby necessitating extreme care in sample preparation (including fixation) and imaging parameters. Also, given the dynamic nature of these bodies, it is difficult to determine the formation and disassembly of these structures in fixed preparations. We outline a simple and rapid method to generate vectors that can be used to express fluorescent Argonaute protein fusions in S2 cells (or other *Drosophila* derived cells) or *Drosophila* embryos suitable for live-cell imaging techniques. Use of fluorescently tagged proteins in living cells or embryonic tissue also enables one to test the effects of dsRNAs, drugs, antibodies, or other reagents on the protein composition, number, motility, or other dynamics of Argonaute containing RNPs.

2. Materials

2.1. Cell Culture

1. *Drosophila* Schneider 2 cell medium (S2) (Invitrogen).
2. Heat inactivated fetal bovine serum (FBS) (Invitrogen).
3. Sterile Hygromycin-B or Blasticidin solution (Invitrogen).
4. Penicillin streptomycin (liquid) (Invitrogen).

2.2. Collecting Drosophila Embryos

1. Apple Juice Plates http://cshprotocols.cshlp.org/cgi/content/full/2009/7/pdb.rec11871.
2. Embryo Collection Cages (Genesee Scientific).
3. Embryo Collection Baskets (Genesee Scientific).
4. Embryo Wash (5).

2.3. Rapid Cloning of Genes into Drosophila Gateway Vectors

1. The cDNAs for all protein coding isoforms of the wild type (or mutant) Argonaute gene of interest (http://www.fruitfly.org/).
2. pENTR™/D-TOPO Cloning Kit (Invitrogen).
3. Platinum Pfx DNA Polymerase (Invitrogen).
4. 10 mM dNTP mix (Invitrogen).
5. ORF specific primers (3 per cDNA) (5′+CACC, 3′+stop codon, and 3′ no stop codon).
6. LB-agar plates (supplemented with 100 μg/ml ampicillin or 60 μg/ml kanamycin).
7. The *Drosophila* Gateway Vector Collection (A resource developed by the Murphy lab for the *Drosophila* research community) http://www.ciwemb.edu/labs/murphy/Gateway%20 vectors.html.
8. Qiagen plasmid Maxi-prep kit (Qiagen).
9. One-shot Top10 competent cells (Invitrogen).
10. Ampicillin, sodium salt (Sigma).
11. Kanamycin sulfate (Sigma).

2.4. Preparing S2 Cells for Imaging

1. pCoHygro or pCoBlast (Invitrogen).
2. 6-well tissue culture dishes (uncoated) (Sarstedt).

3. 8-well chambered #1.5 coverglass (Nunc Lab Tek II #70377-81).
4. Sonicated didecyldimethylammonium bromide (DDAB) solution (250 μg/ml) (6).

2.5. Preparing Embryos for Live-Imaging

1. Embryo Glue (1 m of 3 M double-sided tape soaked in 50 ml of heptane overnight).
2. Halocarbon oil 700 (Sigma, # H8898).
3. #1.5 22 mm square coverglass.
4. Used compact discs (CD)/digital video discs(DVD).
5. Glass micropipettes (Eppendorf).
6. Pressure Injection system (Eppendorf).
7. P-transposase encoding Δ2–3 helper plasmid, obtained from the *Drosophila* Genomics Resource Center https://dgrc.cgb.indiana.edu/product/View?product=1001.

2.6. Confocal Imaging

1. Axiovert200M inverted microscope (Zeiss).
2. 63× or 100× multi-immersion LCI-PlanNeoflaur, or equivalent, lenses (Zeiss).
3. Ultraview spinning disc confocal system (Perkin Elmer Biosciences).
4. Stage Heater (Zeiss).
5. Image acquisition software.
   (a) Volocity 5.2 (Perkin Elmer Biosciences).
   (b) ImageJ with the μManager plugin can also be used for acquisition. http://www.micro-manager.org/ (7).

2.7. Image/Video Post-Processing

1. Volocity 5.2 software (Perkin Elmer Biosciences).
2. Imaris software version 6.2 (Bitplane AG).
3. Autoquant X version 2.1.0.
4. VirtualDub (8).

3. Methods

This method outlines the steps for rapid live-cell visualization of fluorescently tagged Argonaute family proteins in cultured *Drosophila* S2 cells or developing embryos. For simplicity we will only describe how to create protein fusions for visualization of AGO1 and AGO2. This method is equally applicable to dynamic visualization of other P-body components including other Argonaute family proteins, Gawky, Pacman (Pcm), Lsm4 (9), and others.

Cytoplasmic RNP composition and/or number has been shown to be dramatically affected by cellular stress, including heat shock (10), so considerable care must be taken to keep the cell or embryo cultures at consistent temperature and limit other environmental stresses before and during the imaging process. The *Drosophila* community has benefited greatly from the availability of the *Drosophila* Gateway Vector Collection, which allows rapid transfer of an initial entry clone into multiple vectors suitable for expressing fluorescent protein fusions, under the control of inducible or constitutive promoters in either cultured S2 cells or creating transgenic animals.

We have compared images of S2 cells or embryos acquired using multiple imaging techniques including: video microscopy, wide-field microscopy coupled with 4-D deconvolution, and laser-scanning, or spinning disc confocal microscopy.

We have found that a spinning disc confocal system such as the Perkin Elmer Ultraview was superior in terms of the ability to rapidly collect sufficient images at a sufficient vertical (Z) resolution for three dimensional reconstruction of each time-point, while minimizing bleaching (and related cellular stress). In our hands it was possible to use wide field-microscopy with either a cooled CCD (Zeiss Axiovert) or EMCCD (Hamamatsu C9100) camera to obtain sufficient samples for three-dimensional reconstruction using post acquisition processing including deconvolution. Here we describe the method used to capture images with a spinning disc confocal microscope. In each case, images were collected with a similar base microscope stand and lens (Zeiss Axiovert 200M, LCI-PlanNeoflaur 63×/1.3NA) for which either water or glycerine can be used as an immersion medium. Accurate tracking of RNPs containing Argonaute proteins requires recreation of a three-dimensional volume encompassing the entire cell to follow particle movement through different Z-planes. An often overlooked parameter in live-imaging is aberrations induced by mismatch between the immersion medium and the medium the cells/embryos are suspended in. Water is the best match for Schneider Cell Medium while glycerine is the closest match to Halocarbon oil 700.

3.1. Rapid Generation of Plasmids to Express Fluorescently Labelled Argonaute Proteins in S2 Cells or Animals

3.1.1. PCR Cloning of Ago1 and Ago2 into pENTR/D

1. We first isolated plasmid DNA containing cDNAs for *Ago1* and *Ago2* (see Note 1).
2. We next designed PCR primers that spanned the open reading frame (ORF) of interest. Three primers were needed, a 5′ primer that begins at the start of the ORF and two 3′primers: one retaining the endogenous stop codon and another with the stop codon deleted to allow for construction of C-terminal fusions (see Notes 2–4).
3. The PCR mix contained: 200 ng template DNA, 2 μl 10 mM dNTP, 1 ml 50 mM Mg^{2+}, 5 μl 10× Platinum PFX buffer,

specific10 μM 5′ and 3′primers, and 10U of Pfx DNA polymerase. The mixture was then made up to 50 μl using sterile water. This was mixed by gently pipetting the solution up and down before loading into a thermocycler.

4. The PCR conditions used for amplifying the *Ago1* ORF were as follows: pre-denaturing the template at 95°C for 5 min followed by 35 cycles of denaturing the template at 95°C for 1 min, annealing at 55°C for 1 min, and extension at 72°C for 2 min (1 min per kb of target). This was followed by a final extension at 72°C for 10 min. The samples were then held at 4°C after finishing the program. The conditions as described should be modified for templates with different lengths (extended extension time), *Tm* of primers (adjust annealing time) or CG content (adjust extension time).
5. The resulting PCR products were verified by agarose gel electrophoresis and sequencing. Only one prominent band should be evident on a 0.8% (w/v) agarose gel. If multiple bands were observed, the annealing temperature was increased until only a single band was produced. If this cannot be accomplished, then a fragment of the correct size can be purified from the agarose gel although this step may lead to reduced cloning efficiencies.

3.1.2. Directional PCR Cloning of Argonaute Family Genes into pENTR/D TOPO Vector

1. The PCR fragment of the *Argonaute* gene ORF was cloned with pENTR/D directional TOPO Cloning Kit as described in the supplied manual. The fresh PCR product (2 μl) was mixed with 3 μl of water and 1 μl of the prepared pENTR/D vector and incubated at room temperature for 30 min to overnight (see Note 5).
2. Top-10 One Shot chemically competent *Escherichia coli* were transformed as follows: 2 μl of reaction mix were added to one vial of the competent cells followed by gently mixing and incubation on ice for 5–30 min. Then the cells were heat-shocked in a 42°C water-bath for 30 s followed by incubation on ice. We added 250 μl of S.O.C medium to the transformation reaction and incubated it at 37°C for 1 h while shaking. Approximately 50–100 μl of the transformation mixture was spread onto warmed LB-agar plates supplemented with 60 μg/ml of kanamycin and incubated at 37°C overnight (see Note 6).
3. A single bacterial colony was selected and grown overnight in liquid culture (5 ml LB + kanamycin) at 37°C with shaking (250 rpm). Argonaute clones in pENTR/D were verified by DNA sequencing using M13 forward from 5′ end and M13 backward primers from 3′ end before proceeding to the recombination step.

4. To store the transformed bacteria, we supplemented 500 μl of a separate overnight culture with 30% glycerol and flash froze them in liquid nitrogen, before transferring the stocks to a –80°C freezer.

3.1.3. Recombination of Entry Clones with GFP/RFP Expressing Drosophila Gateway Vectors (See Note 7)

1. The sequence verified *Ago1* and *Ago2* pENTR/D cDNAs were recombined into appropriate destination vectors *Drosophila* Gateway Vector Collection. The choice of vector depends on the intended use (see Note 8).
2. For cultured S2 cells, the pARW and pAGW vectors were used to create N-terminal fusions of *Ago1* and *Ago2* expressed under the control of an *Actin* promoter (11) (see Note 9).
3. For expression in developing Actin embryos, the pPRW and pPGW vectors were used as destination vectors. These vectors express transgenes under the control of the UAS promoter and also contain suitable transposase sites for genomic integration (11–13).
4. The recombination reaction was performed using Gateway LR Clonase II enzyme mix according to supplied instructions.
5. The resulting reaction (3 μl) was then used to transform One Shot chemically competent *E. coli* – DH5α, which were then plated on LB-agar plates supplemented with 100 μg/ml ampicillin.
6. The Gateway cloning system is designed such that each resulting colony on the bacterial plates is the product of a positive recombination event. Single colonies were grown in 5 ml LB media supplemented with 100 μg/ml ampicillin at 37°C while shaking overnight.
7. For long-term storage, 500 μl of overnight bacterial culture was supplemented with 30% glycerol and then flash frozen in liquid nitrogen before transferring to a –80°C freezer.

3.1.4. Preparation of Plasmids for Cell or Animal Transfection

Purified cDNA vector for each construct was prepared by growing transformed DH5α bacteria overnight in 100 ml LB supplemented with 100 μg/ml ampicillin. A commercial maxi prep kit was used to purify the plasmids. The purified plasmid DNA was dissolved at a concentration of 1 μg/μl in sterile double-distilled water.

3.2. Creating Transformed S2 Cell Lines Expressing Fluorescent Argonaute Proteins

1. The transformation reaction was made with 200 μl of Didecyldimethylammonium bromide (DDAB) solution (250 μg/ml) and 100 μl serum-free S2 culture medium (v/v = 1:2). These were mixed in a sterile micro-centrifuge tube and left at room temperature for 5 min until the mixture became turbid (Han 1996). Next, 2 μg pARW-*Ago1* and/or pAGW-*Ago2* DNA and 840 μl DDAB-medium were added to the mixture, and incubated at room temperature for an additional 15 min (see Note 10).

2. An aliquot containing approximately 5×10^6 S2 cells was centrifuged in 15-ml conical tube (1,000×*g* for 5 min at room temperature). Most of the medium was aspirated with care taken not to disturb the cell pellet.
3. Cells were washed once with 2 ml serum-free media and then centrifuged in 15-ml conical tube (1,000×*g* for 5 min). Then, S2 cells were resuspended with serum-free S2 medium at 10^6 cells/ml, and transferred into a single T25 flask (see Note 11).
4. The mixture containing prepared cDNA vector with medium and DDAB was added to single flasks of S2 cells, followed by gently swirling to make sure the solution was well mixed. The flasks were returned to the 25°C incubator for 4–6 h.
5. After incubation, the cells were centrifuged (1,000×*g* for 5 min at room temperature), washed, then resuspended in 5 ml of S2 medium supplemented with 1% (v/v) heat-inactivated FBS and premixed penicillin/streptomycin solution (10,000 units of penicillin and 10,000 μg of streptomycin per ml) for at least 24 h.
6. At this point S2 cells were ready for microscopic analyses.

3.3. Generating S2 Cell Lines that Stably Express Fluorescent Argonaute Proteins

1. To create S2 lines stably expressing pAGW-*Ago1* or pARW-*Ago2* (as in Subheading 3.2), we performed transfection as above but also added the selection vector (100 ng pCoHygro or pCoBlast) at a 19:1 (w/w) ratio to expression vector.
2. To select for cells co-transfected with selection and expression vectors, we used S2 cell medium (supplemented with 1% (v/v) FBS) with appropriate selection reagent (300 μg/ml for hygromycin-B or 25 μg/ml for blasticidin).
3. S2 medium including selection reagent was replaced every 4–5 days. When the cell density reached $6–20 \times 10^7$ cells/ml, they were serially passaged at a 1:2 dilution to remove dead cells. Then aliquots of the stably transfected cells were frozen for long-term storage.

3.4. Expressing Fluorescent Argonaute Proteins in Developing Drosophila Embryos

1. Genetic transformation of *Drosophila melanogaster* embryos was performed via P-element mediated insertion of the expression vector into the germline of animals co-expressing P-element transposase from an endogenous or plasmid source (13). Both pPRW-*Ago1* or pPGW-*Ago2* contain flanking P-element insertion sites (see Note 12).
2. The P-element plasmids (pPRW-*Ago1* or pPGW-*Ago2*) were coinjected with a helper plasmid encoding transposase (Δ2–3) into the posterior cytoplasm of 0–1 h old embryos using a glass needle attached to a positive pressure injection apparatus.

3. Injected embryos were allowed to develop into adults which were then mated to w^{1118} flies. The progeny of this mating that expressed the mini-w^{+} marker gene from the pPRW-*Ago1* or pPGW-*Ago2* (red eyed), were made homozygous and retained as transformed stocks.
4. Homozygous flies transformed with pPGW-*Ago2* or pPRW-*Ago2* do not express these transgenes until they are mated to flies expressing GAL4. For this purpose, we used flies expressing GAL4 under the *nanos* (*y^1 w^*; P{GAL4-nos.NGT}40*) (14) or *tubulin* (*y^1 w^*; P{tubP-GAL4}LL7/TM3, Sb^1*) (15) promoters. These combinations allowed for high levels of transgene expression in early embryos.

3.5. Preparing Embryos Expressing GFP-Ago1 or RFP-Ago2 for Live-Imaging

1. Flies were transferred to an embryo collection cage and allowed to acclimatize for 2–3 days. This cage has an open bottom that is covered with a petri dish containing apple juice agar (Fig. 1).
2. To maintain the population of flies in the cage, apple juice agar plates at the bottom of the cage were supplemented with live yeast (1:1 *Saccharomyces cerevisiae* and water) as a food source and replaced every 12 h.
3. Approximately 2–3 h before imaging, a fresh apple juice plate supplemented with yeast paste was placed at the bottom of the cage (Fig. 1). This was left in place for 1 h to allow transgenic flies to lay embryos. After 1 h the plate was removed and replaced.
4. After embryo collection, plates were stored at 25°C until the embryos reached the desired developmental stage (up to 24 h).

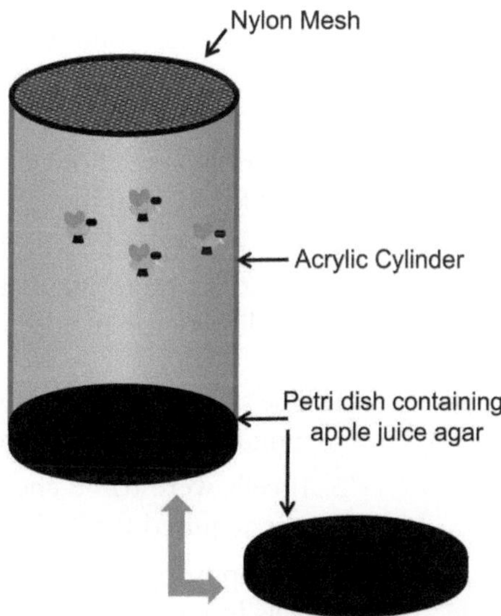

Fig. 1. A diagram of the cage used to raise flies for embryo collection.

5. The embryos were washed from the apple juice plate into an egg collection chamber by rinsing with embryo wash. A fine paintbrush was used to detach embryos adhering to the plate or to the walls of the collection chamber such that the eggs were collected in the mesh at the bottom of the collection chamber.
6. The collected embryos were dechorionated directly in the collection basket by immersion in a 50% (v/v) bleach solution. They were then rinsed with copious amounts of distilled water until no bleach odour was detected.
7. Dechorionated embryos were then lined up end to end on the centre of a cover slip coated with a thin layer of embryo glue.
8. If the embryos were to be injected with fluorescent dsRNA, antibodies or other reagents, they were first partially dehydrated in a sealed plastic container containing Drierite desiccant for 2–4 min. The optimal time depends on the relative humidity of the room and the amount of time dechorionated embryos were exposed to ambient air. This step was omitted if the embryos were not to be injected.
9. Dehydrated embryos on cover slips were then covered in thin layer of Halocarbon 700 oil.
10. The centre 3 cm plastic portion of a compact disc, normally designed to accomodate the disc spindle, was cut out. The coverslip containing the embryos was positioned onto the cut out portion of the compact disc such that the embryos were aligned with the centre of hole. The coverslip was secured in place using embryo glue (see Note 13).
11. Additional halocarbon 700 oil was added to cover the entire spindle hole.

3.6. Preparation of S2 Cell Cultures for Live Imaging

1. Appropriate volumes of the culture of transfected S2 cells were transferred by pipette into Lab-Tek II chambered #1.5 coverglasses (1 ml – 2 well/500 μl – 4 well/250 μl 8 well).
2. The cells were allowed to settle onto the bottom of the chambers for 10 min in a 25°C incubator before imaging.
3. The chambered coverglass dishes were transferred directly to the Zeiss stage incubator set at 25°C and then allowed to equilibrate for an additional 5 min before imaging.

3.7. Imaging Parameters

1. The stage heater insert was mounted on the stage of the spinning disc confocal microscope.
2. If cells were to be chemically treated, a baseline image stack was acquired for 2–3 min before treatment.
3. The appropriate immersion media was applied to the lens. For cells in aqueous media water was best, while for halocarbon 700 oil, glycerol has the closest matching refractive index (16).

4. A small sample of TetraSpeck microspheres, 0.1 μm, fluorescent blue/green/orange/dark red (Invitrogen) was added directly to a region of the preparation near, but not obscuring, the samples of interest. Focusing on the Tetraspeck beads, the collar of the objective was adjusted such that there appeared to be the same pattern of aberration at a focal plane slightly above and below the plane at a point where the beads were focused most clearly.
5. If the images were to be subsequently deconvolved, a single image stack encompassing one or more fluorescent beads ±3 μm was collected from each sample later calibration.
6. The stage heater can also be raised to 36.5°C for induction of transgenes under the control of the heat-shock promoter (the pHW series), (9).
7. The field of view was moved to a cell or embryo to be analyzed. If an X-Y motorized stage was used, multiple cells/embryos were visualized per experiment.
8. Imaging parameters were set such that multiple Z-planes were captured beginning immediately below the cell of interest, and ending just above.
9. For a 63× 1.3 NA objective, the optimal distance between Z-planes should be approximately 120 nm when visualizing live cells expressing GFP in an aqueous medium with water as an immersion medium. For embryos expressing GFP under Halocarbon 700 oil with glycerine as an immersion medium, it should be approximately 155 nm (see Note 14).
10. For rapidly moving RNPs containing Argonaute proteins, we captured a stack of images encompassing the entire cell every 30 s. The electron multiplier setting of the camera was set to 85% of maximum to allow for exposure times of approximately 150–200 ms per Z-plane with the laser intensity minimized (50–70%) to avoid photobleaching the fluorescent proteins.
11. Depending on the signal intensity of the protein being imaged, cells were imaged for up to 1 h. Under optimal conditions, we were able to image S2 cells for 12 h, which allowed for detection of mitotic events. However, for routine determination of events in interphase cells, 10–20 time points were sufficient.

3.8. Image Post-Processing, Movie Production and Image Quantification

3.8.1. Movie Production Using Volocity 5.1 (See Note 16)

1. For publication, movies were made using "extended view" in the Volocity visualization module and exported as uncompressed AVI format (see Note 15).
2. For publication as a figure, a series of still images from the movie were made at an appropriate time spacing (e.g. 2 min). The final time-spacing depended on the event we were trying to show (Fig. 2).

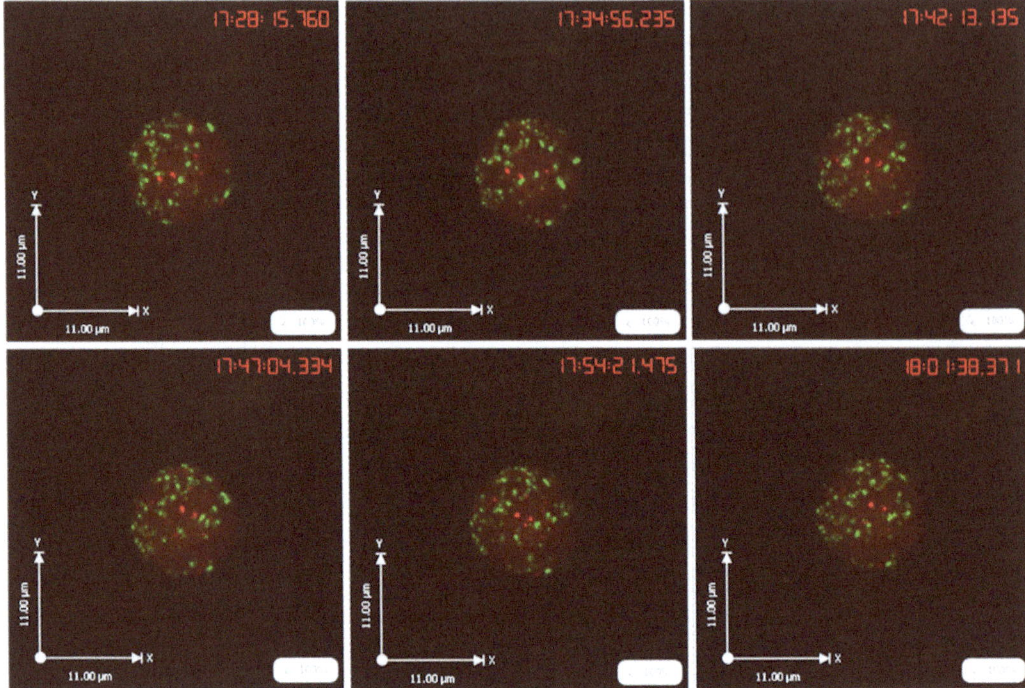

Fig. 2. Individual frames captured from a 4D movie showing the dynamic movement of Argonaute proteins (GFP-Ago1 and RFP-Ago2) in S2 cells. A selection of time points (shown in *red* at *upper right*) was captured from the "extended view" in Volocity using the "snapshot" function. The extended image from each time point was pasted sequentially in the figure to show the movement of GFP-Ago1 and RFP-Ago2 RNP particle relative to each other.

3. Minor corrections to the movie speed (frames per second), cropping, adding titles or basic thresholding, etc., were made using Virtual-Dub 0.96.

3.8.2. Movie Production Using Autoquant AutoDeblur Gold 2.1.1 and Imaris 6.2 (See Note 16)

1. We exported each image series to be analyzed as open microscopy format (OME).
2. The OME format was opened with Imaris 6.2.
3. We selected "Image Processing – Autodeblur …" to transfer files directly from Imaris to Autoquant X.
4. Using Autoquant – We selected "3D deconvolution."
5. For each set of imported images we ensured that the imaging parameters were correct. In particular, we were careful to ensure that the type of image was set to "spinning disc confocal," and the lens type, immersion media and vertical Z-distance matched the actual acquisition parameters that were used.
6. Generally, the default settings were selected for blind deconvolution.
7. After the deconvolving process was complete the image stacks were transferred back to Imaris 6.2.

8. For publication, the image stacks were visualized as a maximum projection using the "Easy 3D" viewer. "Snapshots" were made of sequential timepoints and pasted into a program such as Photoshop (CS), to create figures (Fig. 3a).
9. Alternatively, the "Gallery" function of Easy 3D within Imaris can be used to create a montage of timepoints or Z-slices for publication.

3.8.3. Analysis of RNP Particle Movement Using Imaris 6.2 (See Note 16)

1. We opened the images set in Imaris 6.2.
2. To follow each fluorescent body/particle, we used the "Spots" algorithm (step-by-step wizard). A separate set of tracks was generated for each colour.
3. For this analysis we selected: Region Growing, Region of Interest and Track Spots (over time). Select blue arrow button for next.
4. The region of interest was selected. Multiple regions of interest can be processed at one time. Generally, we selected a region of interest representing one cell. Select blue arrow button for next.
5. The approximate diameter of Ago1 and/or Pcm containing RNPs was entered into the "Estimated diameter" field (~1 μm). Background subtraction was selected. Select blue arrow button for next.
6. We used the threshold filter to eliminate background spots that did not represent Ago1 or Pcm containing RNPs. This varied depending on expression levels. As a general rule, we reduced the threshold until the spots correlated with the brightly staining areas of each image, with no spots being selected outside the region considered to be within the cells. Select blue arrow button for next.
7. We selected "spot region type" as " Local Contrast." Select blue arrow button for next.
8. Tracks representing the mobility of each spot were then displayed. Tracks that are not representative of actual RNP particle movement (i.e. movement between cells) were disconnected and removed manually. Select green arrow button to finish.
9. Multiple tracks for two different signals can be displayed at one time (Fig. 2b). The direction and velocity of each RNP particle can be exported into a spreadsheet program for analysis of speed or direction (Fig. 3b).

3.8.4. Analysis of Colocalizaiton Using Imaris 6.2 (See Note 16)

1. The image stack was opened in Imaris 6.2 as above.
2. For colocalization analysis "View" and then "Coloc" functions were selected.

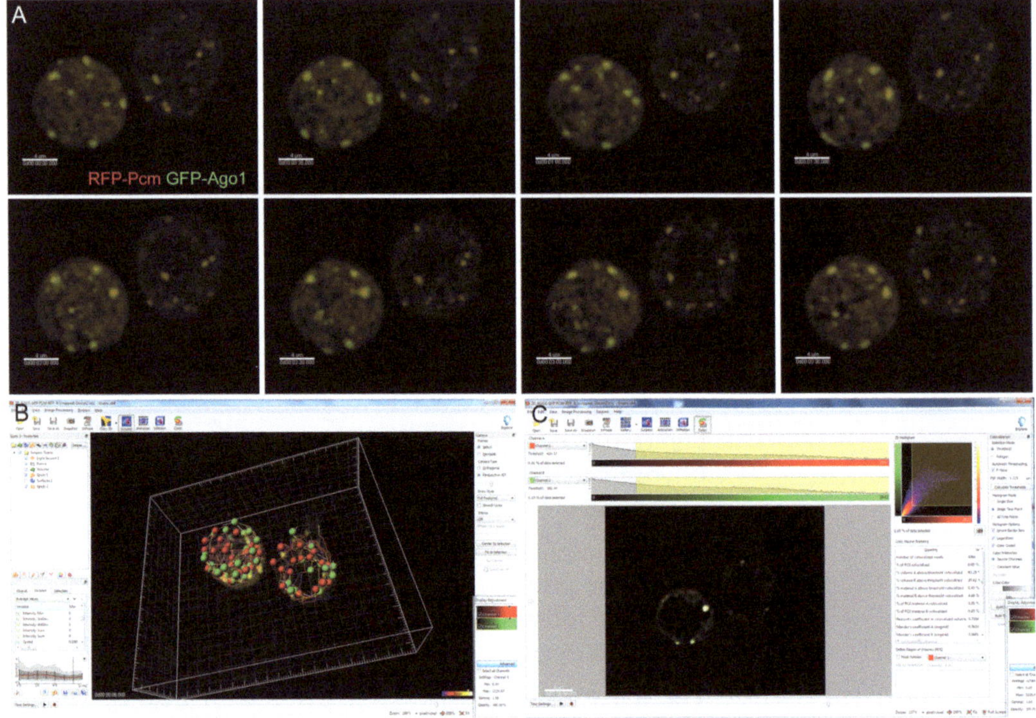

Fig. 3. (**a**) A time-series showing cytoplasmic bodies containing the 3′exonuclease Pacman (RFP-Pcm) and Argonaute1 protein (GFP-Ago1) after deconvolving using Autoquant X2 and visualized using Imaris 6.4. The images appear *yellow* as there is essentially complete co-location between these two protein fusions. (**b**) Visualization of the tracks made by each RNP particle containing either Pcm or Ago using Imaris 6.4.2. (**c**) Co-localization analysis of the larger spots containing Pcm, Ago1 or both using Imaris 6.4.2.

3. From the analysis screen the intensity threshold was adjusted such that they highlighted regions which encompass the AGO1-containing spots.

4. The percent colocalization as well as Pearson's and Mander's coefficients are displayed for each time point (Fig. 3c).

4. Notes

1. Validated template cDNAs for almost all *Drosophila* genes (and their various isoforms), were obtained from the Berkeley *Drosophila* Genome Project (BDGP).

2. It is best if the Tm of primers are close to 60°C, and that the Tm of the primers for both ends are equal.

3. For the pENTR/D-TOPO vector, the sequence 5′CACC 3′ must be included in the 5′ end of the 5′ gene specific primer.

4. If the epitope tag is to be added on to the C-terminus of the protein to be visualized, use the primer that omits the stop codon from the 3′ end of the ORF.
5. pENTR/Directional TOPO vector (Invitrogen) is recommended here. Compared to the classic cloning method, it shortens the time for cloning a gene largely by omitting the need for restriction enzyme cutting, ligation, and moreover, it simplifies selection of positive colonies containing the blunt-end PCR product cloned in the correct orientation. However, clones made in other Gateway entry vectors can be used if they are (1) limited to the ORF, (2) have an insert cloned in the same reading frame of the GFP/RFP destination vectors.
6. Do not vigorously shake the cells as this may interrupt the TOPO-cloning reaction.
7. The Carnegie *Drosophila* Gateway Vector collection comprises 68 Gateway-based vectors containing the specific recombination sites attR1 and attR2 that can be recombined with the pENTR/D-TOPO recombination site and incorporate the cloned cDNA sequence into the "destination" plasmid. Multiple destination plasmids containing different epitope tags are available. Each epitope tag is also paired with an array of different promoter so that the transcription can be triggered under different conditions (i.e. heat-shock or UAS-Gal4 system).
8. An exhaustive list can be found at http://www.ciwemb.edu/labs/murphy/Gateway%20vectors.html (9).
9. When creating fusions of Ago1–Ago3, one must consider two important functional domains: the PAZ domain which binds siRNAs and miRNAs and the PIWI domain which in some Argonaute proteins possess RNase-H-like activity. Another important domain found in mammalian Argonaute proteins is the Mid domain that may be required for binding to the 5′cap of mRNAs (1).
10. *Drosophila* S2 cells are being used for transfection, gene expression, RNAi, and high-resolution live-cell microscopy. The approximate length of the S2 cell cycle in serum supplemented media is 20 h (17).
11. Culturing S2 cells with serum-free media SFX (Thermo Fisher) instead of S2 media induces cell morphology to adopt a more flattened shape, thereby simplifying three-dimensional imaging. However, removing serum from the culture medium does lead to a slower rate of proliferation.
12. Transgenic animals were created by a commercial service (http://www.thebestgene.com).

13. The use of a shallow plastic ring facilitates easy manipulation of the embryos once mounted and also allows for use of an axially presented microinjection needle to deliver dsRNA or other chemicals into the embryos.
14. An online calculator for appropriate ideal Nyquist rate for Z-axis sampling rates can be found at http://support.svi.nl/wiki/NyquistCalculator.
15. The use of compressed video codecs as part of Quicktime (MOV) or Windows Movie (WMV) is not recommended unless required by a specific journal. Compression can often introduce artefacts into your video.
16. An explanation of all of the possible methods of image analysis using Volocity 5.2 and Imaris 6.4 are beyond the scope of this protocol. Please refer to the user guide for these programs for further information.

Acknowledgments

The authors would like to thank Dr. Sarah C. Hughes, Dr. Fred Mast and the other participants in the 14th annual UBC live-imaging course (2009) for advice on confocal microscopy and developing the methodology for live imaging cell culture and the subsequent data analysis. This work was supported by a CIHR grant (84154) to AJS.

References

1. Hutvagner, G., and Simard, M. J. (2008) Argonaute proteins: Key players in RNA silencing. *Nature Reviews Molecular Cell Biology*, 9(1), 22–32.
2. Noma, K., Sugiyama, T., Cam, H., Verdel, A., Zofall, M., Jia, S., Moazed, D., and Grewal, S. I. (2004) RITS acts in cis to promote RNA interference-mediated transcriptional and post-transcriptional silencing. *Nature Genetics*, 36(11), 1141–2.
3. Brennecke, J., Aravin, A. A., Stark, A., Dus, M., Kellis, M., Sachidanandam, R., and Hannon, G. J. (2007) Discrete small RNA-generating loci as master regulators of transposon activity in *Drosophila*. *Cell*, 128(6), 1089–1103.
4. Gunawardane, L. S., Saito, K., Nishida, K. M., Miyoshi, K., Kawamura, Y., Nagami, T., Siomi, H., and Siomi, M. C. (2007) A slicer-mediated mechanism for repeat-associated siRNA 5' end formation in *Drosophila*. *Science* 315(5818), 1587–90.
5. Hughes, S. and Krause, H.M. (1998) Single and double FISH protocols for *Drosophila* . in: Methods in Molecular Biology, vol. 122: Protocols in Confocal Microscopy., ed. Stephen Paddock, (Humana Press), 93–101.
6. Han, K. (1996). An efficient DDAB-mediated transfection of *Drosophila* S2 cells. *Nucleic Acids Research*, 24(21), 4362–63.
7. μManager – developed by Arthur Edelstein and Nico Stuurman in the laboratory of Ron Vale at the University of California San Francisco (http://www.micro-manager.org).
8. VirtualDub – developed by Avery Lee (http://www.virtualdub.org).
9. Schneider, M.D., Najand, N., Chaker, S., Pare, J.M., Haskins, J., Hughes, S.C., Hobman, T.C., Locke, J., and Simmonds, A.J. (2006). Gawky is a component of cytoplasmic mRNA processing bodies required for early *Drosophila* development. *Journal of Cell Biology* 174, 349–58.

10. Kedersha, N., and Anderson, P. (2002) Stress granules: Sites of mRNA triage that regulate mRNA stability and translatability. *Biochemical Society Transactions*, 30(Pt 6), 963–69.
11. The Drosophila Gateway Vector Collection developed by Terrence Murphy (http://www.ciwemb.edu/labs/murphy/Gateway%20 vectors.html).
12. Brand, A.H., and Perrimon, N. (1993) Targeted gene expression as a means of altering cell fates and generating dominant phenotypes. *Development* 118, 401–15.
13. Spradling, A. C., and Rubin, G. M. (1982). Transposition of cloned P elements into Drosophila germ line chromosomes. Science (New York, N.Y.), 218(4570), 341–7.
14. Barrett, K., Leptin, M., Settleman, J. (1997) The Rho GTPase and a putative RhoGEF mediate a signaling pathway for the cell shape changes in *Drosophila* gastrulation. *Cell* 91(7), 905–15.
15. Lee, T., and Luo, L. (1999) Mosaic analysis with a repressible neurotechnique cell marker for studies of gene function in neuronal morphogenesis. *Neuron* 22(3): 451–461.
16. Cavey, M and Lecuit, T. (2008) Imaging Cellular and Molecular Dynamics in Live Embryos Using Fluorescent Proteins in *Drosophila*. Methods in Molecular Biology vol 420: Drosophila, Methods and Protocols (Humana Press), 219–238.
17. Rogers S.L. and Rogers G.C. (2008) Culture of *Drosophila* S2 cells and their use for RNAi-mediated loss-of-function studies and immunofluorescence microscopy *Nature Protocols* 3, 606–611.

Chapter 11

Live Cell Imaging of Argonaute Proteins in Mammalian Cells

Justin M. Pare, Joaquin Lopez-Orozco, and Tom C. Hobman

Abstract

The central effector of mammalian RNA interference (RNAi) is the RNA-induced silencing complex (RISC). Proteins of the Argonaute family are the core components of RISC. Recent work from multiple laboratories has shown that Argonaute family members are associated with at least two types of cytoplasmic RNA granules: GW/Processing bodies and stress granules. These Argonaute-containing granules harbor proteins that function in mRNA degradation and translational repression in response to stress. The known role of Argonaute proteins in miRNA-mediated translational repression and siRNA-directed mRNA cleavage (i.e., Argonaute 2) has prompted speculation that the association of Argonautes with these granules may reflect the activity of RNAi in vivo. Accordingly, studying the dynamic association between Argonautes and RNA granules in living cells will undoubtedly provide insight into the regulatory mechanisms of RNA-based silencing. This chapter describes a method for imaging fluorescently tagged Argonaute proteins in living mammalian cells using spinning disk confocal microscopy.

Key words: Argonaute, Processing bodies, Stress granules, Live cell imaging, Confocal microscopy

1. Introduction

The Argonaute superfamily comprises a group of RNA-binding proteins that form the core of ribonucleoprotein complexes (RNPs) that mediate RNA interference (RNAi) and related gene-silencing pathways (reviewed in refs. 1–3). The basic functions of Argonautes in the canonical RNAi pathway are reasonably well understood. Small-guide RNAs direct Argonaute-containing RNPs to homologous mRNAs, thereby providing the specificity in this gene-silencing pathway (reviewed in refs. 1, 4). The Argonaute superfamily has been divided into two subgroups: Argonaute and Piwi (5–7). Expression of the Piwi subfamily members is restricted to germline tissues and undifferentiated cells, whereas the Argonaute group members are ubiquitous. The two main types of

Tom C. Hobman and Thomas F. Duchaine (eds.), *Argonaute Proteins: Methods and Protocols*,
Methods in Molecular Biology, vol. 725, DOI 10.1007/978-1-61779-046-1_11, © Springer Science+Business Media, LLC 2011

small RNAs that associate with Argonaute subfamily members differ in origin and function. Short-interfering RNAs (siRNAs) and microRNAs (miRNAs) are derived from long double-stranded (ds)RNA and hairpin RNA precursors, respectively. Successive processing of miRNA precursors in the nucleus, by the RNase Drosha (8), and in the cytoplasm, by the RNase Dicer (9), produces mature miRNAs. These mature miRNAs are then transferred to an Argonaute-containing RNP complex termed RNA-induced silencing complex (RISC) (10). It is estimated that expression of more than 50% of all human genes may be controlled posttranscriptionally by miRNA-dependent mechanisms (11). Because of its central role in gene expression, the RNAi apparatus itself is subject to regulation and evidence suggests that this occurs at multiple levels. Notably, localization and presumably the functions of Argonaute proteins are regulated by phosphorylation (12). In the cytoplasm, Argonaute-dependent posttranscriptional gene silencing (PTGS) has been linked to discrete cytoplasmic puncta called GW- or P-bodies (13–15).

1.1. Processing Bodies

GW/Processing bodies (GW/P-bodies) are known to play important roles in mRNA catabolism and contain RNA decapping enzymes and exonucleases, as well as Argonaute proteins. An ongoing matter of debate is whether these structures are required for small RNA-mediated PTGS or whether they simply form as a consequence of silencing. Regardless of which school of thought is ultimately proven correct, it is likely that formation of GW/P-bodies may increase the efficiency or kinetics of PTGS.

1.2. Stress Granules

In addition to P-bodies, Argonautes have been shown to rapidly associate with stress granules (SGs) when cells encounter translational stress (16, 17). Microscopically visible SGs are not present in "unstressed" cells; however, treatment with arsenite or the translational repressor hippuristanol results in rapid formation of SGs (16, 18). SGs contain stalled translation complexes and have been implicated in miRNA-mediated translational repression (19).

Given that interaction of Argonautes with P-bodies and SGs is dynamic, altering the rate of association between Argonautes and these granules is a potential mechanism to modulate PTGS. Indeed, this process seems to be regulated in mammalian cells in that biogenesis of P-bodies is linked to maturation of miRNAs (20).

This chapter describes a protocol for imaging live mammalian cells expressing fluorescently labeled Argonaute proteins and the methods used to induce stress granule (SG) formation during image acquisition. These techniques have been used to visualize the dynamic association of Argonautes with RNA granules such as GW/P-bodies and SGs (17). Also described are the software-based analytical methods for determining the numbers of nascent SGs induced during acquisition as well as for measuring their fluorescent intensities.

2. Materials

2.1. Generating HeLa TREx Cell Line Stably Transfected with pcDNA4/TO/GFP-hAgo2

2.1.1. Construction of pcDNA4/TO/GFP-hAgo2

1. peGFP-hAgo2 (Addgene).
2. Restriction enzymes for vector construction: *Eag*I, *Afl*II, and *Xba*I.
3. PCR primers:

   GFP-hAgo2 FOR 5′-TAC TCT TAA GTC GCC ACC ATG GTG AGC AAG G-3′.

   GFP-hAgo2 REV 5′-ACT TCT AGA TTA AGC AAA GTA CAT GGT GCG C-3′.
4. Long Template PCR System (Roche).
5. TOPO PCR Blunt Kit (Invitrogen).
6. XL10-Gold Ultracompetent cells (Stratagene).
7. pcDNA4/TO plasmid (Invitrogen).
8. T4 DNA ligase and buffer (Invitrogen).
9. Subcloning efficiency DH5α cells (Invitrogen).

2.1.2. Transfection of HeLa TREx Cell Line with pcDNA4/TO/GFP-hAgo2

1. HeLa TREx cell line (Invitrogen).
2. Dulbecco's Modified Eagle's cell culture medium (DMEM) supplemented with l-glutamine (Invitrogen).
3. Penicillin/streptomycin (Invitrogen).
4. Blasticidin (Sigma).
5. Fetal bovine serum (FBS) (Invitrogen).
6. Sterile, autoclaved phosphate-buffered saline (PBS).
7. Lipofectamine 2000 (Invitrogen).
8. OptiMEM (Invitrogen).

2.1.3. Selection of Stably Transfected HeLa TREx Cells

1. Zeocin (Invitrogen).
2. Doxycycline (DOX; Sigma) prepared at stock concentration of 1 mg/mL in sterile ddH_2O.

2.2. Cell Culture and Transfection

2.2.1. Transient Transfection of HeLa Cells

1. HeLa cell line (American Type Culture Collection).
2. Glass-bottom p35 dishes (Mat-Tek).
3. Dulbecco's Modified Eagle's cell culture medium (DMEM) supplemented with l-glutamine (Invitrogen).
4. Penicillin/streptomycin (Invitrogen).
5. Fetal bovine serum (FBS) (Invitrogen).
6. Sterile, autoclaved phosphate-buffered saline (PBS).
7. OptiMEM (Invitrogen).

8. Lipofectamine 2000 (Invitrogen).
9. pcDNA4/TO/GFP-hAgo2 plasmid (described in Subheading 3.1).
10. pDsRed-C1-Monomer-TIA-1 plasmid (Clontech) (see Note 1).

2.3. Preparation of Equipment and Reagents Prior to Imaging

2.3.1. Preparation of Microscope

1. Axiovert 200M microscope (Carl Zeiss) equipped with a spinning disk confocal unit (UltraView ERS). Apochromat 63× objective lens is used for viewing mammalian cells.
2. Objective lens warmer and power supply controller (Bioptechs).
3. Heated stage (Heating Insert P) and power supply (Tempcontrol 37-2 digital 2-channel controller) from Carl Zeiss.
4. Volocity 4D acquisition software (Perkin Elmer).

2.3.2. Preparation of CO_2-Independent Medium

1. CO_2-independent cell culture medium (Invitrogen).
2. Sodium Arsenite (Sigma-Aldrich) 50 mM.
3. Hippuristanol (Pelletier Lab, McGill University), prepared at a stock concentration of 1 mM in anhydrous dimethyl sulfoxide (DMSO).

2.4. Image Analysis and Quantitation

1. Imaris ×64 with MeasurementPro module (Bitplane).

3. Methods

The method described here can be used to visualize components of the RNAi pathway and RNA granules (e.g., GW/P-bodies, SGs) during cellular stress. As an example, we outline the construction of a stably transfected, inducible cell line expressing GFP-tagged human Argonaute 2 (a core component of RISC). We also describe the protocol for transient transfection of this GFP-tagged hAgo2 and TIA-1 (a component of mammalian SGs) fused with DsRed. Finally, the procedure for visualization of these fluorescently tagged proteins in living mammalian cells, coupled with the induction of SGs using oxidative damage (arsenite) or translational inhibition (hippuristanol) is described.

3.1. Generating HeLa TREx Cell Line Stably Transfected with pcDNA4/TO/GFP-hAgo2

3.1.1. Construction of pcDNA4/TO/GFP-hAgo2

1. pEGFP-hAgo2 is linearized by digestion with *Eag*I and then used as the template for a polymerase chain reaction using the Long Template PCR system and custom primers "GFP-hAgo2 FOR" and "GFP-hAgo2 REV."
2. The resulting ~3,400 bp PCR product is subcloned into pCRII-Blunt-TOPO using TOPO PCR Blunt Kit following the manufacturer's protocol. The TOPO reaction is transformed

into XL10-Gold Ultracompentent cells and plated on LB agar plates containing 50 μg/mL kanamycin.

3. Following digestion of pCRII-Blunt/GFP-hAgo2 and pcDNA4/TO with *Afl*II and *Xba*I, the GFP-hAgo2 cassette is ligated into pcDNA4/TO using T4 DNA ligase. Ligation reactions are transformed into Subcloning Efficiency DH5α and plated on LB agar plates containing 100 μg/mL ampicillin.

3.1.2. Transfection of HeLa TREx Cell Line with pcDNA4/TO/GFP-hAgo2

1. HeLa TREx cells are cultured in DMEM containing 10% FBS, penicillin, streptomycin, and 5 μg/mL blasticidin. All cell culture is conducted at 37°C in an incubator with a 5% CO_2 atmosphere.
2. Twenty-four hours prior to transfection, the cells are trypsinized, counted and seeded into 60 mm dishes at a density of 5×10^5cells/dish.
3. Prior to transfection, the normal growth medium is removed, the cells are rinsed once with PBS and the medium is replaced with OptiMEM.
4. For each p60 to be transfected, 2.5 μg of pcDNA4/TO/GFP-hAgo2 is mixed with 65 μL of OptiMEM. Separately, 3.75 μL of Lipofectamine 2000 is mixed with 60 μL of OptiMEM.
5. After 5–10 min at room temperature, the mixtures are combined and incubated at room temperature for an additional 15–30 min. The resulting mixture is then applied to the cells which are then returned to 37°C for up to 4 h. Following transfection, the cells are rinsed with PBS and regular growth medium is added.

3.1.3. Selection of Stably Transfected HeLa TREx Cells

1. At 48 h posttransfection, cells in p60 dishes are trypsinized and reseeded in a 150 mm dish. Growth medium is replaced with DMEM containing 10% FBS, penicillin, streptomycin, blasticidin, and 400 μg/mL *Zeocin*.
2. Medium is changed every 2–3 days until the appearance of distinct colonies is evident (typically 10–14 days).
3. Colonies are trypsinized using autoclaved silicone grease and cloning rings. Cells from each colony are transferred to wells in a 96-well plate. Cells are continually passaged into wells of increasing size (24-, 12-, and finally 6-well plates).
4. Once the cells are confluent in 6-well plates, clonal populations are screened for inducible expression of GFP-hAgo2. After seeding on to sterile coverslips, cells are cultured in the presence of 1 μg/mL DOX for 24 h and then examined by fluorescence microscopy.

3.2. Cell Culture and Transfection

3.2.1. Transient Transfection of HeLa Cells

1. HeLa cells are cultured in DMEM containing 10% FBS, penicillin, and streptomycin.
2. Twenty-four hours prior to transfection, the cells are trypsinized, counted, and seeded at a density of 2.0×10^5 cells per glass-bottom p35.
3. Just prior to transfection, the normal growth medium is removed, the cells are rinsed once with PBS and the medium is replaced with OptiMEM.
4. For each p35 to be transfected, plasmids encoding GFP-hAgo2 and DsRed-TIA (500 ng of each) are mixed with 25 μL of OptiMEM. Separately, 1.5 μL of Lipofectamine 2000 is mixed with 23.5 μL of OptiMEM.
5. After 5–10 min at room temperature, the mixtures are combined and incubated for an additional 15–30 min. The resulting mixture is then applied to the cells and they are returned to 37°C for up to 4 h, after which the cells are rinsed with PBS prior to adding regular growth medium (see Note 2).

3.2.2. Induction of GFP-hAgo2 in Stably Transfected HeLa TREx Cells

1. HeLa TREx cells stably transfected with pcDNA4/TO/GFP-hAgo2 are cultured in DMEM containing 10% FBS, penicillin, streptomycin, 5 μg/mL blasticidin, and 400 μg/mL Zeocin.
2. Twenty-four hours prior to the induction of GFP-hAgo2 expression, the cells are trypsinized, counted, and seeded at a density of 2.0×10^5 cells per into glass-bottom p35 dish.
3. For induction of HeLa TREx cells, growth medium is supplemented with 1 μg/mL DOX for a minimum of 6 h. Continued induction for over 24 h is not generally recommended as expression levels of GFP-hAgo2 can become too high.

3.3. Preparation of Equipment and Reagents Prior to Imaging

3.3.1. Preparation of the Microscope

1. At 24 h posttransfection, and approximately 15–30 min prior to imaging, the objective warmer is applied to objective lens (typically a 63× objective is used for imaging HeLa cells). The standard stage is replaced with the heated stage.
2. The temperature of both the objective warmer and the heated stage is set to 37°C. Allow sufficient time to warm and equilibrate equipment.
3. Parameters of the experiment should be set within the acquisition software to reduce down time once the samples are in place.
4. The interval between time points is set to between 30 and 60 s.
5. Z-stacks are set to contain 10–15 slices and the position of the top and bottom of the stack is set beyond the top and bottom of the cell of interest to ensure the entire cell is captured in the acquisition. Allow the software to determine the appropriate slice thickness (see Note 3).

3.3.2. Preparation of CO_2-Independent Medium

For each p35 to be imaged, a total of 4 mL of pre-warmed CO_2-independent medium containing 10% FBS is prepared. One milliliter is made to contain a 2× concentration of the drug being used to induce cellular stress (e.g., 100 μM arsenite or 2 μM hippuristanol). Two aliquots (1 and 2 mL per dish) of drug-free, CO_2-independent medium are prepared separately. All aliquots of medium are kept at 37°C (see Note 4).

3.4. Live Cell Imaging and Induction of Translational Stress

3.4.1. Imaging Prior to the Induction of Stress

1. The regular growth medium is aspirated and the cells are rinsed with PBS.
2. Following aspiration of the PBS, 1 mL of pre-warmed, drug-free, CO_2-independent medium is added to cells.
3. The p35 is secured in the heated stage and for the pre-treatment imaging, the lid of the stage is left off to prevent excessive movement of the stage during the addition of the arsenite or hippuristanol.
4. Image acquisition is started and 1–4 "unstressed" time points are recorded to provide $t=0$ time points as well as to visualize the cells prior to the addition of the stress.

3.4.2. Imaging During the Induction of Stress

1. Following acquisition of the final "pre-treatment" time point, acquisition is paused and 1 mL of pre-warmed CO_2-independent medium containing a 2× concentration of the stress-inducing drug (arsenite or hippuristanol) is added resulting in a final drug concentration of 1× (e.g., 50 μM arsenite or 1 μM hippuristanol) (see Note 5)
2. The lid of the heated stage (left off in Subheading 3.4.1, step 3) is returned to prevent further heat loss and evaporation.
3. Acquisition is resumed and allowed to continue for a predetermined length of time. SGs induced by arsenite or hippuristanol are typically visible in HeLa cells within 10 min of treatment (Fig. 1, see Note 6).

3.5. Wash Out and Recovery from Cellular Stress

1. Image acquisition is paused and the medium containing arsenite or hippuristanol is removed using a Pasteur pipette connected to a vacuum line. Medium containing arsenite should be disposed of in a manner that is in accordance with local regulations.
2. The cells are carefully rinsed twice with PBS to remove trace amounts of arsenite or hippuristanol.
3. Two milliliters of pre-warmed CO_2-independent medium is added to the dish.
4. Image acquisition is resumed while the cells recover from stress (Fig. 2, see Note 7).

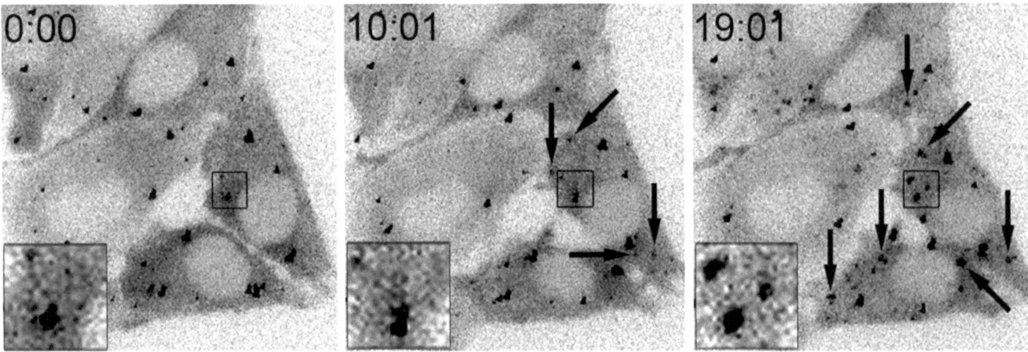

Fig. 1. Induction of SGs in HeLa cells expressing GFP-hAgo2. Live HeLa cells transfected with pcDNA4/TO/GFP-hAgo2 were imaged prior to and following the addition of hippuristanol (final concentration is 1 μM). Z-stacks were acquired every 20 s for 20 min. Time points are shown from before the addition of hippuristanol (0:00) as well as 10:01 and 19:01 min after the addition. *Insets* show enlarged regions that are indicated by *white boxes*. *Arrows* indicate nascent SGs (adapted, with permission, from Pare JM, et al. Hsp90 regulates the function of argonaute 2 and its recruitment to stress granules and P-bodies. *Mol Biol Cell* 2009;20:3273–84).

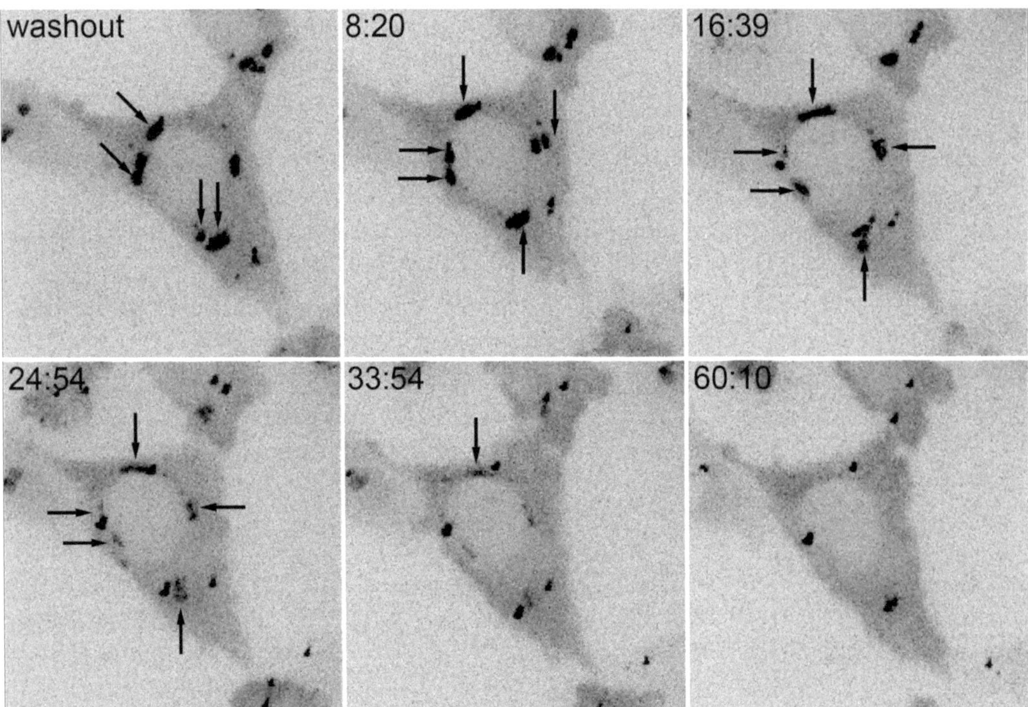

Fig. 2. Dissolution of SGs in live HeLa cells expressing GFP-hAgo2 following wash out of hippuristanol. SG formation was induced with 1 μM hippuristanol for 30 min. Z-stacks were acquired every 45 s for 60 min beginning immediately before and subsequent to washout of hippuristanol. Time points are shown from pre-washout as well as 8:20, 16:39, 24:54, 33:54, and 1:00:10 following washout. *Arrows* indicate SGs.

3.6. Image Analysis and Quantitation

1. Datasets should be exported in LIFF format to preserve *z*-axis, time, and channel data.
2. LIFF file is opened with Imaris image analysis software. Please note that for this section, terms specific for the Imaris image analysis software and the creation of an algorithm designed to measure fluorescent intensity of granules are denoted with quotation marks.
3. A new "Spots" algorithm is started.
4. "Region growing" is selected; "Region of Interest" and "Track (over time)" are left unselected.
5. The source is selected as the channel which corresponds to the protein/fluor of interest.
6. Average diameter of nascent SGs is entered in the "Estimated diameter" field (typically ~1 μm).
7. A filter is added to exclude spots that are *below* a threshold "Quality." This threshold must be set at a value low enough to include faint nascent SGs but not so low as to introduce a significant number of false positives. This value will vary depending on GFP-hAgo2 expression levels.
8. A second filter is used to exclude spots *over* a threshold "Quality." This step is included to filter preexisting GFP-hAgo2-positive GW/P-bodies which are always brighter than nascent SGs.
9. "Absolute Intensity" is selected.
10. "Region volume" levels are manually adjusted until the SGs in the previewed image are filled-in (the previewed image allows you to observe the effects of your adjustments in real-time).
11. Click the "Finish" button to complete the algorithm build (Fig. 3a).
12. Click on the "Export Statistics" button to save the data in Microsoft Excel file format.
13. Average fluorescent intensities from individual time points are grouped within predetermined, standardized bin ranges using statistical analysis software (e.g., Microsoft Excel with Histogram tool in the Data Analysis ToolPak add-in) (Fig. 3b). Counts within bin ranges, for each individual time point, are represented as a line graph with counts (Number of SGs) on the *y*-axis and fluorescent intensities (Bin range values) on the *x*-axis (Fig. 3c).

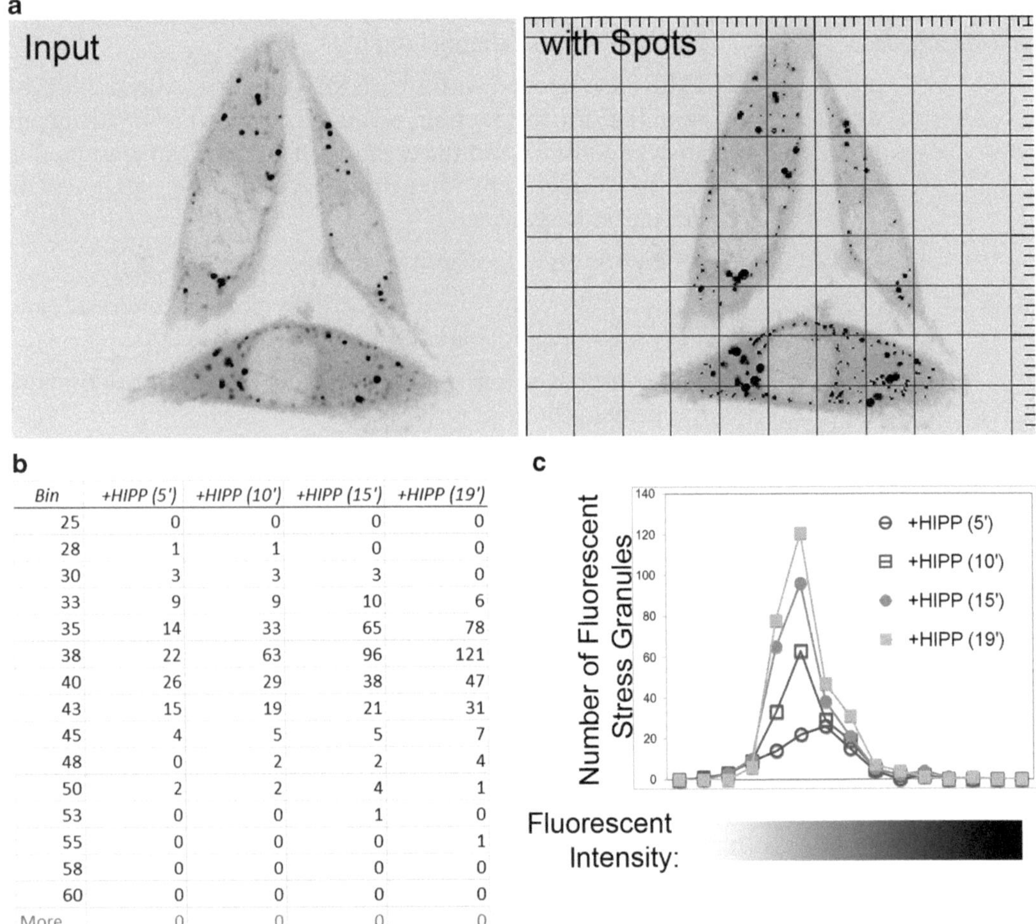

Bin	+HIPP (5')	+HIPP (10')	+HIPP (15')	+HIPP (19')
25	0	0	0	0
28	1	1	0	0
30	3	3	3	0
33	9	9	10	6
35	14	33	65	78
38	22	63	96	121
40	26	29	38	47
43	15	19	21	31
45	4	5	5	7
48	0	2	2	4
50	2	2	4	1
53	0	0	1	0
55	0	0	0	1
58	0	0	0	0
60	0	0	0	0
More	0	0	0	0

Fig. 3. Quantitation of SG number and fluorescent intensity. (**a**) Using Imaris image analysis software (with MeasurementPro module), a "Spots" algorithm was used to identify and measure granules falling within a user-specified range of size and intensity, corresponding to SGs. (**b**) Statistical analysis software was used to generate a histogram, using arbitrary bin ranges, from the data obtained measuring fluorescent intensities of SGs with Imaris. (**c**) Histogram values are represented as a *line* graph where fluorescent intensities (bin ranges) are plotted on the *x*-axis and the number of GFP-positive SGs (counts) on the *y*-axis. Each time point is represented as a *separate line* on the graph (reproduced, with permission, from Pare JM, et al. Hsp90 regulates the function of argonaute 2 and its recruitment to stress granules and P-bodies. *Mol Biol Cell* 2009;20:3273–84).

4. Notes

1. pEGFP-C1-TIA-1, received from JF Caceres (Medical Research Council Human Genetics Unit, Western General Hospital, Edinburgh, Scotland, UK). TIA-1 was subcloned into pDsRed-C1-Monomer (Clontech) in this laboratory.
2. Transfections are typically performed in triplicate; however, consideration must be made when designing live cell imaging experiments as single imaging sessions may last for several hours depending on the length of treatment or washout.

3. Considerations must be made to reduce the amount of radiation the live cells are exposed to minimize cellular damage caused by imaging. To that effect, decisions regarding acquisition setup, specifically Z spacing, are important when designing live cell imaging experiments.

   The ideal setting for Z-stack stepping is 0.2–0.3 μM (resolution of the light microscope) when imaging P-bodies and SGs. In our experience, the height of live HeLa cells ranges from 10 to 15 μm. The overall height, coupled with exposure times and multiple channels, have led us to conclude that this slice thickness is not feasible.

   Two-channel acquisition (DsRed and GFP), with typical exposure times of 200–600 ms and a predefined number of slices ranging from 10 to 15, results in 4–18 s of exposure to the laser light per time point.

4. Smaller volumes of medium can be used if required. This is particularly relevant if using precious reagents such as hippuristanol. Only the cells growing on the glass coverslip in the depressed area of the dish need to be submerged.

5. This method allows for rapid diffusion of the drug resulting in more complete mixing as well as limiting the local concentration of the drug at the site of addition. Additionally, the toxicity of solvents such as DMSO is minimized using this technique.

6. Drift in the *z*-axis is a common issue when imaging live cells over an extended period of time and caution should be taken to ensure that cells of interest remain within the defined range of focus. To correct for drifting, adjustments to the position of the objective lens can be made during live acquisition using the fine focus knob.

7. HeLa cells treated with hippuristanol have been observed to recover, which is to say TIA-positive SGs resolve, in approximately 30 min (Fig. 2). In contrast, longer recovery times are needed when arsenite treatment is used to induce SGs. Furthermore, when arsenite treatment exceeds 20–30 min, the damage may be irreversible (i.e., SGs do not resolve even after extended recovery times).

Acknowledgments

We would like to thank Dr. A Simmonds (University of Alberta) for providing advice on live-cell imaging. This work was funded by grants from the Canadian Institutes of Health Research (CIHR; to T.C.H.). J.M.P is the recipient of a Doctoral Scholarship from CIHR. J.L.O. is the recipient of a studentship from AHFMR. Portions of the introduction to this chapter, as well as Figs. 1 and 3c, have been reproduced from Pare JM, et al. Hsp90 regulates the function of argonaute 2 and its recruitment to stress granules and P-bodies. Mol Biol Cell 2009;20:3273–84.

References

1. Hannon, G.J. (2002) RNA interference. *Nature* **418** 244–51
2. Peters, L. and Meister, G. (2007) Argonaute proteins: mediators of RNA silencing. *Mol. Cell* **26** 611–23
3. Hutvagner, G. and Simard, M. J. (2008) Argonaute proteins: key players in RNA silencing. *Nat. Rev. Mol. Cell Biol.* **9** 22–32
4. Jaronczyk, K., Carmichael, J. B. and Hobman, T. C. (2005) Exploring the functions of RNA interference pathway proteins: some functions are more RISCy than others? *Biochem. J.* **387** 561–71
5. Bohmert, K., Camus, I., Bellini, C., Bouchez, D., Caboche, M. and Benning, C. (1998) AGO1 defines a novel locus of Arabidopsis controlling leaf development. *EMBO J.* **17** 170–80
6. Cox, D.N., Chao, A., Baker, J., Chang, L., Qiao, D. and Lin, H. (1998) A novel class of evolutionarily conserved genes defined by piwi are essential for stem cell self-renewal. *Genes Dev.* **12** 3715–27
7. Carmell, M.A., Xuan, Z., Zhang, M. Q. and Hannon, G. J. (2002) The Argonaute family: tentacles that reach into RNAi, developmental control, stem cell maintenance, and tumorigenesis. *Genes Dev.* **16** 2733–42
8. Lee, Y., Ahn, C., Han, J., Choi, H., Kim, J., Yim, J., Lee, J., Provost, P., Radmark, O., Kim, S. and Kim, V. N. (2003) The nuclear RNase III Drosha initiates microRNA processing. *Nature* **425** 415–9
9. Bernstein, E., Caudy, A. A., Hammond, S. M. and Hannon, G. J. (2001) Role for a bidentate ribonuclease in the initiation step of RNA interference. *Nature* **409** 363–6
10. Hammond, S.M., Bernstein, E., Beach, D. and Hannon, G. J. (2000) An RNA-directed nuclease mediates post-transcriptional gene silencing in Drosophila cells. *Nature* **404** 293–6
11. Lewis, B.P., Burge, C. B. and Bartel, D. P. (2005) Conserved seed pairing, often flanked by adenosines, indicates that thousands of human genes are microRNA targets. *Cell* **120** 15–20
12. Zeng, Y., Sankala, H., Zhang, X. and Graves, P. R. (2008) Phosphorylation of Argonaute 2 at serine-387 facilitates its localization to processing bodies. *Biochem. J.* **413** 429–36
13. Jakymiw, A., Lian, S., Eystathioy, T., Li, S., Satoh, M., Hamel, J. C., Fritzler, M. J. and Chan, E. K. (2005) Disruption of GW bodies impairs mammalian RNA interference. *Nat. Cell Biol.* 7 1267–74
14. Liu, J., Valencia-Sanchez, M. A., Hannon, G. J. and Parker, R. (2005) MicroRNA-dependent localization of targeted mRNAs to mammalian P-bodies. *Nat. Cell Biol.* 7 719–23
15. Sen, G.L. and Blau, H. M. (2005) Argonaute 2/RISC resides in sites of mammalian mRNA decay known as cytoplasmic bodies. *Nat. Cell Biol.* 7 633–6
16. Leung, A.K., Calabrese, J. M. and Sharp, P. A. (2006) Quantitative analysis of Argonaute protein reveals microRNA-dependent localization to stress granules. *Proc. Natl. Acad. Sci. U.S.A.* **103** 18125–30
17. Pare, J.M., Tahbaz, N., Lopez-Orozco, J., LaPointe, P., Lasko, P. and Hobman, T. C. (2009) Hsp90 Regulates the Function of Argonaute 2 and Its Recruitment to Stress Granules and P-bodies. *Mol. Biol. Cell*
18. Bordeleau, M.E., Mori, A., Oberer, M., Lindqvist, L., Chard, L. S., Higa, T., Belsham, G. J., Wagner, G., Tanaka, J. and Pelletier, J. (2006) Functional characterization of IRESes by an inhibitor of the RNA helicase eIF4A. *Nat. Chem. Biol.* **2** 213–20
19. Anderson, P. and Kedersha, N. (2008) Stress granules: the Tao of RNA triage. *Trends Biochem. Sci.* **33** 141–50
20. Pauley, K.M., Eystathioy, T., Jakymiw, A., Hamel, J. C., Fritzler, M. J. and Chan, E. K. (2006) Formation of GW bodies is a consequence of microRNA genesis. *EMBO Rep.* 7 904–10

Chapter 12

Reporter-Based Assays for Analyzing RNA Interference in Mammalian Cells

Lydia V. McClure, Gil Ju Seo, and Christopher S. Sullivan

Abstract

RNA interference (RNAi) is a process whereby small RNAs serve as effectors to direct posttranscriptional regulation of gene expression. The effector small RNAs can arise from various sources including plasmids that express short-hairpin RNAs (shRNAs) or microRNA (miRNAs), or alternatively, from synthetic small-interfering RNAs (siRNAs). These small RNAs enter a protein complex that binds directly to mRNA targets and this results in transcript-specific inhibition of protein expression. Though the key core components of the mammalian RNAi processing and effector complexes have been identified, accessory and regulatory factors are less well-defined. Reporter assays that can quantitatively assess RNAi activity can be used to identify modulators of RNAi. We present two methods to quantitatively analyze RNAi activity that have overlapping and distinct utility. The first method uses an eGFP reporter in transiently transfected cells to identify RNAi modulators. The second method uses cells that express luciferase-based reporters in a stable fashion. This assay can easily be conducted in 96-well plate format. Both methods can be used to identify novel proteins or small molecules that modulate RNAi activity.

Key words: RNA interference, RNAi, microRNA, miRNA, Small hairpin RNA, shRNA, Small interfering RNA, siRNA, microRNA reporter assay, Dual luciferase assay, Enhanced green fluorescent protein, eGFP, Mammalian cells, Flow cytometry, Nodamura, Virus, B2

1. Introduction

RNA interference (RNAi) is an evolutionarily conserved process that plays a role in regulating gene expression. In some organisms, components of the RNAi machinery regulate endogenous gene expression via microRNAs (miRNAs) and other effector classes of small RNAs (1). Additionally, RNAi functions to prevent the expression of harmful genetic elements such as transposons and viruses (2, 3). RNAi is clearly an antiviral defense in some organisms such as plants and insects (4), while in mammalian

Tom C. Hobman and Thomas F. Duchaine (eds.), *Argonaute Proteins: Methods and Protocols*, Methods in Molecular Biology, vol. 725, DOI 10.1007/978-1-61779-046-1_12, © Springer Science+Business Media, LLC 2011

cells, it remains an unresolved question as to whether RNAi plays any role in antiviral defense (5). Elegant biochemical and molecular studies have elucidated the key core components of the RNAi response. But, an understanding of the factors that regulate the RNAi response is less well defined. Furthermore, many questions still remain about the relative contribution of individual RNAi components (such as the various Argonautes) to the degree and type of RNAi response that is initiated. As the link between some cancers and miRNAs becomes more evident (6), identifying small molecules that globally manipulate RNAi activity could have therapeutic or experimental value. Reporter assays that can quantitatively assess RNAi activity can be used to better understand the effects of cellular changes (such as viral infection, changes in physiology, and stress) on the global regulation of RNAi. Such reporters can also be used to identify novel regulators (gene products or synthetic molecules) of the RNAi response or for structure/function analysis of known components of the RNAi machinery.

We present here two different reporter assays that can be used to quantitatively assess RNAi activity in mammalian cells. One assay utilizes a modified enhanced green fluorescent protein (eGFP) as a reporter and the other employs modified *Renilla* luciferase. Both systems have overlapping and distinct utility depending on the design of the screen to be employed. Using the eGFP reporter, we were able to identify the Nodamura Viral Protein B2 as a negative regulator of RNAi that can function in mammalian cells (7). Our use of a modified version of the screen with different initiators of the RNAi response (siRNAs versus plasmid-expressed shRNAs, see Note 1) led to an understanding of the mechanism of how the B2 proteins from Nodamura and other Nodaviruses inhibit RNAi. We present in detail the protocols for our eGFP and luciferase-based RNAi activity screens.

2. Materials

2.1. RNAi Screen Using eGFP Reporters and Flow Cytometry

2.1.1. Construction of RNAi eGFP Reporter and NovB2 Expression Vectors

1. Taq polymerase (Promega, Madison, WI).
2. Restriction enzymes (New England Biosciences (NEB), Ipswich, MA).
3. Custom DNA oligonucleotides (Invitrogen, Carlsbad, CA).
4. pcDNA 3.1 expression vector (Invitrogen, Carlsbad, CA).
5. Destabilized eGFP reporter vector pD2eGFP (BD Biosciences Clontech, San Jose, CA).
6. pDsRed-Express-DR (BD Biosciences Clontech, San Jose, CA).

2.1.2. Cell Culture and shRNA Vector Transfection

1. Cell culture media: Dulbecco's Modified Eagle's Medium (DMEM) (Gibco, Bethesda, MD) with 10% fetal bovine serum (FBS, HyClone, Ogden, UT) and 100 U/ml penicillin and 100 mg/ml streptomycin (Sigma, St. Louis, MO). Complete medium is stored at 4°C for up to 5 weeks and is pre-warmed at 37°C before being applied to cells.
2. 293 and 293T human embryonic kidney cells (HEK 293 and HEK 293 T, American Type Culture Collection (ATCC, Manassas, VA) maintained as recommended by ATCC.
3. Transfection reagent Fugene 6 (Roche).
4. Custom eGFP reporter vectors.
5. Custom dsRED expression vector.
6. Vectors expressing shRNAs against eGFP or irrelevant mRNA (Firefly luciferase) (InvivoGen, San Diego, CA).
7. Custom pcDNA3.1 puro NoVB2 expression vector.

2.1.3. Flow Cytometry and Analysis

1. FACSCalibur flow cytometer (Becton Dickinson).
2. Magnesium-free and magnesium-containing phosphate-buffered saline (PBS) at pH 7.4 (Sigma, St. Louis, MO).
3. Trypsin/EDTA solution (Gibco, Bethesda, MD).
4. Falcon flow cytometry tubes (Becton Dickson, Franklin Lakes, NJ).

2.1.4. SiRNA Transfection

1. Anti-eGFP siRNA (target sequence: 5′ AAGCTGACCCTG-AAGTTCATC 3′) (20 μM, 1,000× stock) (Dharmacon Thermo Fisher Scientific, Lafayette, CO) (see Note 2).
2. Irrelevant control siRNA (target sequence anti-Luciferase GL2; target sequence: 5′ CGTACGCGGAATACTTCGA 3′) (Dharmacon Thermo Fisher Scientific, Lafayette, CO).
3. Transfection reagent Lipofectamine 2000 (Invitrogen, Carlsbad, CA).

2.2. Measurement of RNAi Activity Using a Luciferase-Based Reporter Assay

2.2.1. Plasmid Construction for miRNA Expression Vectors and Luciferase Reporters

1. pcDNA3.1 hygromycin and pcDNA3.1 puromycin (a custom modified version of pcDNA3.1 neomycin, (7)) expression vectors (Invitrogen, Carlsbad, CA).
2. JCV viral genome template (NCBI accession number NC_001699).
3. Destabilized *Renilla* luciferase (pGL4.84hRlucCP/Puro) (Promega, Madison, WI).
4. Destabilized Firefly luciferase (pGL4.22luc2CP/Puro) (Promega, Madison, WI).
5. Topo TA cloning kit (Invitrogen, Carlsbad, CA).
6. Luria Broth Base (Miller's LB Broth) (Sigma, St. Louis, MO).

7. Ampicillin (Fisher Scientific, Fair Lawn, NJ): prepared fresh or stored as single use aliquots at –20°C.
8. Phusion® High-Fidelity DNA Polymerase and Taq Polymerase (New England Biosciences (NEB), Ipswich, MA).
9. Custom PCR primers (Invitrogen, Carlsbad, CA).
10. DNA clean and concentration 5 spin columns (Zymo Research, Orange, CA)
11. Agarose gel and PCR product purification kits (Qiagen, Valencia, CA).
12. Restriction enzymes (New England Biosciences (NEB), Ipswich, MA).
13. T4 DNA ligase (New England Biosciences (NEB), Ipswich, MA).

2.2.2. Establishment of Stable Luciferase Reporter Cell Lines

1. HEK 293 cells (American Type Culture Collection (ATCC), Manassas, VA) maintained as recommended by ATCC.
2. Transfection reagent Lipofectamine 2000 (Invitrogen, Carlsbad, CA).
3. Puromycin and Hygromycin B are each dissolved in tissue-culture grade water at 10 and 100 mg/ml, respectively. These stock solutions are stable for up to a year at –20°C (InvivoGen, San Diego, CA).
4. Growth medium is prepared with Dulbecco's Modified Eagle's Medium (DMEM, Gibco, Bethesda, MD) with 10% FBS (HyClone, Ogden, UT) and with 100 U/ml penicillin and 100 mg/ml streptomycin (Sigma, St. Louis, MO). Complete media is kept at 4°C and warmed to 37°C before use.
5. Pyrex cloning cylinders (Fisher Scientific, Fair Lawn, NJ).

2.2.3. Transfection with Control Antisense Inhibitor of RNAi

1. 2′-*O*-methylated oligonucleotides (Dharmacon Thermo Fisher Scientific, Lafayette, CO) used as antisense miRNA inhibitors are stored as a 100 μM stock solution at ≤ –20°C.
2. Transfection reagent Lipofectamine 2000 (Invitrogen, Carlsbad, CA).

2.2.4. Dual Luciferase Assay

1. Luminoskan Ascent Luminometer (ThermoScientific, Waltham, MA).
2. 96-Well Polystyrene Microplates, solid bottom, white (Greiner Bio-One, Monroe, NC).
3. Dual luciferase assay kit from Promega (includes Passive Lysis Buffer, LAR II substrate, Stop & Glo buffer and Stop & Glo substrate (Promega, Madison, WI)). All reagents must be at room temperature before use and should be thawed immediately prior to running the experiment.

3. Methods

Either of the two different reporter assays described here can be used to measure RNAi in mammalian cells. The first uses eGFP and is particularly useful for transient transfection assays. This protocol assays individual cells, first gating on those that were co-transfected with the dsRED Express marker. In this way, the assay can be performed on cells with a low-transfection efficiency, since only transfected cells will be scored. The second assay uses luciferase as a reporter protein. This assay is particularly amenable to large-scale screens in 96-well plate format because it uses cells that stably express the reporters and gives consistent results that are reproducible throughout multiple cellular passages.

3.1. RNAi Screen Using eGFP Reporter and Flow Cytometry

Proteins that can function as inhibitors of RNAi have been identified in viruses that infect plants, insects, and mammals. We used this eGFP reporter assay to identify the Nodamura virus (NoV) B2 protein as an inhibitor of mammalian RNAi. B2 was found to function by inhibiting multiple steps in the RNAi response (7). This was determined by using either pre- or post-dicer products as triggers of RNAi and assaying for differences in the degree of inhibition (see Note 1). NoV B2 is a useful positive control when screening to identify new modulators of the mammalian RNAi response. This method describes: (1) the construction of the eGFP reporter vector and the NovB2 positive control vector, (2) mammalian cell transfection, (3) flow cytometry analysis, and (4) an alternative protocol using siRNA transfection instead of the shRNA-encoding vector to induce RNAi.

3.1.1. Construction of RNAi eGFP Reporter and NovB2 Expression Vectors

1. eGFP Reporter: the destabilized eGFP gene is shuttled from the pD2eGFP vector into the pcDNA3.1 expression vector employing standard unidirectional restriction enzyme/ligation cloning. To increase the dynamic range of the assays, eGFP reporters are generated that contain multiple binding sites complementary to the shRNA of interest (see Note 3). In this case, an shRNA expression vector expressing a small RNA that recognizes a 21 nucleotide region within the eGFP mRNA is used (CAAGCUGACCCUGAAGUUCA). An additional six binding sites are engineered into the 3′ UTR by ligating long DNA oligonucleotides together and filling in with Taq polymerase: GFPutrNotF, CGCA**GCGGCCGC**GCAAGCTGACCCTGAAGTTCAGCAAGCTGACCCTGAAGTTCAGCAAGCTGACCCTGAAGTTCAGCAAGCTGACCCTGAAGTTCAGAAT; GFPutrApaR, GGCT**GGGCCC**GAATTCGTCGGCGGGGTGCTTCACGTACACCTTGGGTCGGCGGGGTGCTTCACGTACACCTTGGGTCGGCGGGGTGCTTCACGTACACCT. Because of the repeated nature of the two oligonucleotides

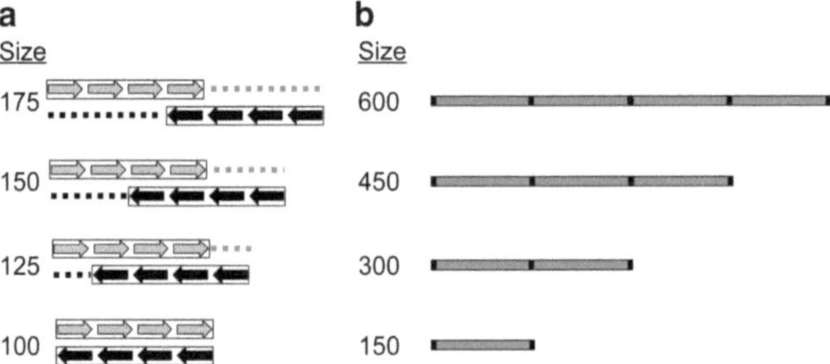

Fig. 1. Two different strategies to engineer concatameric 3′ UTRs of RNAi reporters that contain multiple binding sites for experimental shRNAs or miRNAs. (**a**) For the eGFP reporters, oligonucleotides (100 nucleotides in length) each containing four copies of the eGFP shRNA target binding site were annealed and filled in with Taq polymerase. The dashed line represents the "filled-in" product. Each arrow represents a copy of the shRNA target site. The gray arrows depict the forward strand and the black arrows depict the reverse strand. Gel fractionation of the annealed products allows for the isolation of longer concatamers for subsequent cloning. In our hands, we were only able to isolate annealed products with a maximum of six repeated binding sites. However, theoretically, annealed products containing up to seven repeats of the shRNA target binding site are possible (~175 bp final product). (**b**) For the luciferase reporters, we PCR-amplified a region of the JCV genome that was complementary to the 5p and 3p miRNAs, which are derived from the JCV pre-miRNA. The PCR products had nonpalindromic BanI restriction sites engineered onto the ends so that after BanI digestion, the PCR products could be unidirectionally ligated into concatamers. Gel fractionation was utilized to isolate concatamers that contained four copies of the JCV miRNA binding sites. The black boxes represent the locations of the BanI sites, the gray boxes represent the ~150 nucleotides of JCV genome that contains a region that is complementary to both the JCV 5′ and 3′ miRNAs. Size indicates approximate number of nucleotides of each concatamer.

that are being annealed, double-stranded products of various lengths are derived from this reaction (Fig. 1a). Therefore, size fractionation is carried out by agarose gel electrophoresis to select those inserts containing the most binding sites for the shRNA of interest. The purified double-stranded DNA insert is then cloned into the 3′ UTR of eGFP using *Not*I and *Apa*I restriction enzymes and DNA ligase via standard cloning procedures (see Note 3).

2. DsRed transfection marker: destabilized dsRed Express is used during flow cytometry to gate on cells that were efficiently transfected with the eGFP RNAi reporter. The DsRed cassette from pDsRed-Express-DR is subcloned into pcDNA3.1 using standard unidirectional restriction site subcloning procedures.
3. The NoVB2 expression vector: the gene that encodes NoV B2 is cloned into pcDNA3.1Puro, a modified pcDNA3.1 vector in which the neomycin resistance cassette was replaced with a Puromycin resistance gene. The gene that encodes NoV B2 is synthesized using PCR and overlapping, inverse orientation ~75 nucleotide long oligonucleotides as template.

The external primers include flanking *Kpn*I and *Xba*I restriction enzyme sites (7). The internal four oligonucleotides are mixed with the external outermost two flanking oligonucleotides at a 1–10 ratio. PCR is carried out using Taq polymerase under standard PCR conditions.

3.1.2. Cell Culture and shRNA Vector Transfection

1. Experiments are performed using HEK 293 T cells maintained under standard conditions in the growth medium of DMEM with 10% FBS, 100 U/ml penicillin, and 100 mg/ml streptomycin in an incubator at 37°C with 5% CO_2.
2. One to two days before the assay, a T75 flask of HEK293 T cells is trypsinized and vigorously resuspended in growth medium. Cells are brought up to a total volume of ~60 and 2 ml are transferred into each well of a six well dish (see Note 4).
3. Cells at a final density of 60–80% confluency are co-transfected with plasmids expressing fluorescent reporter plasmids (eGFP reporter plasmid pcDNA3.1 dseGFP and transfection marker pcDNA3.1 DsRed-Express), candidate RNAi modulator plasmid (or positive control pcDNA3.1puro NopVB2), and short-hairpin RNAs (see Note 5). Vectors expressing shRNAs against eGFP and an irrelevant target (Firefly luciferase) are driven by the H1 polymerase III promoter called psiRNA-hH1 (InvivoGen, San Diego, CA) (see Note 6).
4. Fugene6 is used as the transfection reagent (see Note 5) and reactions are set up in master mixes as follows: 1×: 2 ml medium, 4 μl Fugene reagent, 1.11 μg of total DNA: 5 ng eGFP reporter vector (pcDNA3.1 dseGFP), 5 ng DsRed tranfection control vector (pcDNA3.1 pDsRed-Express-DR), 100 ng shRNA vector, and 1 μg of candidate RNAi modulator vector (or control parental vector, pcDNA3.1 puro) (see Note 7).

3.1.3. Flow Cytometry and Analysis

1. Transfected cells are washed with magnesium-free PBS and trypsinized with trypsin/EDTA solution. Cells are then pelleted, washed one time in PBS-containing magnesium, resuspended in PBS-containing magnesium at 10^5–10^7 cells/ml (see Note 8), and transferred into a 5 ml flow cytometry tube.
2. All experiments were conducted using a Becton Dickinson FACSCalibur flow cytometer.

   If necessary, set up compensation parameters on the flow cytometer to minimize DsRed signal in the eGFP detection range (see Note 9).
3. Gate on living cells.
4. Of the living cells, gate on highest dsRed-expressing (top ~15%) and plot eGFP signal from these cells as a histogram (Fig. 2a).

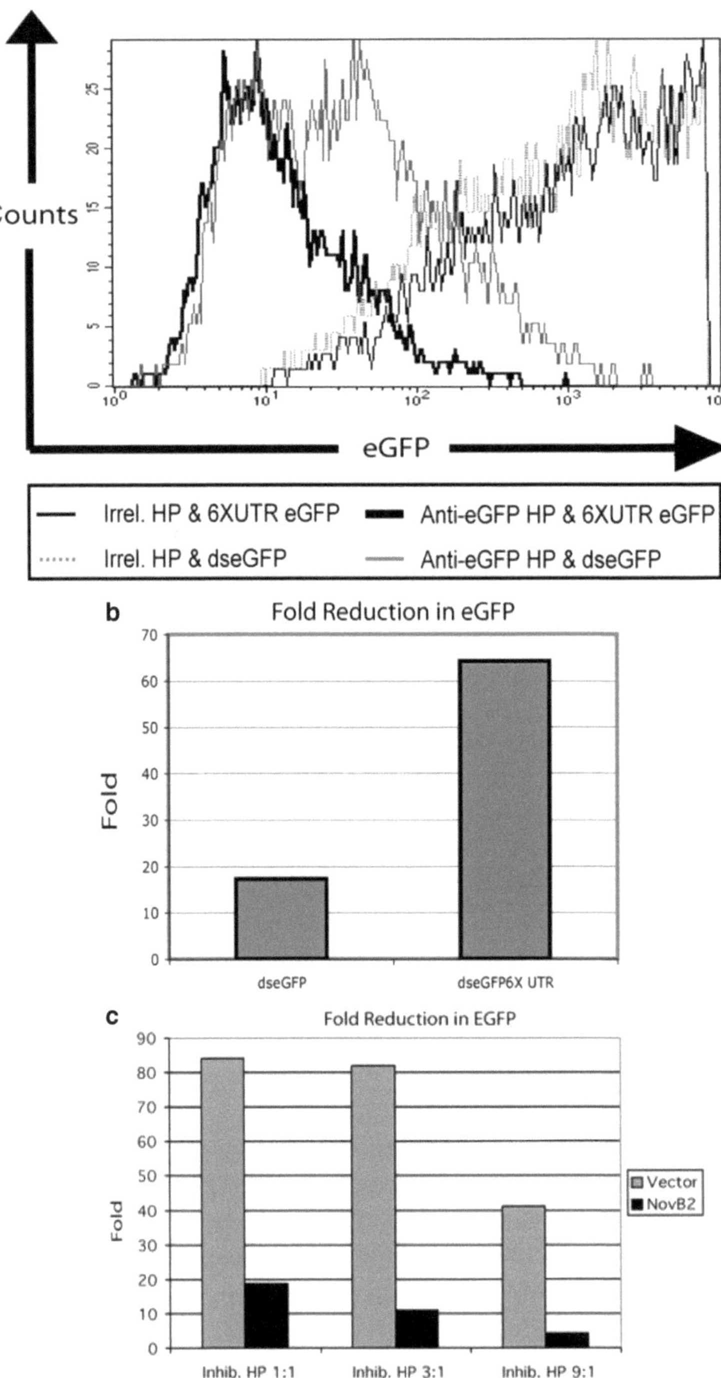

Fig. 2. eGFP-based reporter assay for modulators of RNAi in mammalian cells. (**a**) Histogram plot showing specific reduction in eGFP levels in HEK 293T cells transfected with anti-eGFP shRNA-expressing vector (anti-eGFP HP). Using a reporter engineered to have six additional shRNA target sites in its 3′ UTR increases the dynamic range of the assay. The dseGFP construct contains a single-shRNA binding site within the mRNA of eGFP. 6× UTR eGFP refers to a vector that expresses dseGFP with a 3′ UTR containing six additional shRNA binding sites. A destabilized eGFP is used for all vectors to more closely reflect mRNA levels. The expression of eGFP by the anti-eGFP shRNA is compared with an irrelevant shRNA control vector (Irrel. HP dseGFP) that is targeted against luciferase. (**b**) Bar graph plot showing fold reduction in eGFP levels of each reporter (vector expressing eGFP alone or vector that additionally expresses six extra binding sites in the 3′ UTR). (**c**) Co-transfection of NovB2 protein reduces RNAi activity (adapted from ref. 7).

5. Determine the geometric mean of eGFP fluorescence in these cells. Fold reduction in signal (a measure of RNAi activity) is plotted as the ratio of geometric means of irrelevant shRNA vector-transfected cells to eGFP shRNA-transfected cells (Fig. 2b). The positive control RNAi inhibitor NoVB2 should induce an approximately four- to eightfold higher eGFP geometric mean in the reporter cells transfected with the anti-eGFP shRNA (four- to eightfold less reduction in eGFP levels, Fig. 2c).

3.1.4. Variation of Protocol: Using Synthetic siRNAs to Trigger RNAi

Instead of using shRNA vectors to drive RNAi, an alternative protocol can be employed that utilizes siRNAs (see Note 1). This allows rapid determination of whether an RNAi modulator is affecting RNAi processes pre- or post-Dicer cleavage (since Dicer activity is more important for shRNA activity than for siRNA activity).

1. Using the Lipofectamine 2000 transfection reagent, transfect cells with the eGFP reporter plasmid, DsRed transfection marker plasmid, candidate RNAi modulator, and siRNAs (either anti-eGFP or irrelevant control siRNAs) (see Note 5). 5 ng each of the eGFP reporter and DsRed transfection marker are co-transfected with a final concentration of 20 nM siRNA. Add 10 ng total of vector mix (5 ng each of eGFP reporter vector and DsRed vector, see Note 7), along with 1 μg of candidate RNAi modulator vector (or control parental vector, pcDNA3.1 puro) and 20 pmol (1 μl of 20 μM, 1,000× stock) of siRNA into 50 μl of serum-free medium. In a separate tube, add 1.5 μl of Lipofectamine 2000 to 50 μl of serum-free medium. Wait 5 min, mix contents of both tubes and add to 1 ml of medium on cells at 50–60% confluency in a six-well dish.
2. Allow transfection to proceed for 36–48 h and proceed with flow cytometry as described above in Subheading 3.1.3.

3.1.5. Interpretation of Results

RNAi activity is measured as the reduction in geometric mean fluorescent intensity of eGFP signal in anti-eGFP-shRNA-treated cells. The fold reduction in the geometric mean fluorescent intensity is plotted as a ratio of eGFP levels from cells expressing the anti-eGFP shRNA versus an irrelevant shRNA (Fig. 2b). Candidate inhibitors of RNAi activity score positive when less reduction in eGFP signal is observed (Fig. 2c).

3.2. Measuring RNAi Activity Using a Luciferase-Based Reporter

We have previously developed expression vectors and reporters based on the viral miRNAs encoded by JCV virus (8). We utilized these reagents to develop an assay to systematically screen numerous candidate modulators of the mammalian RNAi pathway in a single experiment format (e.g., drug libraries and plasmid libraries). The dual luciferase-based assay is a commonly used method to quantify relative luciferase enzyme activity, which generally directly reflects the quantity of luciferase protein present in the cell. Luciferase-based assays are useful because they can be easily scaled up to a 96-well format. This method describes (1) the construction of miRNA expression vectors, (2) the design of sensitive luciferase-based reporters containing concatamers of a sequence targeted by exogenous viral miRNAs (see Note 10, Fig. 1b), (3) the establishment of stable reporter cell lines expressing both reporters and miRNA expression construct using two rounds of transfection, each with different drug selection, and (4) the dual luciferase assay to measure RNAi activity.

3.2.1. Plasmid Design for miRNA Expression Vectors and Luciferase Reporters

1. The JCV miRNA is PCR-amplified from a portion of the JCV viral genome. Primers are designed to amplify the pre-miRNA with approximately 1,000 bp of flanking sequence on each side of the pre-miRNA. Restriction enzyme digestion/cloning using *Bam*HI and *Xho*I is used to subclone the PCR-purified pre-miRNA cDNA into pcDNA3.1 hygromycin, which drives the expression of the pri-miRNA using the HCMV immediate early promoter (see Notes 11 and 12).
2. *Renilla* luciferase reporters are constructed with concatameric miRNA binding sites in their 3′ UTR using PCR-based methods (see Notes 13 and 14). Two plasmids, pGL4.84hRlucCP/Puro and pGL4.22luc2CP/Puro are used as PCR template. *Kpn*I/*Xho*I restriction enzyme sites are engineered into the PCR products through gene-specific primers and are used to subclone each luciferase gene into the pcDNA3.1puromycin expression vector. The primers are as follows: *Renilla* luciferase gene forward primer, ATT**<u>GGTACC</u>**ATGGCTTCCAAGGTGTACGACCC; Firefly luciferase gene forward primer, GCTGGTACCATG GAAGATGCCAAAAACATTAAG; and the shared Firefly and *Renilla* luciferase gene reverse primer, AAT**<u>CTCGAG</u>** TTAGACGTTGATCCTGGCGCTGGC.
3. Four concatameric JCV miRNA binding sites are engineered by amplifying a portion of the JCV genome complementary to the miRNA. A plasmid containing JCV genomic DNA serves as the PCR template to amplify a ~150 bp region, a portion of which is perfectly complementary to both the 5p and 3p miRNAs. PCR primers contain the nonpalindromic *Ban*I restriction enzyme site (this allows for unidirectional concatamer formation, see Fig. 1b). Standard PCR conditions are used with Phusion polymerase and 25 amplification cycles. The primers used were as follows: JCV TAg gene forward primer, GAAGGCACCAGACCCATTCTTGACTTTCCT and JCV TAg gene reverse primer, GCA**<u>GGTGCC</u>**ACAGAT GTGAAAAGTGCAGTT (see Note 15).

4. The PCR products are purified using Zymo spin columns, digested with *Ban*I, purified using Zymo spin columns, and ligated with T4 DNA ligase overnight at 16°C. Multicopy concatamer PCR products are purified on a 1% agarose gel. A band corresponding in size to four concatamers has an approximate size of 600 bp and is cut from the gel. The DNA is purified from the gel slice using the Qiagen gel elution kit. *Taq* polymerase is then used to fill in the overhanging nucleotides and "Taq-on" overhanging A nucleotides. The resulting products are cloned into the pCR2.1 vector using the TOPO TA cloning kit. The vectors are then confirmed by sequencing analysis using the M13 forward or reverse primers.
5. PCR is used to subclone the concatamers from pCR2.1 into the 3′ UTR region of pCDNA3.1dsRluc. The primers are as follows: JCV TAg concatamer forward primer, ATG **CTCGAG**CGGCCGCCAGTGTGATGGATA, and JCV TAg concatamer reverse primer, GCA**TCTAGA**GTAACGGCC GCCAGTGTGCTG (both primers correspond to pCR2.1 vector-specific sequences). Restriction enzyme digested PCR products are shuttled into the *Xho*I/*Xba*I sites of pCDNA3.1dsRluc vectors.

3.2.2. Establishing Reporter Stable Cell Lines

1. First, clonal cell lines are derived that express the luciferase constructs in a stable manner. HEK 293 cells are plated into 10 cm culture dishes in growth medium (see Notes 16 and 17). Cells at ~20% confluency are co-transfected with the pcDNA 3.1dsRlucJCVTAg plasmid as well as pcDNA3.1dsFFluc. 5 μg each of the *Renilla* and Firefly reporter plasmids are diluted in 500 μl of DMEM without serum or antibiotics. In a separate tube, 20 μl of Lipofectamine 2000 is diluted in 500 μl of DMEM without serum or antibiotics. After 5 min, the two tubes are mixed and incubated for 20 min at room temperature. The mixture is then added dropwise into the 10 cm culture dish.
2. After 24 h, the transfection media is replaced with medium containing 3 μg/ml of puromycin and stable cell lines are selected. Cell death should be seen by 7 days and drug resistant cells will form colonies after 2–3 weeks.
3. Colony selection can be accomplished either by using Pyrex® cloning cylinders or by directly picking using a pipette tip. For the latter, the dish is held up to a light source to see dense regions of cells (colonies) whose location are then marked on the dish with a colored marker pen. Next, the entire dish is bathed in a dilute trypsin solution and a light microscope is used to locate the "marked" colonies on the dish. Then a pipette tip of a p1000 pipettor is used to scrape and pipette up and down locally on top of a single colony (see Note 18).

The selected colony is transferred to a 96-well plate from the pipettor and, once confluent, expanded to a single well of a 24-well dish.

4. After validation of luciferase activity, the colonies expressing the highest luciferase activity are selected and used in the second clonal selection step to drive expression of the miRNA (described below).
5. The second clonal selection is used to drive the expression of the JCV miRNA within the cell lines already expressing the luciferase reporters. The JCV miRNA expression vectors are transfected into the clonal cell lines as described above (Subheading 3.2.2). Stable cells are selected in medium containing 100 μg/ml of hygromycin B. Cell death begins approximately 7 days after treatment, as with puromycin treatment. Normally, drug-resistant colonies form after 2–3 weeks of hygromycin B treatment. Colonies with at least a sixfold repression of the *Renilla* activity are chosen as reporter cell lines for further experiments (see Note 19).

3.2.3. Seeding of Cells for Reporter Assay

1. Pre-seeded reporter cells (stably expressing the luciferase reporter, control luciferase, and miRNA (from Subheading 3.2.2)) are allowed to reach ~80% confluency in a T75 flask. Next, remove the medium and gently rinse the cells with 5 ml of tissue-culture grade D-PBS. One milliliter of trypsin–EDTA is added and the flask is tapped vigorously for approximately 30 s to detach the cells. The cells and trypsin–EDTA mixture is pipetted up and down approximately ten times. The trypsin reaction is then quenched by resuspending cells in 9 ml of growth medium using a serological pipette. Centrifuge the cells at 3,300 × *g* for 3 min. Discard the supernatant and resuspend the cell pellet in 10 ml of growth medium by gently pipetting approximately ten times up and down.
2. A single T75 flask is enough to seed approximately four 96-well plates. For one well of a 96-well plate, seed 25 μl of cell resuspension and 75 μl of growth medium. For one entire 96-well plate, add 2.5 ml of cell resuspension to 7.5 ml of growth medium, gently mix, and then add 100 μl to each well.

3.2.4. Treatment with Control Inhibitor

1. Antisense oligonucleotides complementary to the JCV miRNA are used as a positive control to inhibit RNAi activity (see Note 20). Combined transfection of the two RNA inhibitor oligonucleotides, each antisense to either the 5p or 3p JCV miRNAs, specifically blocks the activity of the JCV miRNAs within the RISC complex. As a negative control, we use an antisense oligonucleotide to a different,

irrelevant viral miRNA. The antisense oligonucleotides contain 2′-*O*-methylated nucleotides to increase RNA stability and binding affinity. Furthermore, it has been shown that longer antisense inhibitors with specific secondary structure have greater efficacy at inhibiting RISC (9). These inhibitors are designed with a partial short hairpin on each side of the single-stranded region that is complementary to the miRNA of interest (9). The oligonucleotides that serve as the irrelevant control or that block the JCV miRNA are as follows: irrelevant control antisense inhibitor, AGAAGAGAGAAAUCUCU UCUUGGCCACUCGGGGGACAACACUAAUCG CCAACAGACAUCUUCUCUUUCGAGAGAAGA; JCV 3p miRNA antisense inhibitor, AGAAGAGAGAAAUCUCUU CUCAGAAGACUCUGGACAUGGAUCAAGCACUG AAUCACAAUCUUCUCUUUCGAGAGAAGA; and JCV 5p miRNA antisense inhibitor, AGAAGAGAGAAAUCUC UUCUCUGAAUCACAAUCACAAUGCUUUUCCCAG GUCUCAUCUUCUCUUUCGAGAGAAGA.

2. RNA inhibitors are transfected using Lipofectamine 2000. One day after seeding, cells are transfected with a mixture of Lipofectamine 2000 and antisense inhibitors. A mastermix is made and distributed into each well for a final 1× concentration of: 0.25 μl of Lipofectamine 2000, 50 nM of each of the 5p and 3p inhibitors in 50 μl of DMEM (without antibiotics or serum). Inhibition of RNAi activity is proportional to the amount of inhibitor oligonucleotide included in the transfection. The effect reaches a plateau at 50 nM. Therefore, 100 nM of total antisense oligonucleotide (50 nM each of the 5p and 3p JCV inhibitors) should be used to maximize the inhibition of RNAi. Allow 24 h after inhibitor transfection before assaying RNAi activity.

3.2.5. Dual Luciferase Assay

The dual luciferase assay is performed using a Luminoskan Ascent Luminometer that is pre-warmed for at least 10 min prior to use (see Note 21). Remove the growth medium cautiously with an aspirator and then add 20 μl of room temperature passive lysis buffer. Incubate the cells for 15 min at room temperature to ensure lysis. Cell lysates are transferred from 96-well tissue culture plates to 96-well, white, solid bottom plates for use with the luminometer (see Note 22). Program the luminometer to perform a 2 s pre-measurement delay followed by a 5–10 s measurement period for each reporter assay. Add 50 μl of LAR II to read Firefly luciferase enzyme activity and record the results. Then add 50 μl of Stop & Glo to read *Renilla* luciferase enzyme activity. Normalize all treated cells to an untreated control well and average the replicates (see Note 23).

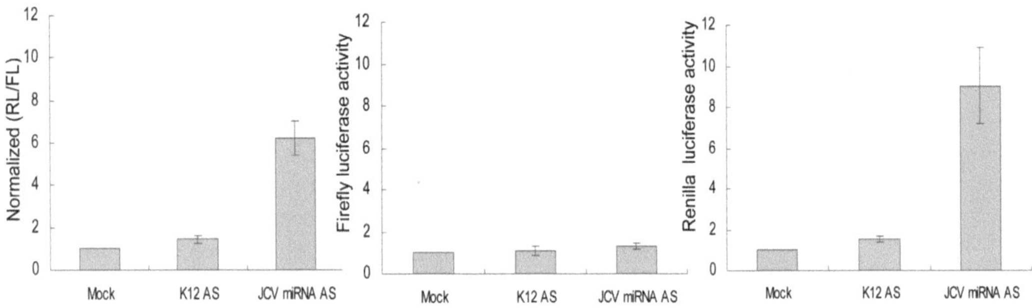

Fig. 3. Luciferase-based assay for RNAi activity. Stable cell lines were generated that express an exogenous viral miRNA that targets a Renilla luciferase (RL) reporter transcript. The cells also express Firefly luciferase (FL) as a control that is not affected by RNAi. Negative modulators of RNAi activity will result in increased Renilla luciferase activity only. The right panel shows the Renilla luciferase activity levels after treatment with a specific antisense oligonucleotide inhibitor of the viral miRNA (JCV miRNA AS), an irrelevant oligonucleotide (K12 AS), or no oligonucleotide (mock). Note: all luciferase results are normalized to the mock treatment (transfection reagent alone without any AS oligonucleotide). The center panel shows Firefly luciferase activity levels under the same treatment conditions. The left panel depicts the Renilla luciferase activities normalized to Firefly luciferase levels. Note the ~6-fold increase in the ratio of Renilla to Firefly luciferase levels that is indicative of RNAi inhibition.

3.2.6. Interpretation of Results

RNAi activity is measured in stable reporter lines after treatment with miRNA inhibitors such as the positive control antisense 2′-*O*-methylated inhibitor oligonucleotides or candidate modulators. *Renilla* and Firefly luciferase levels are individually normalized to mock-transfected cells (mock). Next, *Renilla* luciferase levels are normalized to Firefly luciferase levels (Fig. 3). Cell lines expressing the exogenous viral miRNA typically show a six- to tenfold reduction in *Renilla* activity compared with cells expressing the luciferase reporters only (data not shown). Treatment with the antisense inhibitors increases the normalized *Renilla* activity in these cells approximately sixfold (Fig. 3).

4. Notes

1. Using synthetic siRNAs instead of Dicer-dependent small RNAs derived from shRNA or miRNA-encoding vectors can help to identify which component of the RNAi machinery a particular modulator is affecting. In addition, we have used synthetic hairpins that mimic the post-microprocessor, pre-Dicer pre-miRNA substrates with success.
2. We routinely use siRNAs from (Dharmacon Thermo Fisher Scientific, Lafayette, CO), but have used siRNAs from several other companies with success.
3. Using multiple binding sites of the anti-eGFP sequence increases the dynamic range of the experiment (10). A reporter with only a single binding site displays an approximately 18-fold

reduction in eGFP levels, whereas a reporter containing at least four additional binding sites was repressed by 60-fold (7).

4. The actual dilution will require some optimization to get 60–80% confluency and will depend on how many days there are between plating the cells and conducting the experiment.

5. An important negative control in this experiment is the irrelevant shRNA. To confirm that any modulator of RNAi is specifically regulating a siRNA–mRNA target interaction, the irrelevant shRNA is co-transfected with the vectors and treated with the same potential RNAi modulator. If an siRNA is to be used instead of an shRNA-expressing vector, Lipofectamine 2000 or other transfection reagents capable of combined DNA and small RNA transfection must be used.

6. We have used shRNA vectors driven by either the H1 or U6 pol III promoters with similar results. We therefore anticipate many of the available pol III-driven shRNA vectors will work in this assay. The protocol described for the eGFP reporter system utilizes a pol III-driven shRNA. The protocol could easily be adapted to a pol II-driven vector that expresses exogenous miRNAs or shRNA engineered into pre-miRNA-like hairpins.

7. Using a master mix is key to reproducibility. We recommend that the eGFP and DsRed vectors be pre-mixed and the same volume administered to each experimental transfection.

8. Cells are trypsinized and vigorously pipetted up and down (~20×) to make a suspension of single cells. To prevent clumping and clogs during flow cytometry, cells are resuspended in 10^5–10^7 cells/ml (approximately 300 μl to 1 ml per six well).

9. For proper compensation on the flow cytometer, it is imperative to always include four control wells: (1) mock-transfected cells; (2) cells transfected with both eGFP and DsRed; (3) eGFP alone-transfected cells; and (4) DsRed alone-transfected cells. Ensure that no signal from either the red or green channel bleeds into the incorrect emission detection range.

10. Viral miRNAs are particularly useful for generating RNAi reporter cell lines because they are efficiently processed by host cells but are not endogenously encoded by the host cells.

11. The PCR method we used for generating the JCV miRNA-expressing vector utilized a "touchdown" protocol whereby the initial annealing temperature starts higher (60°C) and decreases by 0.5°C increments through the first ten rounds of PCR ending at 55°C for the remaining cycles.

12. We used high-fidelity Phusion polymerase to reduce the chance of PCR-induced mutations.

13. We design our concatameric miRNA binding sites to have perfect complementarity to the miRNAs. This results in

robust cleavage of the reporter transcript and allows independent confirmation of any effect observed by Northern blot analysis. However, we have also successfully utilized concatameric reporters with bulges engineered in the middle of the complementary site to mimic miRNA-mediated translation repression (10).

14. Both the Firefly and *Renilla* luciferase reporters contain a destabilizing PEST sequence (BD Biosciences Clontech, San Jose, CA). The shorter half-life of these reporters allows for luciferase protein levels that more closely mirror changes in mRNA levels. Additionally, since both inserts have the same 3′ end that codes for the PEST sequence, the same reverse primer can be used for PCR amplification of each insert.
15. The process of generating concatamers of miRNA binding sites should be initiated with abundant amounts of PCR product, as the loss of DNA can be a problem during the ligation and subsequent elution steps of this protocol.
16. Treat the tissue culture dishes with care as the HEK 293 and 293T cells are only weakly attached to the dish.
17. Because both the Firefly and *Renilla* plasmids utilize the same selection marker (Puromycin), each colony should be carefully checked to ensure the expression of both constructs.
18. When picking colonies, work rapidly to avoid over-trypsinization.
19. Expression of miRNAs (or shRNAs) should be confirmed by small RNA Northern blot analysis.
20. Antisense oligonucleotides are useful for ensuring that reporter cells derived from the second clonal selection (for the miRNA-expressing vector) have reduced *Renilla* luciferase levels due to RNAi activity triggered by the JCV miRNA.
21. The software package for the 96-well luminometer contains automated dual luciferase protocols. Be sure to include control wells in the appropriate location that contain blank wells and substrate only (these controls are critical for automated background subtraction).
22. 96-Well white polystyrene luminometer assay plates are reusable. After use, rinse plates well with deionized water followed by 70% ethanol. Allow to air-dry completely before future use.
23. Each assay condition should be performed in triplicate (at a minimum) and for future normalization, data should be saved in a format compatible with the Microsoft Excel spreadsheet software.

Acknowledgments

Work in the Sullivan Lab involving the protocols presented in this chapter is supported by NIH grant R01AI077746-01 and a fellowship from the UT Austin Institute for Cellular and Molecular Biology.

References

1. Bartel, D.P. (2009) *MicroRNAs: target recognition and regulatory functions.* Cell, **136**(2): p. 215–33.
2. Carthew, R.W. and E.J. Sontheimer (2009), *Origins and Mechanisms of miRNAs and siRNAs.* Cell, **136**(4): p. 642–55.
3. Ghildiyal, M. and P.D. Zamore (2009) *Small silencing RNAs: an expanding universe.* Nat Rev Genet. **10**(2): p. 94–108.
4. Ding, S.W. and O. Voinnet (2007) *Antiviral immunity directed by small RNAs.* Cell, **130**(3): p. 413–26.
5. Cullen, B.R. (2006) *Is RNA interference involved in intrinsic antiviral immunity in mammals?* Nat Immunol. 7(6): p. 563–7.
6. Croce, C.M. (2009) *Causes and consequences of microRNA dysregulation in cancer.* Nat Rev Genet. **10**(10): p. 704–14.
7. Sullivan, C.S. and D. Ganem (2005) *A virus-encoded inhibitor that blocks RNA interference in mammalian cells.* J Virol. **79**(12): p. 7371–9.
8. Seo, G.J., et al. (2008) *Evolutionarily conserved function of a viral microRNA.* J Virol. **82**(20): p. 9823–8.
9. Vermeulen, A., et al. (2007) *Double-stranded regions are essential design components of potent inhibitors of RISC function.* RNA. **13**(5): p. 723–30.
10. Doench, J.G., C.P. Petersen, and P.A. Sharp (2003) *siRNAs can function as miRNAs.* Genes Dev. **17**(4): p. 438–42.

Chapter 13

Artificial Tethering of Argonaute Proteins for Studying their Role in Translational Repression of Target mRNAs

Stephanie Eckhardt, Emilia Szostak, Zhaolin Yang, and Ramesh Pillai

Abstract

Small RNAs such as microRNAs (miRNAs) and small-interfering RNAs (siRNAs) associate with members of the RNA-binding Argonaute family proteins. Together they participate in transcriptional and posttranscriptional gene silencing mechanisms. The fate of the target mRNA is determined, in part, by the degree of complementarity with the small RNA. To examine the exact role of the Argonaute protein in the silencing complex, human Argonautes were artificially recruited to reporter mRNAs in a small RNA-independent manner by the BoxB-N-peptide tethering system. Tethering of Argonaute proteins to a reporter mRNA leads to the inhibition of translation, mimicking the repression seen with miRNAs. Similar tethering experiments were performed with fly and fission yeast Argonaute proteins and other components of the small RNP (ribonucleoprotein) complex, uncovering their specific roles in the silencing complexes containing them.

Key words: Tethering, BoxB, N-peptide, Argonaute, Translational repression, miRNA

1. Introduction

Small RNAs such as microRNAs (miRNAs) and small-interfering RNAs (siRNAs) participate in a variety of gene regulatory pathways in most eukaryotes studied (1, 2). Both classes of RNAs function as part of RNA–protein complexes which contain a member of the Argonaute family as a central component (3). The Argonaute (Ago) proteins recruit additional interacting factors to assemble a silencing complex called RNA-induced silencing complex (RISC) containing siRNAs or an miRNA ribonucleoprotein (miRNP), when miRNA is the guide RNA (4). In most eukaryotes, the role of small RNAs in posttranscriptional silencing by either target mRNA degradation or repression of

Tom C. Hobman and Thomas F. Duchaine (eds.), *Argonaute Proteins: Methods and Protocols*, Methods in Molecular Biology, vol. 725, DOI 10.1007/978-1-61779-046-1_13, © Springer Science+Business Media, LLC 2011

translation is well described. In the fission yeast, *Schizosaccharomyces pombe*, siRNAs associate with Ago1 to form an RNA-induced transcriptional silencing (RITS) complex that participates in the assembly of heterochromatin at various genomic loci (5). Although transcriptional silencing by small RNAs in higher eukaryotes is reported, it is presently unclear because this is a wide-spread phenomenon (6).

Initial biochemical studies indicated that the RISC and miRNP are distinct complexes, yet were shown to share several key components (7). Also, it was revealed that the degree of complementarity between the small RNA in these complexes and the target has a decisive role in determining the fate of the target mRNA. Perfectly complementary targets are degraded by the endonucleolytic activity (slicer activity) of the Ago protein (8, 9). In contrast, partially complementary targets are silenced by translational repression, with very little change in target RNA levels. In other instances, partially complementary targets are also subject to mRNA decay but via a pathway that is mechanistically distinct from slicer-mediated RNA cleavage (10). To appreciate the contribution of Ago proteins in the silencing process, they were artificially tethered to a reporter mRNA by an N-peptide–BoxB interaction system (11). This allows a small RNA-independent recruitment of Ago proteins to a target mRNA.

The first 22 amino acids (aa) of the transcriptional antitermination protein N of the bacteriophage lambda (λN peptide) specifically recognize a short 19-nucleotide BoxB hairpin (12). N-peptide fusions of a protein of interest can be artificially recruited to a target RNA bearing one or more BoxB hairpins, allowing an independent examination of its functions (13). Similarly, human Argonaute proteins were modified by the addition of the N-peptide followed by a haemaglutinin (HA) tag at the N-terminus of the protein (Fig. 1). The HA-tag allows tracking the expression of the fusion protein by western blot analysis. A luciferase reporter mRNA was modified by the insertion of five BoxB hairpins in the 3′-untranslated region (UTR) of the RNA. Co-transfection of plasmids expressing the 5BoxB reporter mRNA and N-HA-Ago2 fusion protein into HeLa cell cultures results in the repression of the reporter mRNA (Fig. 2). Presence of multiple BoxB hairpins leads to increased repression, which is very similar to the observation that multiple miRNA-binding sites on a target mRNA contributes to efficient repression. The experimental setup also allows mapping of functional domains in the Ago2 protein and the determination that all human Ago proteins mediate the repression of the reporter mRNA (Fig. 3). The observation that silencing is without any change in reporter mRNA levels led to the conclusion that the default effect of Argonaute proteins is the repression of translation of the target mRNA (Fig. 4). Sucrose gradient analysis revealed

a

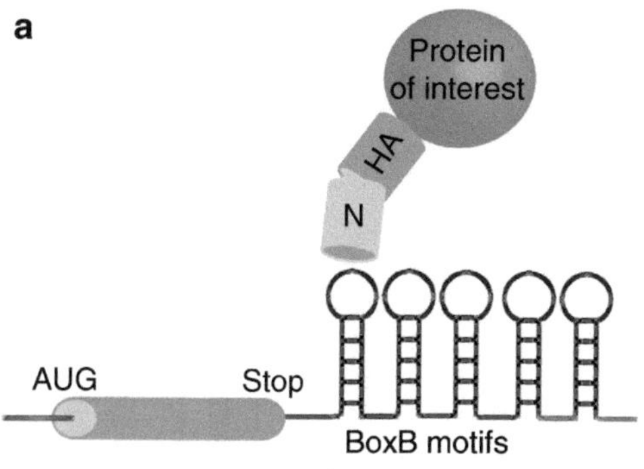

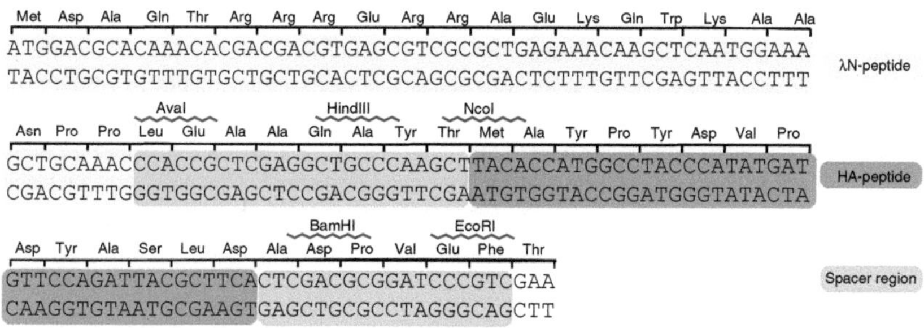

c

TAATTCTAGATAAGTCCAACTACTAAACTGGGGATTCCTGGGCCCTGAAGAAGGGCCCCTCGACTAAGTCCAACTAC

TAAACTGGGCCCTGAAGAAGGGCCCATATAGGGCCCTGAAGAAGGGCCCTATCGAGGATATTATCTCGACTAAGTCC

AACTACTAAACTGGGCCCTGAAGAAGGGCCCATATAGGGCCCTGAAGAAGGGCCCTATCGAGGATATTATCTCGAG

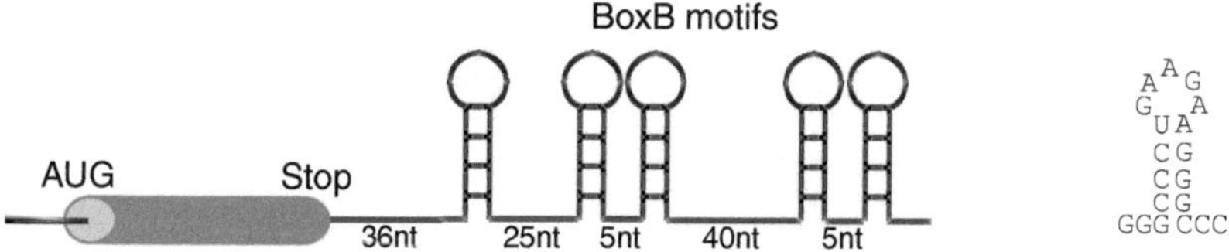

Fig. 1. The Argonaute tethering experiment. (**a**) A schematic representation of the *Renilla* luciferase reporter mRNA with 5BoxB hairpins inserted into the 3′-untranlsated region. The N-HA-tagged Argonaute protein (shown as protein of interest) is depicted as binding the BoxB hairpin; several of the five hairpins might be bound by the protein in the actual experiment. (**b**) The coding sequence for the N-HA peptides in the pCIneo-N-HA vector (11) and the *Eco* RI site used for cloning the Argonaute proteins are shown. (**c**) Sequence context of the five BoxB hairpins (highlighted) in the 3′-UTR of the RL-5BoxB mRNA. The stop codon (TAA) of *Renilla* luciferase is indicated.

that translational inhibition of the RL-5BoxB reporter mRNA is at the initiation step of translation (14).

A similar approach was used by others in tethering the *S. pombe* Ago protein to nascent transcripts of a reporter, leading to the recruitment of the RITS complex and the formation of

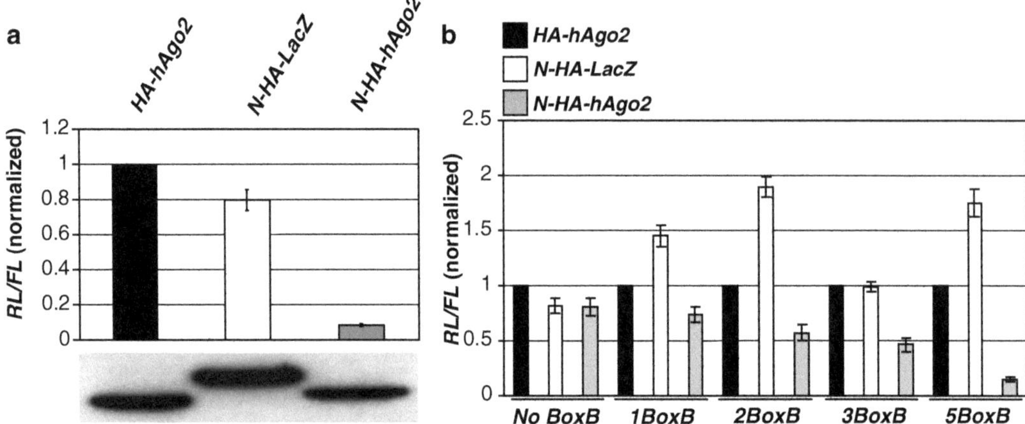

Fig. 2. (**a**) RL activity detected in extracts from HeLa cells expressing the indicated fusion proteins. Cells were cotransfected with constructs expressing the RL-5BoxB reporter, fire fly (FL) reference and indicated fusion proteins. Histograms represent normalized mean values (±SD) of RL/FL activities from a minimum of three experiments. RL activity values seen in the presence of HA-hAgo2 were set as 1. Expression levels of fusion proteins, as determined by western analysis using anti-HA antibody, are shown below the histogram. Generally, the N-HA-LacZ protein is expressed at a ~tenfold higher level than N-HA-hAgo2 or HA-hAgo2 and for this reason, ten times less of the N-HA-LacZ-expressing extract was loaded. Increased RL activity in extracts expressing N-HA-LacZ is likely due to the effect of the protein on the stability of mRNA reporters containing BoxB sequences. (**b**) Activity of reporter RL mRNAs containing different numbers of BoxB hairpins (reproduced from ref. (11) with permission from Cold Spring Harbour Laboratory Press).

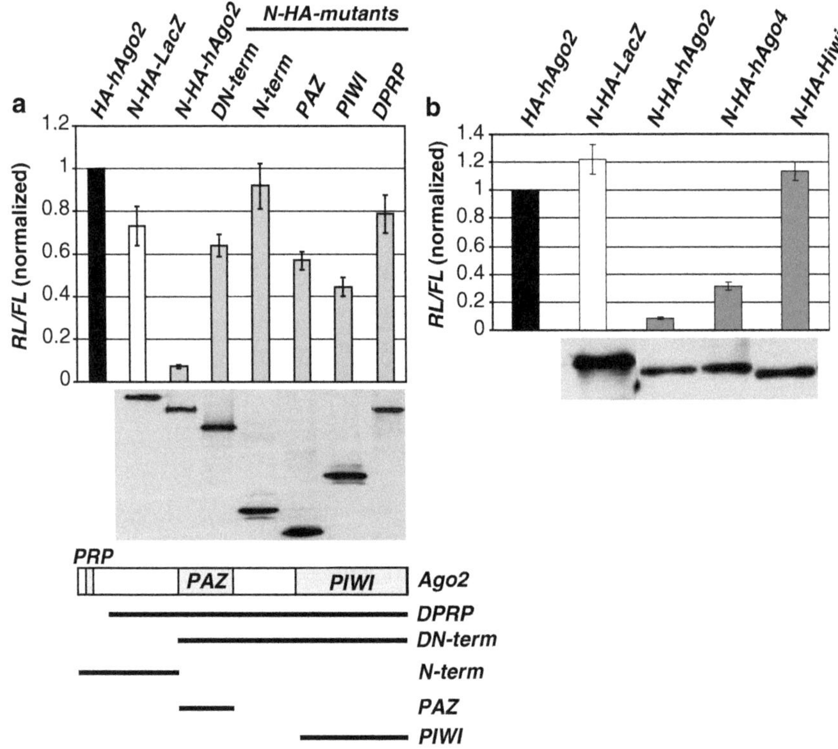

Fig. 3. Tethered hAgo2 and hAgo4, but not hAgo2 mutants and Hiwi, induce repression of the RL reporter. (**a**) RL activity in extracts from HeLa cells cotransfected with plasmids expressing RL-5BoxB reporter and N-HA-tagged mutant hAgo2 fusions. Western analysis of fusion protein expression is shown below the histograms. Schematic representation of hAgo2 and its deletion mutants is shown in the *lower panel*. (**b**) Tethering of hAgo4 but not the piwi protein Hiwi represses translation (reproduced from ref. (11) with permission from Cold Spring Harbour Laboratory Press).

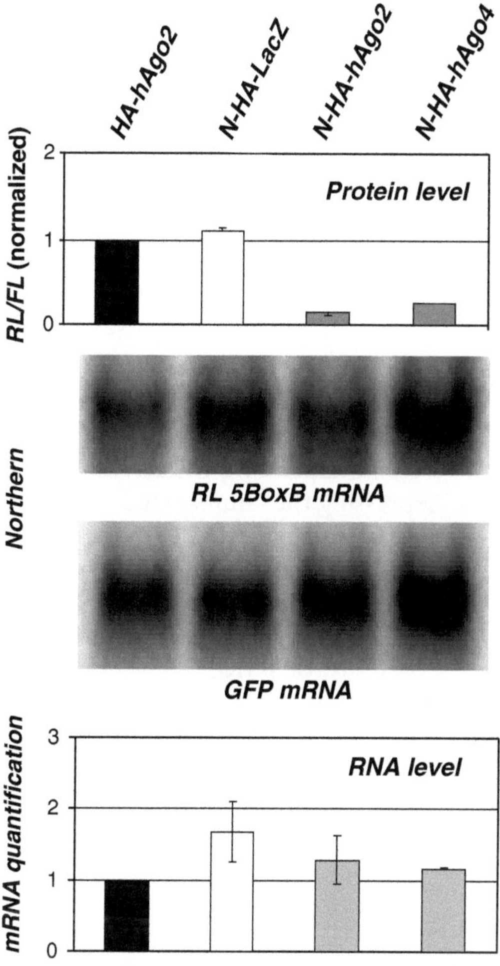

Fig. 4. Repression by N-HA-hAgo2 and N-HA-hAgo4 occurs without changes in reporter mRNA level. Northern blot analysis (*middle panels*) was performed with total RNA isolated from transfected cells, using probes specific for RL-5BoxB mRNA and green fluorescence protein (GFP) mRNA, expressed from the cotransfected plasmid. The RL activity in extracts from the same transfected cells is shown in the *upper panel*. Phosphor Storage screen scan quantification of the RL-5BoxB mRNA, normalized to GFP mRNA, is shown in the *lower panel*; values are means (±SD) from three independent experiments. The RL-5BoxB mRNA level in cells cotransfected with HA-hAgo2 is set to 1 (reproduced from ref. (11) with permission from Cold Spring Harbour Laboratory Press).

heterochromatin on the target loci (15). Tethering experiments with Ago-interacting GW182 protein in fly and human cells revealed the essential role of Ago proteins in recruiting GW182 to mediate translational repression and mRNA decay (16, 17). Thus, the RNA tethering assay can be used to dissect the molecular mechanism of individual components in the multiprotein small RNA complex that acts on RNA targets.

This chapter describes the protocols used in the tethering experiments with human Ago proteins as reported (11).

2. Materials

2.1. Preparation of Reporter and Protein Expression Constructs

1. The mammalian expression vector pCIneo (Promega) was modified by the insertion of annealed oligonucleotides coding for the 22 aa λN-peptide, followed by the 14 aa HA tag or carrying HA tag alone, giving rise to pCIneo-NHA and pCIneo-HA constructs (Fig. 1b).
2. The full-length human Argonaute proteins or their deletion versions are cloned downstream of the N-HA tag in the *Eco*RI/*Not*I sites of pCIneo-N-HA construct. This results in the expression of N-HA-tagged Argonaute protein from the cytomegalovirus (CMV) promoter.
3. Vectors expressing the *Rotylenchulus reniformis* (phRL-TK; expressing the "humanized" *Renilla* luciferase) and *Photinus pyralis* or fire fly (pGL3 Promoter) luciferases are commercially available (Promega). Annealed oligonucleotides encoding five separate units of the BoxB hairpin sequences (5BoxB) were inserted downstream of the stop codon in the *Xba*I site of phRL-TK vector resulting in the phRL-TK-5BoxB construct (Fig. 1c). To achieve higher expression levels of the reporter required for Northern blot analysis, cassettes encoding the RL-5BoxB were subcloned into the pCIneo backbone to obtain pCMV-RL-5BoxB construct. All modified constructs are available from the authors (11).

2.2. Mammalian Cell Culture, Transfections, and Luciferase Assays

1. Dulbecco's modified Eagle's medium (DMEM; Invitrogen) and DMEM supplemented with l-glutamine (Invitrogen), pencillin/streptamycin (Invitrogen), and 10% foetal bovine serum (Invitrogen). Media is warmed up to 37°C before use.
2. Exponentially growing HeLa cell cultures are seeded into 6-well tissue culture plates (Costar) and incubated at 37°C.
3. Transfection reagents – PLUS™ reagent (Invitrogen) and Lipofectamine (Invitrogen).
4. Sterile, autoclaved 1× PBS (phosphate-buffered saline): 137 mM NaCl, 2.7 mM KCl, 4.3 mM $Na_2HPO_4{\cdot}7H_2O$, and 1.4 mM KH_2PO_4 (pH 7.3).
5. Dual-Luciferase Reporter Assay kit (Promega) and luminometer (Centro LB 960, Berthold Technologies).

2.3. SDS–Polyacrylamide Gel Electrophoresis

1. Hoeffer SE-260 gel system (GE Health) and Hamilton needle (Hamilton)
2. Buffer A (4×) (resolving gel buffer): For 1 L, dissolve 181.64 g Tris base (MW 121.1) in water, adjust to pH 8.8. Add 20 mL 20% SDS (sodium dodecyl sulphate).

3. Buffer B (4×) (stacking gel buffer): For 500 ml, dissolve 30.28 g Tris base (MW 121.1) in water, adjust to pH 6.8. Add 10 mL 20% SDS.
4. Running buffer (1×): 0.025 M Tris base, 0.19 M Glycine, and 0.1% SDS (make a 10× stock).
5. SDS–PAGE sample loading buffer (2×): 100 mM Tris–HCl pH 6.8, 4% (w/v) SDS, 0.2% (w/v) bromophenol blue, 20% (v/v) glycerol, and 10% (v/v) 2-mercaptoethanol.
6. Thirty percent (30%) acrylamide/bis-acrylamide solution (37.5:1) (National diagnostics) (see Note 1).
7. APS (Ammonium persulphate): prepare a 10% solution in water, keep at 4°C (see Note 2) and TEMED (tetramethylethylenediamine).
8. Pre-stained protein molecular weight markers (Fermentas).

2.4. Western Blot Analysis

1. Transfer buffer (1×): 25 mM Tris base, 190 mM glycine, and 20% methanol.
2. Reinforced nitrocellulose membrane (Whatman OPTITRAN BA-S 85) and 3MM chromatography paper (Whatman).
3. Transblot Semidry transfer apparatus (Biorad).
4. Blocking buffer (1×): 5% non-fat dry milk powder in PBS-Tween (1× PBS with 0.1% Tween-20).
5. Primary antibodies: Anti-HA rabbit polyclonal (Y-11, Santa Cruz, sc-805); rabbit IgG horse radish peroxidase (HRP) conjugate (GE Health).
6. ECL Plus Western Blotting Detection system (GE Health). Hyperfilm ECL (GE Health) and film exposure Hypercassette (GE Health).
7. Developing machine (AGFA Curix 60).

2.5. Northern Blot Analysis

1. Absolutely RNA® RT-PCR miniprep kit (Stratagene).
2. NanoDrop spectrophotometer (NanoDrop Technologies, Inc.).
3. DEPC (diethyl pyrocarbonate)-treated water: mix 1 mL of DEPC with 1 L of double-distilled water and incubate overnight by vigorous shaking. Autoclave to remove DEPC and then aliquot.
4. Water-saturated phenol (acidic phenol), chloroform, and absolute ethanol.
5. Reagents for denaturing agarose gels: agarose and 37% formaldehyde.
6. Running buffer for denaturing gel (10×): 0.2 M MOPS, pH 7.0, 20 mM sodium acetate, and 10 mM EDTA (see Note 3).

7. RNA sample loading buffer (2×): To obtain 3 mL, mix 1.5 mL formamide, 300 μL of 10× MOPS, 525 μL of 37% formaldehyde, 15 μL of ethidium bromide (10 mg/mL stock), and a speck each of bromophenol blue and xylene cyanol. Make up with water. Freeze in aliquots at –20°C.
8. Nylon membrane (Hybond N+; GE Health), UV crosslinker (Stratalinker with 254 nm UV bulbs, Stratagene), and hybridization oven.
9. Gel-extraction kit (QIAgen), Random Primed DNA labelling kit (Roche) and microspin G-25 spin columns (GE Health). Radioactive nucleoside triphosphate [α-^{32}P] dCTP, 3,000 Ci/mmol (Perkin Elmer).
10. Hybridization buffer (1×; Church buffer): 0.25 M $NaHPO_4$, 0.25 M NaH_2PO_4, 1 mM EDTA, 1% BSA (bovine serum albumin), and 7% SDS.
11. Prepare 20× SSC (sodium chloride/sodium citrate): 3 M NaCl, 0.3 M sodium citrate, adjust pH to 7.0 with 1 M HCl.
12. Northern wash buffer: 2× SSC with 0.1% SDS (low stringency wash) and 0.2× SSC with 0.1% SDS (high stringency wash). Buffers are warmed up to 65°C before use.
13. Typhoon Scanner (GE Health) or other phosphor imaging apparatus. Image Quant Analyses software. Alternatively, Northern blots can be exposed to X-ray film.

3. Methods

3.1. Transfection of HeLa Cell Cultures with Reporters

1. Seed HeLa cells into required number of 6-well plates at a dilution so that they are 60–70% confluent at the time of use on or at the next day.
2. For each well, prepare a mixture of the following plasmids in a 1.5 mL eppendorf tube: 100 ng phRL-TK-5BoxB, 100 ng of pGL3 Promoter (for normalization of transfections), and 500 ng of plasmid expressing either HA or N-HA Argonaute protein fusions. When analysis by Northern blotting is required, prepare the following: 100 ng pCMV-RL-5BoxB, 100 ng of pGL3 Promoter 500 ng of plasmid expressing either HA or N-HA Argonaute protein fusions, and 75 ng of a GFP expression plasmid for the normalization of RNA levels. The strong CMV promoter gives considerably higher expression levels than the weaker Herpes Simplex Virus Thymidine Kinase (HSV-TK) promoter, required for the detection of the RL-5BoxB mRNA by Northern blotting.

3. The DNA mixture is diluted in DMEM to a final volume of 100 μL.
4. Add 4 μL of PLUS™ Reagent and mix by pipetting, incubate for 15 min (min) at room temperature.
5. In the meantime, mix 4 μL of Lipofectamine Reagent with 96 μL of DMEM media. Incubate for 15 min at room temperature.
6. Combine the DNA-Plus mix with the Lipofectamine solution and mix well by pipetting. Incubate a further 20 min at room temperature.
7. Each transfection in an experimental series is performed in triplicates, so a master-mix for three transfections can be prepared (steps 2–6). The experiment series should be reproduced at least three times.
8. During the incubation period, wash the cells in the 6-well plates: remove culture media by aspiration and wash once with 2 mL of DMEM media. Remove the wash and replace with 800 μL of DMEM. Leave cells in the 37°C incubator till required.
9. Remove cells from the incubator and gently pipette the transfection mix from step 5 into the wells. Return cells to incubator.
10. Replace the media after 4 h of incubation with fresh culture media containing supplements.
11. Change media once more after 24 h.

3.2. Luciferase Assays

1. Approximately 24 h posttransfection, remove media by aspiration and wash cells with 2 mL of 1× PBS. Remove as much of PBS as possible to prevent dilution of the cell extract.
2. Add 300 μL of passive lysis buffer (PLB) provided in the Dual-Luciferase Reporter Assay kit to the cells in each well of the 6-well plate. Incubate for 15 min at room temperature with gentle shaking on a tilting platform.
3. Luciferase measurements are made with a luminometer using 10 μL of the lysate from each well. The following amounts of the luciferase assay reagents are used: 50 μL of LAR II and 25 μL of Stop and Glo. A detailed protocol is provided by the manufacturer of the kit. Due to the high level of expression of hRL from the CMV promoter, the lysate might need to be diluted with 1× PLB to stay within the detection range of the luminometer.
4. Analysis of the luciferase measurements: the value obtained from RL measurements is divided by the fire fly measurements from the same sample. This normalized value for the triplicates of each transfection is then used to calculate the

mean and the standard deviation (SD). The mean obtained in transfections with plasmid expressing HA-Argonaute fusion is set to 1 (Fig. 2a).

3.3. SDS–PAGE

To detect the expression of the HA or N-HA-tagged Argonaute proteins or their deletions in the transfection experiments, western blot analysis is performed using anti-HA antibodies: The samples are first resolved by 10% SDS–PAGE and then analysed by western blotting.

1. The 10% SDS–PAGE is performed with the Hoeffer SE-260 gel system (or equivalent apparatus).
2. Glass plates are scrubbed clean with soap and water, and allowed to drain in a vertical position. Clean them with 70% ethanol applied to a tissue paper to remove contaminants that can trap air bubbles while pouring the gel.
3. Set up the glass plate and aluminium plate separated by 1.5 mm thick spacers on the gel caster. If doing it in the first time, check whether the set up is watertight by filling the space between the plates with water. When satisfied, pour off the water into the sink and start to prepare the gel afterwards.
4. Insert the comb into the space between the plates, and using a waterproof marker pen, mark a level indicating 1 cm below the end of the teeth of the comb. This will be the level to which the resolving gel will be poured. Remove the comb.
5. Prepare the following solutions without TEMED. Resolving gel solution: 2.5 mL Buffer 4×A, 3.3 mL 30% acrylamide solution, 4.2 mL water, and 66 μL 10% APS. Stacking gel solution: 1.9 mL Buffer 4×B, 1.5 mL 30% acrylamide solution, 2.25 mL water, and 50 μL 10% APS.
6. Add 20 μL of TEMED to the resolving gel solution mix and pour into the space between the plates till the level marked in step 4. This procedure has to be done as rapidly as possible to avoid the solution gelling before it can be poured. Since polymerization is accelerated by higher temperatures, it is helpful to briefly place the solution (before adding TEMED) on ice. Overlay the gel solution with 70% ethanol to a height of about 1 mm. This removes any potential air bubbles on the surface and also accelerates the polymerization process. Allow to set for 10 min.
7. Pour off the 70% ethanol and soak up any remaining liquid with a tissue paper.
8. Add 20 μL TEMED to the stacking gel solution and fill up the space on top of the resolving gel. Quickly insert the comb to form the wells. Let the gel sit for another 10 min (see Note 4).

9. Remove the comb, fix the gel to the gel apparatus and then fill up the top and bottom tanks with 1× SDS–PAGE running buffer.
10. Rinse the wells by pipetting running buffer into them using a 18-guage needle attached to a 5 mL syringe.
11. Mix 10 μL of the lysate prepared for the luciferase assays with sample loading buffer. Prepare a similar mixture for the pre-stained marker. Heat samples for 1 min at 95°C and load into the wells using a Hamilton needle (Hamilton) or a pipette tip.
12. Connect the cables to a power supply source and run the gel at 15 mA into the stacking gel and then at 30 mA till the dye front is at the bottom of the gel. The migration of the gel can also be monitored by the separation of the pre-stained marker.
13. When finished, switch off the power supply and disconnect the cables before opening the gel apparatus.

3.4. Western Blotting

1. After the samples have been resolved by SDS–PAGE, the plates are removed and the stacking gel is cut off. The resolving gel is then transferred to a container and soaked in transfer buffer.
2. During this time, cut four pieces of Whatman 3MM chromatography paper having same dimensions as the resolving gel and prepare a piece of reinforced nitrocellulose membrane of same size and cut on one corner to mark orientation of the gel.
3. Soak two 3MM paper sections in transfer buffer and place on the semi-dry transfer blotting apparatus. Use a 10 mL plastic pipette to roll out the air bubbles from underneath. Place the nitrocellulose membrane and then the gel on top, followed by two more sheets of soaked 3MM paper. Each time roll out any air bubbles that can cause inefficient transfer (Fig. 5a). Place the lid of the apparatus and attach cables. Run the transfer overnight at 5 V (see Note 5).
4. Switch off the power supply and disconnect the cables. The stack is slowly disassembled and the membrane removed. The transfer of the pre-stained marker will serve to indicate that the transfer worked properly. Since the Argonaute proteins are ~90 kDa, observe whether the molecular weight marker larger than 90 kDa is clearly visible.
5. Block the membrane in blocking buffer on a tilting shaker for 30 min at room temperature.
6. Discard the blocking buffer and replace with the minimum amount of blocking buffer required to cover the membrane (approximately 1–5 mL depending on the container used to hold the membrane).

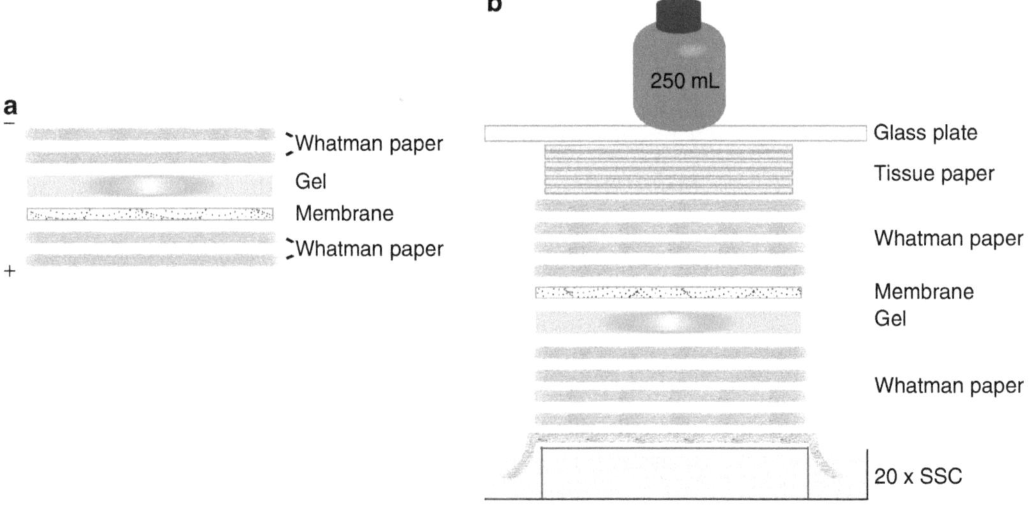

Fig. 5. (**a**) Assembly setup for western blot transfer. (**b**) Assembly setup for Northern blot transfer. Please see Subheadings 3.4 and 3.5 for details.

7. Add the primary antibody at the required dilution and continue incubation with shaking for 1 h. Remove antibody solution and wash five times 5 min each with abundant amounts of 1× PBS-Tween.
8. Incubate the membrane with blocking buffer containing diluted secondary antibody – anti-rabbit IgG-HRP conjugate for 1 h. Wash as in step 7.
9. Prepare the ECL detection reagent as per manufacturer's instructions. Remove the membrane from the wash buffer, place on a sheet of SaranWrap and uniformly apply the ECL detection solution for 1 min. Quickly absorb the excess ECL liquid with a tissue paper and wrap the membrane with the SaranWrap.
10. Place the membrane in the exposure cassette and proceed to the dark room. Remove one X-ray Hyperfilm, cut or fold a corner to indicate orientation and place on the membrane. After about 2 min, develop it in the developer machine. During this time, replace a new film in the cassette to evaluate a second, longer exposure. The anti-HA reactive proteins are visible as dark bands on the film. Overlay on the membrane to mark out the pre-stained marker positions.

3.5. RNA Extraction and Northern Blotting

1. Total RNA from transfected HeLa cells can be prepared using the Absolutely RNA RT-PCR miniprep kit following the manufacturer's instructions. This kit allows DNase-treatment of the RNA bound on the purification columns, which is an advantage. Efficient removal of the transfected plasmid DNA is very important for the subsequent detection of the expressed reporter mRNAs. Measure the concentration of the purified

RNA using a NanoDrop spectrophotometer. Take extreme precaution while working with RNA (see Note 6).

2. The RNA samples are resolved in a formaldehyde denaturing gel and then transferred to a positively charged Nylon membrane (Hybond N+).

3. Clean the agarose gel tray, comb and gel running apparatus with soap and water. Allow them to dry and then set them up in the casting stand in the fume hood. Prepare formaldehyde denaturing 1% agarose gel by boiling 1 g agarose in 72 mL of DEPC-treated, double-distilled water. Allow to cool to 60°C in a waterbath (see Note 7).

4. In the fume hood, add 10 mL of 10× MOPS and 18 mL of 37% formaldehyde. Mix and pour into the setup for agarose gel. Place the comb and allow gel to solidify for approximately 30 min.

5. In the meantime, prepare 1× MOPS running buffer by diluting the 10× buffer in autoclaved double-distilled water. Pour sufficient buffer into the gel tank. Remove the comb from the gel and place the gel into the apparatus, submerged in the buffer.

6. Mix the RNA samples with equal volume of 2× RNA loading buffer. Heat the mixture for 5 min at 65°C (see Note 8).

7. Load the samples into the gel with a pipette tip and run at a constant 5 V/cm of gel length. When the dye front is 2–3 cm away from the bottom of the gel, turn off the power supply. Disconnect all cables and remove gel for examination on a UV-transilluminator. The 28S and 18S ribosomal RNA (rRNA) bands should be clearly visible on all the lanes. Observe whether the intensity of the 28S rRNA band is approximately twice that of the 18S rRNA band. This indicates whether the RNA is of good quality. Degraded RNA samples will show either equal intensities for the two bands or a smeary pattern. To prevent RNase contamination during the imaging process, always place the gel on the transilluminator with a clean SaranWrap plastic sheet between them.

8. Place the gel in a container and wash with abundant amounts of autoclaved double-distilled water to remove the formaldehyde. Change the solution twice. Dispose formaldehyde-containing solutions into dedicated waste containers.

9. Prepare a transfer stack by placing a long strip of Whatman 3MM paper of the same width as the gel on a raised platform. The two ends of the strip should be in contact with the transfer buffer (20× SSC) contained in a large tray at the bottom. The strip is soaked with 20× SSC and air bubbles rolled out with a 10-mL plastic pipette. Build the stack with four sheets of 3MM paper soaked in 20× SSC, place the gel, the nylon membrane, followed by four more sheets of 3MM paper.

Each of these is soaked in transfer buffer and any air bubbles are rolled out. Place a stack of dry tissue paper and top it off with a glass plate. Weight down the stack with something modestly heavy (~300 g; for example, a glass bottle of water with 250 mL water). Let the RNA transfer to the membrane by passive transfer overnight (Fig. 5b).

10. On the next day, remove the stack and carefully extract the membrane. Place the membrane on a dry tissue paper and fix the RNA to the membrane by a short UV-crosslinking exposure in a UV-crosslinker (Stratalinker). Push the AUTO CROSS-LINK button on a Model 1800 UV crosslinker from Stratagene. This will deliver energy of 120,000 μJ from the 254 nm UV lamps to form covalent bonds between uracils of the RNA and amino groups of the nylon membrane (see Note 9).

3.6. Northern Analysis Using Radioactive Probes

The expression of reporter mRNA from the transfected luciferase plasmids can be studied by hybridizing the RNA bound to the nylon membrane with specific radioactive probes complementary to the mRNA of interest. Working with radioactivity requires specialized laboratory equipment and training. Please consult the authorities at your institution before embarking on experiments involving radioactivity.

1. The coding region for the *Renilla* luciferase is released from the phRL-TK plasmid by restriction digestion, gel eluted, purified using a gel-extraction kit. The purified RL DNA fragment is then used for the preparation of radioactive probes using the random-primed DNA labelling kit using [α-^{32}P] dCTP as per manufacturer's instructions.
2. Labelled single-stranded DNA probes are separated from free unincorporated deoxyribonucleoside triphosphates by the use of Microspin G-25 columns. Follow manufacturer's protocol. Store the radioactive probe at –20°C until use.
3. Place the nylon membrane into a hybridization bottle and add sufficient hybridization buffer prewarmed to 65°C. Prehybridize for 1 h at 65°C in a rotating hybridization oven. This allows efficient blocking of the membrane to prevent background signals.
4. Denature radioactive probe at 95°C for 1 min and immediately place on ice.
5. Remove the hybridization solution and replace with fresh hybridization solution. Add probe at one million counts per minute (cpm)/mL. Hybridize at 65°C overnight.
6. On the following day, remove radioactive hybridization solution into proper waste container. Wash twice with low-stringency wash buffer (2× SSC containing 0.1% SDS), each for 10 min. Wash once with high-stringency wash buffer (0.1× SSC containing 0.1% SDS) for 5 min. Check for the presence of counts on the membrane using a Geiger-Muller counter.

7. Seal the membrane in plastic sheeting and expose to a storage phosphor screen for 5 h. Expose the screen on a Typhoon Scanner to detect the signals arising from radioactive probe hybridized to the RL mRNA present in the different samples on the membrane. Alternatively, Northern blots can be exposed to X-ray film.
8. The membrane can be stripped to remove the radioactive probe for RL and reprobed with a probe for GFP to allow normalization of transfection efficiency.
9. Signals are quantified using the software provided with the Typhoon Scanner.

4. Notes

1. Wear gloves while handling this reagent as it is a neurotoxin.
2. Ammonium persulfate: Renew the stock every 3 months.
3. Running buffer for denaturing gel (10×): Store in dark up to 3 months at 4°C.
4. At this step, the gel can be stored for later use by wrapping in tissue paper soaked with 1× running buffer. Then place the gel into a plastic packet or wrap with SaranWrap. Store at 4°C.
5. Increasing the voltage can allow reduced transfer times. However, better transfer efficiency is observed with longer duration.
6. Always wear gloves and use RNase-free plasticware (tips, reaction tubes, etc.).
7. Formaldehyde is toxic and has to be handled with gloves and only under the fume hood.
8. Be carefully as the loading buffer contains ethidium bromide, which is a carcinogen and has to be handled with proper gloves. Do not dilute the loading buffer more than 50%, to maintain the required denaturing conditions for the sample.
9. The UV-crosslinked membrane can be stored dry between two Whatman 3MM sheets until further use.

Acknowledgments

The protocols were originally developed with the active help and advice of Caroline G. Artus and Witold Filipowicz at the Friedrich Miescher Institute for Biomedical Research, Basel, Switzerland. Research in R.S.P's group is supported by the European Molecular Biology Laboratory.

References

1. Filipowicz, W., Bhattacharyya, S.N. and Sonenberg, N. (2008) Mechanisms of post-transcriptional regulation by microRNAs: are the answers in sight? *Nat Rev Genet*, **9**, 102–114.
2. Bartel, D.P. (2004) MicroRNAs: genomics, biogenesis, mechanism, and function. *Cell*, **116**, 281–297.
3. Carmell, M.A., Xuan, Z., Zhang, M.Q. and Hannon, G.J. (2002) The Argonaute family: tentacles that reach into RNAi, developmental control, stem cell maintenance, and tumorigenesis. *Genes Dev*, **16**, 2733–2742.
4. Meister, G. and Tuschl, T. (2004) Mechanisms of gene silencing by double-stranded RNA. *Nature*, **431**, 343–349.
5. Verdel, A., Jia, S., Gerber, S., Sugiyama, T., Gygi, S., Grewal, S.I. and Moazed, D. (2004) RNAi-mediated targeting of heterochromatin by the RITS complex. *Science*, **303**, 672–676.
6. Kim, D.H., Villeneuve, L.M., Morris, K.V. and Rossi, J.J. (2006) Argonaute-1 directs siRNA-mediated transcriptional gene silencing in human cells. *Nat Struct Mol Biol*, **13**, 793–797.
7. Siomi, H. and Siomi, M.C. (2009) On the road to reading the RNA-interference code. *Nature*, **457**, 396–404.
8. Meister, G., Landthaler, M., Patkaniowska, A., Dorsett, Y., Teng, G. and Tuschl, T. (2004) Human Argonaute2 mediates RNA cleavage targeted by miRNAs and siRNAs. *Mol Cell*, **15**, 185–197.
9. Liu, J., Carmell, M.A., Rivas, F.V., Marsden, C.G., Thomson, J.M., Song, J.J., Hammond, S.M., Joshua-Tor, L. and Hannon, G.J. (2004) Argonaute2 is the catalytic engine of mammalian RNAi. *Science*, **305**, 1437–1441.
10. Eulalio, A., Huntzinger, E. and Izaurralde, E. (2008) Getting to the root of miRNA-mediated gene silencing. *Cell*, **132**, 9–14.
11. Pillai, R.S., Artus, C.G. and Filipowicz, W. (2004) Tethering of human Ago proteins to mRNA mimics the miRNA-mediated repression of protein synthesis. *Rna*, **10**, 1518–1525.
12. Legault, P., Li, J., Mogridge, J., Kay, L.E. and Greenblatt, J. (1998) NMR structure of the bacteriophage lambda N peptide/boxB RNA complex: recognition of a GNRA fold by an arginine-rich motif. *Cell*, **93**, 289–299.
13. Baron-Benhamou, J., Gehring, N.H., Kulozik, A.E. and Hentze, M.W. (2004) Using the lambdaN peptide to tether proteins to RNAs. *Methods Mol Biol*, **257**, 135–154.
14. Pillai, R.S., Bhattacharyya, S.N., Artus, C.G., Zoller, T., Cougot, N., Basyuk, E., Bertrand, E. and Filipowicz, W. (2005) Inhibition of translational initiation by Let-7 MicroRNA in human cells. *Science*, **309**, 1573–1576.
15. Buhler, M., Verdel, A. and Moazed, D. (2006) Tethering RITS to a nascent transcript initiates RNAi- and heterochromatin-dependent gene silencing. *Cell*, **125**, 873–886.
16. Eulalio, A., Huntzinger, E. and Izaurralde, E. (2008) GW182 interaction with Argonaute is essential for miRNA-mediated translational repression and mRNA decay. *Nat Struct Mol Biol*, **15**, 346–353.
17. Chekulaeva, M., Filipowicz, W. and Parker, R. (2009) Multiple independent domains of dGW182 function in miRNA-mediated repression in Drosophila. *Rna*, **15**, 794–803.

Chapter 14

An Efficient System for Let-7 MicroRNA and GW182 Protein-Mediated Deadenylation In Vitro

Marc R. Fabian, Yuri V. Svitkin, and Nahum Sonenberg

Abstract

Experiments with cell cultures have been useful in analyzing microRNA action. However, miRNA-mediated effects are often assayed many hours or days after miRNA target recognition. Consequently, this has made it difficult to analyze early events of miRNA-mediated repression. The development of cell-free systems that recapitulate miRNA action in vitro has been instrumental in dissecting the molecular mechanisms of miRNA action. Here we describe such a system, derived from mouse Krebs II ascites carcinoma cells, termed Krebs cell-free system. As an example, the protocol for assaying let-7 and GW182 (TNRC6) protein-mediated deadenylation of mRNA in vitro is described.

Key words: microRNA, GW182, Argonaute, Deadenylation, let-7, Krebs-2 cell-free system

1. Introduction

Micro (mi)RNAs are small (~21 nucleotide) noncoding RNAs that extensively control genome expression of numerous species including plants, insects, and animals. In general, miRNAs hybridize to partially complementary sequences in the 3′ untranslated regions of target mRNAs. Once bound to a target mRNA, miRNAs repress protein synthesis by inhibiting translation and/or by mRNA deadenylation and decay (Fig. 1) (1, 2). These effects are not elicited by the miRNA itself, but rather by a complex of proteins, collectively referred to as the RNA induced silencing complex (RISC). Base-pairing of miRNA to the mRNA target sequence is a prerequisite for recruitment of the RISC to the target mRNA. Protein components of the miRNA (mi)RISC, such as Argonaute (AGO) and GW182 (TNRC6), are critical for effecting repression (1–3) (Note 1). Despite all that has been

Tom C. Hobman and Thomas F. Duchaine (eds.), *Argonaute Proteins: Methods and Protocols*,
Methods in Molecular Biology, vol. 725, DOI 10.1007/978-1-61779-046-1_14, © Springer Science+Business Media, LLC 2011

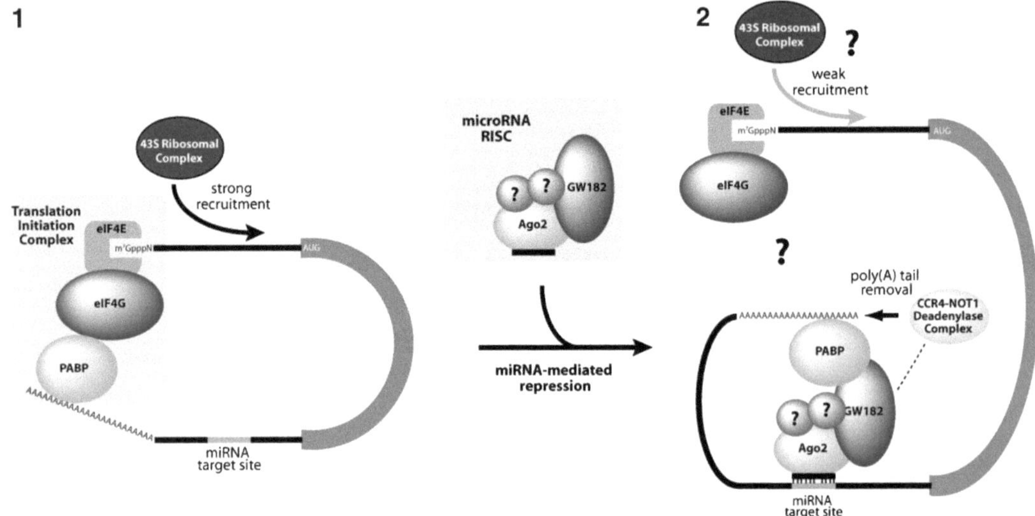

Fig. 1. Model for miRNA-mediated repression. (**1**) mRNA circularization via eIF4G-PABP interaction stimulates cap-dependent translation (strong 43 S ribosomal complex recruitment) by enhancing eIF4E's binding to the mRNA 5′ cap structure. (**2**) When miRISC binds to its target site in the 3′UTR, this allows GW182 to interact with PABP thereby inhibiting its interaction with eIF4G. The result of this is repression of cap-dependent translation by decreasing eIF4E's binding to the 5′ cap structure (weaker binding). Moreover, sequestering the poly(A) tail into the vicinity of CAF1 and CCR4 deadenylases facilitates deadenylation of the mRNA.

discovered about miRNAs over the past few years, the mechanisms of miRNA action are still not fully understood.

Cell-free extracts have been essential for elucidating a multitude of molecular mechanisms in vitro (e.g. mRNA splicing, translation, mRNA turnover, viral replication) (4–7). Several groups have recently described cell-free assays that recapitulate miRNA repression in vitro (8–11). Assaying miRNA activity using in vitro systems offers several advantages over cell culture-based assays. For instance, miRNA-mediated repression can be analyzed minutes, rather than several hours or days, following miRNA target recognition. Moreover, proteins can be added to or depleted from in vitro extracts with relative ease, without perturbing other systems. This is in contrast to cell-based experiments where adding or depleting proteins can have unforeseen consequences unrelated to the mechanism that is being studied.

In this chapter, we outline procedures that can be used with Krebs extract to analyze let-7 miRNA-mediated deadenylation, as well as GW182 protein-mediated deadenylation in vitro (Note 2).

2. Materials

Detailed protocols for preparation, nuclease-treatment, and supplementation of Krebs extract has been previously published in Methods in Molecular Biology (7) (Note 3).

2.1. Preparation of 3′UTR DNA Templates for In Vitro Transcription

1. Plasmids encoding *Renilla* luciferase containing wild-type or mutant let-7a miRNA target sites (RL-6xBpA and RL-6xBMUT-pA, respectively) (Note 4) or containing five BoxB stem–loops (RL-5BoxBpA) (8, 12, 13).
2. 10 μM forward oligonucleotide containing a T7 promoter sequence (5′-GGCGCCTAATACGACTCACTATAGGGGT AAGTACATCAAGAGCTTCG-3′).
3. 10 μM reverse oligonucleotide (5′-GGTGACACTATAGA ATAGGGCCC-3′).
4. Titanium Taq DNA Polymerase (Clontech).
5. 10× Titanium Taq DNA Polymerase Buffer (Clontech).
6. 10 mM dNTP mix (dATP, dTTP, dCTP, and dGTP).
7. Qiaquick Gel Extraction Kit (Qiagen).

2.2. Preparation of Radiolabeled In Vitro Transcripts

1. Plasmids encoding *Renilla* luciferase containing wild-type or mutant let-7a miRNA target sites (RL-6xBpA and RL-6xBMUT-pA, respectively) (Note 4) or containing five BoxB stem–loops (RL-5BoxBpA) (8, 12, 13).
2. PCR products containing a promoter for T7 RNA polymerase followed by 6 wild-type (6xB-3′UTR) or mutated (MUT-3′UTR) let-7a target sites or 5 BoxB stem–loops (5BoxB-3′UTR) followed by a 98-nt poly(A) sequence (12, 13).
3. T7 MAXIscript in vitro transcription kit (Ambion); store at −20°C.
4. G(5′)ppp(5′)A RNA Cap Structure Analog (New England Biolabs); store at −20°C.
5. UTP-α^{32}P (800 Ci/mmol; 20 mCi/mL) (Perkin Elmer); store at 4°C.
6. Mini quick-spin RNA columns (Roche); store at 4°C.
7. Oligo d$(T)_{15}$ (Integrated DNA Technologies).
8. 10× RNaseH Buffer (New England Biolabs): 50 mM Tris–HCl, 75 mM KCl, 3 mM $MgCl_2$, 10 mM Dithiothreitol, pH 8.3 at 25°C. Store at −20°C.
9. RNaseH (New England Biolabs).

2.3. Preparation of Recombinant GW182 (TNRC6C) Proteins

1. Plasmids for bacterial expression (pGEX6p-1; GE Healthcare) of wild-type [GST-TNRC6C (1382-1690) and GST-λNHA-TNRC6C (1382-1690)] or mutated [GST-λNHA-TNRC6C (1382-1690) MUT] C-terminal TNRC6C fragments (Note 5) (12) (Fig. 3a).
2. Glutathione-Sepharose 4B (GE Healthcare).
3. L-Glutathione (Sigma); store at 4°C.
4. FLAG M2-agarose (Sigma); store at −20°C.
5. 3× FLAG peptide (Abgent).

2.4. Assaying miRNA-Mediated Deadenylation in Krebs Extract

1. S7 (micrococcal)-nuclease treated supplemented Krebs extract.
2. Radiolabeled in vitro transcripts.
3. Buffer D: 25 mM HEPES-KOH, pH 7.3; 50 mM KCl; 75 mM KOAc; 2 mM $MgCl_2$.
4. Anti-let-7a 2′-*O*-methyl oligonucleotide (Dharmacon).
5. Anti-miR122 2′-*O*-methyl oligonucleotide (Dharmacon) (Note 6).
6. Recombinant GST-TNRC6C (1382-1690), GST-λNHA-TNRC6C (1382-1690), and GST-λNHA-TNRC6C (1382-1690) MUT protein.
7. TRIzol (Invitrogen); store at 4°C.
8. Chloroform.
9. 2-propanol.
10. Glycoblue (Ambion); store at –20°C.
11. 70% ethanol.

2.5. Denaturing Polyacrylamide Gel Electrophoresis

1. 40% acrylamide/bis solution (19:1). Store at 4°C. Caution, unpolymerized acrylamide is a neurotoxin that can be absorbed through the skin (avoid contact and wear gloves).
2. 10× TBE (Tris, borate, EDTA) running buffer: 108 g Tris base, 55 g boric acid, 40 ml of 0.5 M EDTA, pH 8.0, in 1 L of water. Store at room temperature.
3. 10% (w/v) Ammonium persulfate (APS): prepare 10% solution in water; store at 4°C. Make fresh every 2 weeks.
4. Urea (electrophoresis grade).
5. 2× RNA loading dye (Fermentas).
6. *N,N,N,N′*-Tetramethyl-ethylenediamine (TEMED). Store at 4°C.
7. Mini-PROTEAN 3 Vertical Electrophoresis system (BioRad).
8. Gel Dryer (BioRad).
9. Typhoon Phosphorimager (GE Healthcare).

3. Methods

3.1. Let-7a-Mediated Deadenylation Conditions

1. Dispense 7 μL aliquots of supplemented Krebs extract into precooled plastic tubes containing 1.0 μL (0.1 ng/μL) of radiolabeled RNA containing wild-type (6xB-3′UTR) or mutant (6xBMUT-3′UTR) let-7a target sites. 1.0 μL (100 nM stock concentration) of anti-let-7 2′-*O*-methylated oligonucleotides is added to reactions as a negative control to block

let-7-mediated deadenylation. Final reaction volumes are brought up to 10 μL with Buffer D. Master mixes for each reaction can be made for the analysis of multiple time points during deadenylation reactions (Fig. 2b) (Note 7).

2. Mix each reaction with gentle agitation so as to avoid frothing and incubate the tubes at 30°C for 3–4 h.
3. At desired time points, remove 10 μL of deadenylation reaction and mix in a microfuge tube with 1 mL of TRIzol. Vortex the tube for 30 s.

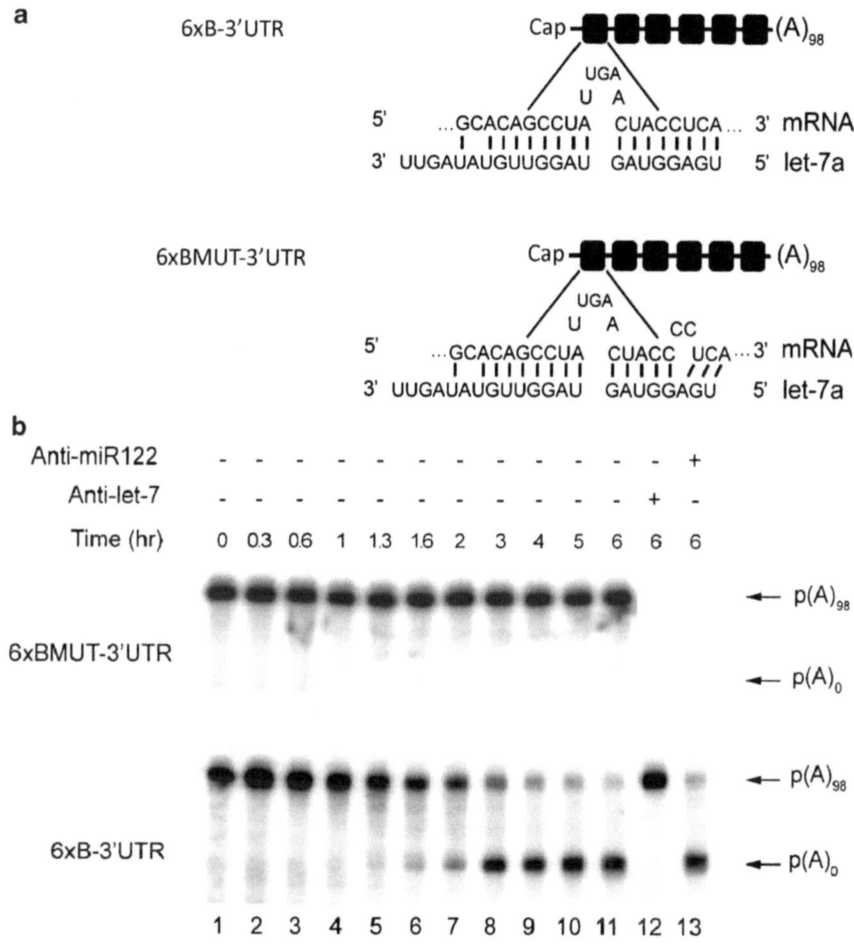

Fig. 2. Example of let-7 miRNA-mediated deadenylation in Krebs extract (13). (**a**) Schematic representation of 6xB-3′UTR reporter RNAs. Sequences of the let-7-binding sites (6xB-3′UTR) and mutated seed sites (6xBMUT-3′UTR) are shown below the drawings. (**b**) Time course of 6xB-3′UTR (lanes 1–13) and 6xBMUT-3′UTR (lanes 1–11) RNA deadenylation in a Krebs extract as determined by autoradiography. 6xB-3′UTR reporter RNAs were incubated in the presence or absence of 10 nM 2′-O-Me oligonucleotide (either anti-let-7a or anti-miR122), and their stabilities were monitored by autoradiography. Polyadenylated and deadenylated RNAs are marked with arrows on the right of the figure as $p(A)_{98}$ and $p(A)_0$, respectively.

4. Add 200 μL of chloroform to tubes with TRIzol and vortex for 10 s.
5. Centrifuge at 20,000*g* at 4°C for 15 min.
6. Pipette 520 μL of aqueous layer of TRIzol extractions and dispense into a tube with 1 μL of Glycoblue and 750 μL of 2-propanol. Invert tubes a few times to mix solutions.
7. Place tubes at −80°C for 1 h or at −20°C overnight to precipitate RNA.
8. After precipitation, pellet RNA by centrifugation at 20,000*g* at 4°C for 15 min.
9. Remove supernatant and add 1 mL of 70% ethanol. Centrifuge at 20,000*g* at 4°C for 10 min.
10. Air-dry RNA pellet and resuspend in 8 μL of RNAse-free water. Store RNA at −80°C until it is used for the electrophoresis step (Note 8).

3.2. Tethered GW182 (TNRC6C) Protein-Mediated Deadenylation Conditions

miRISC-mediated deadenylation can be observed in Krebs extract independent of the let-7 miRNA. This can be accomplished by utilizing the lambda(λ)N peptide-boxB RNA stem–loop tethering strategy and the miRISC protein GW182 paralog TNRC6C (14). Briefly, incubating a recombinant C-terminal fragment of TNRC6C (residues 1382-1690) fused to the λN peptide in supplemented Krebs extract can drive the deadenylation of a radiolabeled polyadenylated RNA containing 5 BoxB RNA stem–loops (Fig. 3).

1. Dispense 7 μL aliquots of supplemented Krebs extract into precooled plastic tubes containing 1.0 μL (0.1 ng/μL) of radiolabeled 3′UTR-5BoxB RNA and 1–2 μL (60 ng/uL) of GST-HA-TNRC6C (1382-1690), GST-λNHA-TNRC6C (1382-1690WT) or GST-λNHA-TNRC6C (1382-1690MUT) recombinant protein (Note 5). Final reaction volumes are brought up to 10 μL with Buffer D. Master mixes for each reaction can be made for the analysis of multiple time points during deadenylation reactions (Fig. 3b).
2. Mix each reaction mixture with gentle agitation so as to avoid frothing and incubate the tubes at 30°C for 3 h.
3. At desired time points, remove 10 μL of the deadenylation reaction and place in a 1.5-mL microfuge tube with 1 mL of TRIzol. Vortex the tube for 30 s. Proceed with purification as outlined in Subheading 3.3, steps 4–10.

3.3. Preparation of Non-Adenylated Radiolabeled RNA

It is useful to produce an aliquot of radiolabled RNA lacking a poly(A) tail to demonstrate what the migration of a deadenylated

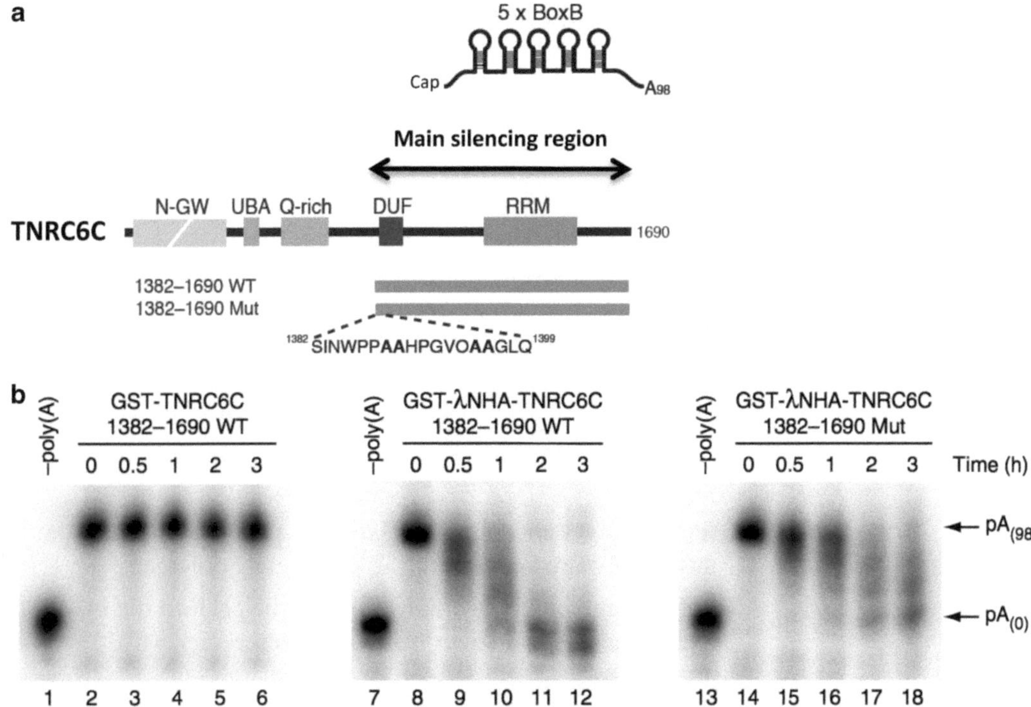

Fig. 3. An example of GW182 (TNRC6) protein-mediated deadenylation in Krebs extract (12). (**a**) Schematic representation of the 5BoxB-3′UTR RNA and TNRC6C C-terminal wild-type (WT) and mutant (MUT) protein fragments used in tethering experiments in Krebs extracts. The region, including GW-rich, ubiquitin-associated (UBA), and glutamine-rich (Q-rich) domains, is responsible for targeting GW182 proteins to cellular processing bodies. The C-terminal part of GW182 proteins (the main silencing region), containing DUF (domain of unknown function) motifs and RNA recognition motifs (RRM), is a major effector domain, mediating translational repression and deadenylation of mRNA. (**b**) Time course of 5BoxB-3′UTR RNA deadenylation in a Krebs extract as determined by autoradiography in the presence of various TNRC6C C-terminal protein fragments. Poly(A)$^-$ RNA was prepared in vitro by treating 5BoxB-3′UTR RNA with oligo d(T) and RNaseH. Polyadenylated and deadenylated RNAs are marked with arrows on the right of the figure as $p(A)_{98}$ and $p(A)_0$, respectively.

RNA looks like following denaturing PAGE (Fig. 3b, lanes 1, 7 and 10). This is accomplished by hybridizing $d(T)_{15}$ oligonucleotide to the poly(A) tail of radiolabeled RNAs and subsequently treating the duplex with RNaseH, an enzyme that cleaves RNA/DNA duplexes.

1. Incubate 1 μL (0.1 ng/μL) of radiolabeled RNA with 2 μL (10 uM stock concentration) of oligo $d(T)_{15}$ and 3 μL of RNAse-free water in a microfuge tube.
2. Heat mixture to 90°C for 30 s and then place tube directly on ice for 1 min.
3. Quick-spin the tube to bring the reaction to the bottom of the microfuge tube.
4. Add 2 μL of 10× RNaseH buffer, 1 μL of RNaseH, and 11 μL of RNAse-free water, and incubate the reaction at 37°C for 30 min.

5. Following the incubation, add 1 mL of TRIZOL, vortex for 30 s and proceed with purification as outlined in Subheading 3.3, steps 4–10.

3.4. Polyacrylamide Gel and Sample Preparation

These instructions are for a Mini-PROTEAN 3 BioRad vertical polyacrylamide apparatus.

1. Clean a set of glass plates with soap, then with 2-propanol.
2. Prepare a 0.5-mm thick, 4.5% polyacrlamide gel by mixing 6 g Urea, 1.2 mL of 10× TBE, 1.4 mL of 40% acrylamide (19:1 acrlyamide:bis), and 5 mL in a 15-mL sterile plastic tube. Heat tube at 37°C for ~10 min to dissolve urea.
3. Before pouring the gel, add 80 µL of 10% (w/v) APS and 15 µL of TEMED to the acrylamide solution and invert tube to mix.
4. Pour the gel and place the comb in the gel. The gel should take approximately 20 min to completely polymerize.
5. Once completely polymerized, place the gel into the gel apparatus and fill the inner and outer reservoirs with 1× TBE buffer.
6. Prepare RNA samples by mixing 4 µL of purified radiolabeled RNA with 4 µL of 2× RNA loading buffer, heat at 80°C for 2 min and then place on ice for 2 min.
7. Remove the comb from gel and flush wells with running buffer to remove any diffused urea.
8. Load samples and run gel at 150 V for 100 min at room temperature (Note 9).
9. When the electrophoresis run is completed, remove the glass plates from the apparatus and plate horizontally. Gently, pry apart the glass plates using a spatula.
10. Take a sheet of 3MM paper that is slightly larger than the size of the gel and place it over the exposed gel. Starting from one corner of the gel, gently peel back the 3MM paper with the attached gel.
11. Once removed from the glass plate, place a sheet of Saran wrap over the gel and place it in a gel drier for 2 h at 80°C.
12. Once dry, expose the gel to X-ray film or phosphorimager screen.

3.5. Supporting Protocols

1. DNA manipulations were performed using standard methods. Luciferase-encoding plasmids (8, 12, 13) were linearized with *ApaI* and transcribed with T7 RNA polymerase. To generate polyadenylated 3′UTR transcripts, luciferase-encoding plasmids were used as templates for PCR reactions with a forward primer containing a promoter for T7 RNA polymerase.

PCR products were purified, digested with ApaI, and transcribed with T7 RNA polymerase.

2. Syntheses of RNA transcripts were performed with the MAXIscript T7 kit (Ambion) according to the manufacturer's instructions. The integrity of in vitro synthesized RNA was verified by denaturing PAGE and by subsequent autoradiography.
3. Quantification of radiolabeled transcripts was performed using trichloroacetic acid (TCA) precipitation of incorporated α^{32}P-UTP and subsequent counting by scintillation.
4. The C-terminal FLAG-tagged GST fusion proteins, GST-HA-TNRC6C(1382-1690WT),GST-λNHA-TNRC6C(1382-1690WT) and GST-λNHA-TNRC6C(1382-1690MUT), were expressed in *Escherichia coli* BL21 (DE3) cells and purified on glutathione–Sepharose resin (GE Healthcare) according to the manufacturer's recommendations. GST-purified proteins were subsequently purified on FLAG M2 Agarose (Sigma) and eluted with 3× FLAG peptide (Abgent).

4. Notes

1. AGO proteins function by binding directly to the miRNA and to GW182 proteins; however GW182 proteins are the major effectors of miRNA-mediated repression. Tethering experiments demonstrate that GW182 proteins can cause repression in the absence of miRNA-associated AGO proteins in *Drosophila melanogaster* (15). Furthermore, tethering GW182 C-terminal fragments that cannot bind AGO can still effect both translational repression and deadenylation in human and insect cells (12, 16–19).
2. Let-7a is highly expressed in Krebs extract; however, other miRNAs may be expressed at high levels as well. This may allow for reporter RNAs containing alternative miRNA target sites to be assayed in Krebs extract. Levels of other specific miRNAs should be analyzed via Northern Blot analysis and/or RT-qPCR.
3. There is variation in translation efficiency between batches of extracts, and also between batches in efficiency of let-7-mediated translational repression and deadenylation.
4. RL-6xBMUT-pA contains two additional C's in each let-7 target site "seed region" (nucleotides 2–7 that are critical for miRNA–mRNA duplex formation). These mutations reduce let-7 miRNA hybridization and subsequent repression.
5. GST-HA-TNRC6C (1382-1690WT) lacks the λN peptide and therefore cannot bind to BoxB stem–loops. This makes

this protein an ideal negative control for tethering experiments. GST-λNHA-TNRC6C (1382-1690MUT) contains EF and WK residues in the DUF motif mutated all to alanine (Fig. 3b). This mutation abolishes the interaction of the DUF domain with the C-terminus of mammalian poly(A) binding protein, and interferes with GW182-mediated deadenylation (12).

6. miR122 is a liver-specific miRNA that is not present in Krebs extract (20). This makes the anti-miR122 2′-*O*-methyl oligonucleotide an excellent parallel control for the addition of 2′-*O*-methyl oligonucleotide to Krebs extract. Unlike anti-let-7 2′-*O*-methyl oligonucleotide, which efficiently blocks let-7-mediated deadenylation and translational repression, addition of miR122-2′-*O*-methyl oligonucleotide has no effect (8, 13).
7. While the let-7 family of miRNAs is relatively abundant in Krebs extract [~150 pM (8), it can be easily saturated by adding too much target RNA to the extract. Therefore, it is extremely important to accurately quantify the amount of reporter RNA being added to the extract.
8. It is extremely important to remove all residual 70% ethanol from the pellet after the final spin. One option to guarantee this is to remove much 70% ethanol after the final spin, and then recentrifuge the tube at 20,000*g* at 4°C to spin down residual ethanol. This residual amount can now be easily removed.
9. RNA samples should be loaded with gel-loading tips with the end of the tip close to the bottom of the well. This ensures that the RNA sample remains tightly packed in the well and ultimately results in a much better image following migration. Moreover, samples must be loaded relatively quickly onto acrylamide gels (e.g. 10 samples on a 10-well acrylamide gel should take no more than 2 min to load) in order to ensure the quality of the final image. If too much time is taken during loading, RNA bands will appear as compressed ovals following electrophoresis rather than as tight thin bands. If more than one gel is being run, load one gel and run it for 5 min at 150 V to allow these samples to enter the first gel. Once this short electrophoresis is complete, the second gel can be loaded.

Acknowledgments

This work was supported by a grant from the Canadian Institutes of Health Research (N.S.) and by a postdoctoral fellowship from the Canadian Cancer Society (M.R.F.).

References

1. Filipowicz, W., Bhattacharyya, S. N., and Sonenberg, N. (2008) Mechanisms of post-transcriptional regulation by microRNAs: are the answers in sight?, *Nat Rev Genet* *9*, 102–114.
2. Eulalio, A., Huntzinger, E., and Izaurralde, E. (2008) Getting to the root of miRNA-mediated gene silencing, *Cell* *132*, 9–14.
3. Eulalio, A., Tritschler, F., and Izaurralde, E. (2009) The GW182 protein family in animal cells: new insights into domains required for miRNA-mediated gene silencing, *RNA* *15*, 1433–1442.
4. Padgett, R. A., Hardy, S. F., and Sharp, P. A. (1983) Splicing of adenovirus RNA in a cell-free transcription system, *Proc Natl Acad Sci U S A* *80*, 5230–5234.
5. Sokoloski, K., Anderson, J. R., and Wilusz, J. (2008) Development of an in vitro mRNA decay system in insect cells, *Methods Mol Biol* *419*, 277–288.
6. Panavas, T., Pogany, J., and Nagy, P. D. (2002) Analysis of minimal promoter sequences for plus-strand synthesis by the Cucumber necrosis virus RNA-dependent RNA polymerase, *Virology* *296*, 263–274.
7. Svitkin, Y. V., and Sonenberg, N. (2004) An efficient system for cap- and poly(A)-dependent translation in vitro, *Methods Mol Biol* *257*, 155–170.
8. Mathonnet, G., Fabian, M. R., Svitkin, Y. V., Parsyan, A., Huck, L., Murata, T., Biffo, S., Merrick, W. C., Darzynkiewicz, E., Pillai, R. S., Filipowicz, W., Duchaine, T. F., and Sonenberg, N. (2007) MicroRNA inhibition of translation initiation in vitro by targeting the cap-binding complex eIF4F, *Science* *317*, 1764–1767.
9. Wang, B., Love, T. M., Call, M. E., Doench, J. G., and Novina, C. D. (2006) Recapitulation of short RNA-directed translational gene silencing in vitro, *Mol Cell* *22*, 553–560.
10. Wakiyama, M., Takimoto, K., Ohara, O., and Yokoyama, S. (2007) Let-7 microRNA-mediated mRNA deadenylation and translational repression in a mammalian cell-free system, *Genes Dev* *21*, 1857–1862.
11. Thermann, R., and Hentze, M. W. (2007) Drosophila miR2 induces pseudo-polysomes and inhibits translation initiation, *Nature* *447*, 875–878.
12. Jinek, M., Fabian, M. R., Coyle, S., Sonenberg, N., and Doudna, J. A. (2010) Structural insights into the human GW182-PABC interaction in microRNA-mediated deadenylation, *Nat Struct Mol Biol* *17*, 238–240.
13. Fabian, M. R., Mathonnet, G., Sundermeier, T., Mathys, H., Zipprich, J. T., Svitkin, Y. V., Rivas, F., Jinek, M., Wohlschlegel, J., Doudna, J. A., Chen, C. Y., Shyu, A. B., Yates, J. R., 3rd, Hannon, G. J., Filipowicz, W., Duchaine, T. F., and Sonenberg, N. (2009) Mammalian miRNA RISC Recruits CAF1 and PABP to Affect PABP-Dependent Deadenylation, *Mol Cell* *35*, 868–880.
14. Baron-Benhamou, J., Gehring, N. H., Kulozik, A. E., and Hentze, M. W. (2004) Using the λN Peptide to Tether Proteins to RNAs, *Methods Mol Biol* *257*, 135–153.
15. Eulalio, A., Huntzinger, E., and Izaurralde, E. (2008) GW182 interaction with Argonaute is essential for miRNA-mediated translational repression and mRNA decay, *Nat Struct Mol Biol* *15*, 346–353.
16. Zipprich, J. T., Bhattacharyya, S., Mathys, H., and Filipowicz, W. (2009) Importance of the C-terminal domain of the human GW182 protein TNRC6C for translational repression, *RNA* *15*, 781–793.
17. Lazzaretti, D., Tournier, I., and Izaurralde, E. (2009) The C-terminal domains of human TNRC6A, TNRC6B, and TNRC6C silence bound transcripts independently of Argonaute proteins, *RNA* *15*, 1059–1066.
18. Eulalio, A., Helms, S., Fritzsch, C., Fauser, M., and Izaurralde, E. (2009) A C-terminal silencing domain in GW182 is essential for miRNA function, *RNA* *15*, 1067–1077.
19. Chekulaeva, M., Filipowicz, W., and Parker, R. (2009) Multiple independent domains of dGW182 function in miRNA-mediated repression in Drosophila, *RNA* *15*, 794–803.
20. Lagos-Quintana, M., Rauhut, R., Yalcin, A., Meyer, J., Lendeckel, W., and Tuschl, T. (2002) Identification of tissue-specific microRNAs from mouse, *Curr Biol* *12*, 735–739.

Chapter 15

Cell-Free microRNA-Mediated Translation Repression in *Caenorhabditis elegans*

Edlyn Wu and Thomas F. Duchaine

Abstract

In vitro recapitulation has recently led to significant advances in the understanding of the molecular functions of microRNAs. Cell-free systems allow a direct perspective on the different steps involved, and provide the experimenter with the opportunity to directly interfere with, or alter the implicated factors. In this chapter, we describe a cell-free translation system based on *Caenorhabditis elegans* embryo, which faithfully recapitulates miRNA-mediated translation repression. Because of the genetic and transgenic flexibility of this animal model, such a system provides a unique experimental resource to study the mechanism and the functions of miRNAs, the Argonautes, and the RISC.

Key words: microRNAs, Translation repression, Caenorhabditis elegans, Gene regulation, Embryo, *In vitro* translation

1. Introduction

MicroRNAs (miRNAs), when embedded within the RNA Induced Silencing Complex (RISC), base pair with their messenger RNA (mRNA) targets to subdue gene expression. Ambros and colleagues reported in 1999 that this gene repression occurs at posttranscriptional levels (1). More than a decade has passed since this publication, and the details of the mechanism of action of miRNAs at the molecular level are still not fully understood. Even since the identification of the Argonaute proteins as the core component of the RISC, the molecular basis for miRNA-mediated silencing has proven hard to refine (2). This is likely because the mechanism(s) is complex, but possibly also because the predominant mechanism involved may be different in distinct developmental, or cellular contexts where miRNAs were studied.

Tom C. Hobman and Thomas F. Duchaine (eds.), *Argonaute Proteins: Methods and Protocols*,
Methods in Molecular Biology, vol. 725, DOI 10.1007/978-1-61779-046-1_15, © Springer Science+Business Media, LLC 2011

In the vast majority of mRNA::miRNA targeting events in animals, base pairing is incomplete and does not activate the *Slicer* activity of the Argonautes (3). What happens then, to the expression of an mRNA target, and to its integrity, once it is targeted by the miRISC? Just like with other fundamental mechanisms of gene expression and regulation, elucidation of the underlying mechanisms only became possible when recapitulation was achieved, in cell culture, and *in vitro*. mRNA reporter systems based on transfection or transgenic expression, for example, provided much insight on the mechanism. This strategy, however, bears some significant limitations stemming from the fact that the reporter activity is examined several hours, if not days after transfection. This severely impinges on the possible insight on the nature, or the order of the very first events following mRNA::miRNA recognition. In addition, with such designs, it is difficult to gain a direct and unambiguous view on the relative contribution of the different steps involved. Thus, on these aspects in particular, *in vitro* reconstitution systems are irreplaceable in providing a *direct* perspective on the complexity of the mechanism of miRNA-mediated silencing. A number of cell-free miRNA-mediated silencing systems have recently emerged, derived from *Drosophila* embryo and cultured cells, rabbit reticulocyte, or mouse, and human cell cultures (for examples see 4–6). Most recently, we developed an *in vitro* translation system from *Caenorhabditis elegans (C. elegans)* embryo, which critically relies on both the 5′ cap and 3′ poly(A) tail determinants to initiate translation on exogenously-provided transcripts. We further showed that this system faithfully recapitulates a miRNA-mediated silencing response based on endogenous miRNAs, and requires ALG-1 and ALG-2, the two Argonautes dedicated to miRNA-mediated silencing in *C. elegans* (7).

In this chapter, we provide a detailed method to prepare this translation-competent extract derived from *C. elegans* embryos, and to assay for miRNA-mediated silencing. Specifically, this chapter describes the protocols for the preparation of *C. elegans* embryos, the preparation of the translation extracts, the design and the preparation of the mRNA reporters for miRNAs, and the translation repression assay itself. Finally we describe an alternative method that is based on 2′-*O*-methylated, sequence specific inhibitors.

2. Materials

2.1. Preparation of C. elegans Embryo, and Extracts

1. Agar/NGM 150-mm plates for *C. elegans* cultures (8).
2. OP50 paste, as a food supply for the large-scale *C. elegans* cultures (8).
3. Bleaching solution: 0.25 M potassium hydroxide (KOH) and 0.6% Sodium hypochlorite (Fisher Scientific).

4. 1× M9 saline: 3 g Anhydrous potassium phosphate monobasic (KH_2PO_4), 6 g Sodium phosphate (dibasic) anhydrous (Na_2HPO_4), 5 g Sodium chloride (NaCl), and 1 mL 1 M magnesium sulfate ($MgSO_4$); add water to 1 L. Sterilize by autoclaving (8).
5. Sephadex G-25 Superfine beads (Amersham Bioscience): Suspend the contents of the container in 500 mL of nuclease-free water. Sterilize by autoclaving.
6. 10 mL Poly-Prep Chromatography Columns (Bio-Rad).
7. 15 mL Dounce glass homogenizer with pestle "tight-fitting" (Kontes).
8. 1 M Dithiothreitol (DTT); store at −20°C.
9. Buffer A: 10 mM N-2-hydroxyethylpiperazine-*N*′-2-ethanesulfonic acid (HEPES)-KOH pH 7.4, 15 mM Potassium chloride (KCl), 1.8 mM Magnesium acetate ($Mg(OAc)_2$), 2 mM DTT. Prepare fresh and keep on ice.
10. Buffer B: 30 mM HEPES–KOH pH 7.4, 100 mM Potassium acetate (KOAc), 1.8 mM $Mg(OAc)_2$, and 2 mM DTT. Prepare fresh and keep on ice.
11. Protein Assay Dye Reagent Concentrate (Bio-Rad).

2.2. Preparation of RNA Substrate

1. pCI neo plasmid (Promega): To be used as a backbone into which the miRNA-response elements (miR-complementary sites) are cloned.
2. The *Renilla* Luciferase 6× target mRNAs (RL 6xmiR-52 and 6xmiR-52 mut): encode the *Renilla* luciferase coding sequence and six copies of a target site (Note 1).
3. The 6× target was synthesized as a miniGene (IDT) and was purchased as an insert in the pIDT Smart vector in its XbaI and NotI sites: 5′-GCGGCCGCGAATTCATTAACACCCGTACATTTTCCGTGCTATTAACACCCGTACATTTTCCGTGCTCAATTCATTAACACCCGTACATTTTCCGTGCTATTAACACCCGTACATTTTCCGTGCTATTAACACCCGTACATTTTCCGTGCTCAATCACCCGTACATTTTCCGTGCTTCTAGA-3′ (RL 6xmiR-52 wild-type) and 5′-GCGGCCGCGAATTCATTAACGTTTGTACATTTTCCGTGCTATTAACGTTTGTACATTTTCCGTGCTCAATTCATTAACGTTTGTACATTTTCCGTGCTATTAACGTTTGTACATTTTCCGTGCTATTAACGTTTGTACATTTTCCGTGCTCAATCGTTTGTACATTTTCCGTGCTTCTAGA-3′ (RL 6xmiR-52 mut). The 6× target cassette is digested with XbaI and NotI and subcloned into pCI neo to the RL open reading frame. The poly(A) tail is prepared by annealing oligonucleotides containing a stretch of 90 adenines,

and compatible ends for annealing into pCI neo RL using the HpaI and MfeI sites (Note 2).

4. MEGAscript T7 Transcription Kit (Ambion).
5. m^7(3′-*O*-methyl)G(5′)ppp(5′)G anti-reverse cap analog (ARCA) (Ambion) or ApppG (New England Biolabs).
6. Premixed Phenol:Chloroform:isoamyl alcohol (25:24:1, Bishop).
7. 3 M Sodium acetate.
8. 100% Ethanol.
9. Sephadex RNA mini Quick Spin columns (Roche Applied Science).
10. 4% Polyacrylamide (19:1 acrylamide/bisacrylamide)-8 M urea denaturing gel.
11. Gel loading buffer II (Ambion).
12. RiboRuler High Range RNA Ladder (Fermentas).

2.3. Translation Conditions

1. 2.5 mM Spermidine (Sigma Aldrich), store at −80°C.
2. Total L-amino acid mix: prepare 1 mM of each amino acid from stock commercial powders (Sigma Aldrich and/or Bioshop). Alternatively, this mixture can also be prepared using an amino acid powder kit (Sigma Aldrich). Store 1 mL aliquots at −80°C.
3. 1 M HEPES-KOH pH 7.5.
4. 10 mM $Mg(OAc)_2$, sterilize by filtration.
5. 2 M KOAc, store at −80°C.
6. 5 μg/μL calf-liver tRNA (Novagen, Note 3).
7. RiboLock RNase inhibitor (Fermentas): 40 U/μL.
8. 1 M Creatine phosphate (Roche Applied Science).
9. 3 μg/μL Creatine phosphokinase (from Rabbit skeletal muscle, Calbiochem).
10. 40 mM ATP: Dilute 100 mM ATP stock in sterile water, store at −80°C.
11. 10 mM GTP: Dilute 100 mM ATP stock in RNase free water, store at −80°C.
12. Master mix: 0.1 mM Spermidine, 60 μM amino acids, 24 mM HEPES-KOH (pH 7.5), 1.28 mM $Mg(OAc)_2$, 25 mM KOAc, 0.1 μg/μL Calf-liver tRNA, 0.096 U/μL RiboLock RNase Inhibitor (Fermentas), 16.8 mM Creatine phosphate, 81.6 ng/μL Creatine phosphokinase, 0.8 mM ATP, 0.2 mM GTP (see Table 1 and Note 4).
13. Dual-Luciferase Reporter Assay (Promega).

Table 1
***In vitro* translation mix preparations**

Reagent	Volume (μL)	Final added concentration
A		
2.5 mM Spermidine	0.5	0.1 mM
1 mM Amino acid mix	0.75	0.06 mM
1 M HEPES-KOH (pH 7.5)	0.3	24 mM
10 mM $Mg(OAc)_2$	1.6	1.28 mM
2 M KOAc	0.156	25 mM
5 μg/μL calf-liver tRNA	0.25	0.1 μg/μL
RNase Inhibitor (40 U/μL)	0.03	0.096 U/μL
1 M Creatine Phosphate	0.21	16.8 mM
3 μg/μL Creatine Phosphokinase	0.34	81.6 ng/μL
40 mM ATP		0.8 mM ATP
10 mM GTP mix	0.25	0.2 mM GTP
Extract	5	n/a
Total master mix volume	9.386	n/a
B		
no 2′-O-Me		
Master mix	9.386	n/a
mRNA	1	1 nM
RNase-free water	2.114	n/a
Total reaction volume	12.5	n/a
with 2′-O-Me		
Master mix	9.386	n/a
mRNA	1	1 nM
625 nM 2′-O-Me	1	50 nM
RNase-free water	1.114	n/a
Total reaction volume	12.5	n/a

Reaction mixes assembly for a 12.5 μL translation reaction in the absence (A) or presence (B) of 2′-*O*-Me inhibitors

14. GloMax 20/20 Luminometer (Promega, Note 5).
15. The 2′-*O*-methylated oligonucleotides (Dharmacon) were designed as antisense oligonucleotides to the mature miRNAs according to Wormbase registry (www.wormbase.org). Oligonucleotides were resuspended in water to a concentration of 100 ng/μL. In this chapter we used the miR-52 2′-*O*-Me (α-miR-52) sequence: 5′-UUAAUAGCACGGAAACA UAUGUACGGGUGUUAAU-3′; miR-1 2′-*O*-Me (α-miR-1) sequence: 5′-UCUUCCUCCAUACUUCUUUACAUUCC AACCUU-3′.

3. Methods

The following method starts with the harvest of large-scale cultures of *C. elegans* gravid adults (animals with rows of embryo in their uteri), and extends until the actual miRNA-mediated translation repression assays. In our lab, we often use variations of this method to take advantage of the genetic flexibility of *C. elegans.* Extracts can be generated from viable mutant strains, or after growing the animals on a bacterial strain (usually HT115), which drives the overexpression of dsRNA to silence target genes (9).

3.1. Culture and Harvest of C. elegans Embryos

1. Harvest embryos from large-scale cultures of *C. elegans* on large NGM-Agar plates, and using OP50 as food. For a suitable scale of preparation (a typical batch), harvest embryos from 30×150 mm plates each containing approximately 50,000 synchronous animals each (1,500,000 animals total).
2. Harvest gravid adults in 1× M9, and distribute equally in 15-mL Falcon table-top centrifuge tubes. We usually pool the animals from two or three plates per tube for the hypochlorite step.
3. Treat the adults with freshly prepared hypochlorite solution. Animal suspensions are treated for 2 min with mild, but constant hand agitation followed by 20 s centrifugations in a table-top centrifuge at 680×*g* and then remove all the supernatant.
4. Add hypochlorite solution to the animal pellet, and repeat step 3 until the suspension is completely devoid of adult cuticles. Monitor the progress of the treatment under a dissection microscope. Complete dissolution of cuticles typically requires three to four suspension–centrifugation cycles, and leaves a small, beige embryo pellet of approximately 1/5th to 1/10th the initial animal volume. Following the final centrifugation, carefully remove all of the supernatants.
5. Completely resuspend the pellet of embryo in M9 saline, and centrifuge again in a table-top centrifuge.
6. For the second wash, add 1 mL of 1 M HEPES-KOH pH 7.5, and complete to a final volume of 15 mL with M9 in the Falcon tube.
7. Proceed to two additional washes with 15 mL of M9 saline.
8. Finally, wash the embryonic pellet three more times in RNase-free water to completely remove the sodium ions (which are known to inhibit protein synthesis when present at high concentration). We usually pool all the embryo pellets in a single Falcon tube at this step.
9. After the final centrifugation, carefully remove all the residual supernatant. Typically this results in a pellet of 500 μL to 1 mL of stacked embryos.

10. Flash-freeze in a 15-mL Falcon tube by immersion in liquid nitrogen.

Following this step, embryos may be stored at −80°C for at least 2 years.

3.2. Preparation of C. elegans Embryonic Extract

A broad diversity of methods is available for the preparation of *in vitro* translation systems that are derived from tissues or cell cultures from various species. Most of these methods were not directly adaptable to *C. elegans* extracts. In fact, a large number of parameters had to be tuned before we obtained robust and reproducible translation initiation in *C. elegans* extracts, and often even the slightest deviations greatly affected the recovered activity. For example, in some systems micrococcal nuclease treatment (to remove the endogenous mRNAs) is required for the translation of exogenous mRNAs (10). Such a treatment kills translation initiation in the *C. elegans* embryonic extract. Among the other critical parameters are the monovalent, and divalent ion concentrations, the temperature, and the presence of 5′-cap and 3′-poly(A) tail on the reporter mRNA.

A flow chart illustrating the preparation of the extracts is shown in Fig. 1. Every step of the extract preparation should be conducted at 4°C, or in a cold room.

1. Rapidly thaw the embryonic pellets in hand and keep it on ice until used.
2. Resuspend the embryonic pellet in 0.3 volumes of Buffer A (Note 6).

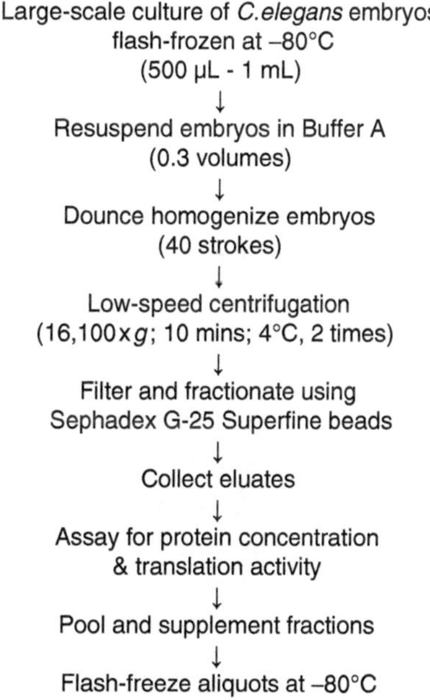

Fig. 1. Flow chart of the procedure for the preparation of the *C. elegans* embryonic extracts.

3. Transfer the slurry to a clean, prechilled Dounce homogenizer. Break the embryos with 40 strokes (total), by series of ten strokes to allow cooling between the series (Note 7).
4. Confirm the lysis of the embryo by visual inspection of 0.5 μL aliquots on a glass slide using a dissection microscope.
5. Transfer the slurry to an RNase-free microcentrifuge tube. Centrifuge the slurry at 16,100 × *g* for 10 min at 4°C.
6. Recover the supernatant and centrifuge once more in the same conditions. Retain a 2 μL aliquot of the resulting supernatant to monitor the dilution of the extract during the fractionation step (Note 8).
7. Fractionate the extract by size-exclusion chromatography using Sephadex G-25 Superfine beads (Note 9). For this, wash the beads three times with Buffer B by suspension–centrifugation in a 15-mL Falcon tube and using a table-top centrifuge at 680 × *g* and 4°C. Beads should make up approximately four times the volume of the extract supernatant.
8. Settle the beads into a 10-mL Poly-Prep chromatography column, and allow Buffer B to flow through until it reaches the surface of the matrix.
9. Load the supernatant onto the column slowly, and directly onto the matrix (drop-wise). Allow the supernatant to completely enter the matrix by gravity.
10. Load the column with 1 extract volume of Buffer B. Discard the dead volume. Start collecting fractions in 1.5 mL RNase-free microcentrifuge tubes as soon as tint of yellow is visible in the eluate (Note 10).
11. Once the flow from the first elution volume stops, add 0.3 volumes of Buffer B and collect fractions. Repeat the elution five to six times or until the eluate appears completely clear.
12. Remove 2 μL aliquots from each fraction, and assess their protein concentration by Bradford assay (Note 11).
13. Save a small aliquot (5 μL) of each fraction to test for translation activity using luciferase reporters (see Subheading 3.4) (Note 12).
14. Supplement the fractions that are active for translation by following Table 1a. For this, the most active fractions can be pooled (Note 13).
15. Aliquot the supplemented fractions and flash freeze as aliquots in liquid nitrogen. The extract remains active for at least 2 years when stored at –80°C.

3.3. Preparation of the RNA Substrate

1. Transcribe RL 6xmiR-52, and RL 6xmiR-52 mut at 30°C for 4 h with the ARCA cap analog using the MEGAscript kit (see Note 14).
2. Following transcription, add 1 μl of DNase Turbo I, and digest the template DNA for 15 min at 37°C.
3. Adjust the volume of the reaction to 70 μl with RNase-free water.
4. Purify the RNA by phenol/chloroform extraction. For this, add 1:1 volume of phenol/chloroform/isoamyl alcohol and vortex for 15 s.
5. Centrifuge for 30 s in a table-top centrifuge at 16,100 × *g*.
6. To remove any residual, unincorporated nucleotides, transfer the aqueous phase to a Sephadex RNA Mini Quick Spin column, and proceed to filtration according to the supplier's instructions.
7. Quantify the recovered RNA, and monitor the size and quality of the transcript on a 4% polyacrylamide–urea denaturing gel and by Ethidium bromide staining. A single band should be visible. Store the RNA as aliquots at −80°C.

3.4. miRNA-Mediated Translation Repression

To assay for miRNA activity, we use a Luciferase reporter mRNA that is fused to a 3′ UTR encoding six copies of a miRNA-binding site (RL 6xmiR-52) (Fig. 2a). Our data and other published reports indicate that translation repression increases with additional copies of miRNA-complementary sites (11, 7).

In the first protocol, we determine the repressive effect of a specific miRNA by comparing the translation of RL 6xmiR-52, with a reporter bearing six copies of binding sites bearing a mutation within the seed complementary sequence (positions complementary to nts 2–4 of miR-52; RL 6xmiR-52 mut).

Note that for our typical experiments, we use a final concentration of 1 nM of reporter mRNA. This is far less than the miR-52 concentration in the extract, but still allows for sufficient sensitivity to detect the translation of the reporter. The investigator is encouraged to determine the precise concentration of their favorite miRNA by qRT-PCR in the embryonic extract, prior to a translation repression assay.

1. Thaw the frozen extract, and assemble the translation reactions in microcentrifuge tubes on ice. For convenience, the master mix content is also outlined in Table 1b (no 2′-*O*-Me).
2. For every 1× reaction, add 2.114 μl of water to the master mix.
3. Dispense 11.5 μL of the master mix (completed with water) to 1 μL of mRNA per tube (1 nM final mRNA concentration). We usually work with duplicates of each time-point, and carry parallel reactions for the RL 6xmiR-52 and RL 6xmiR-52 mut mRNAs.

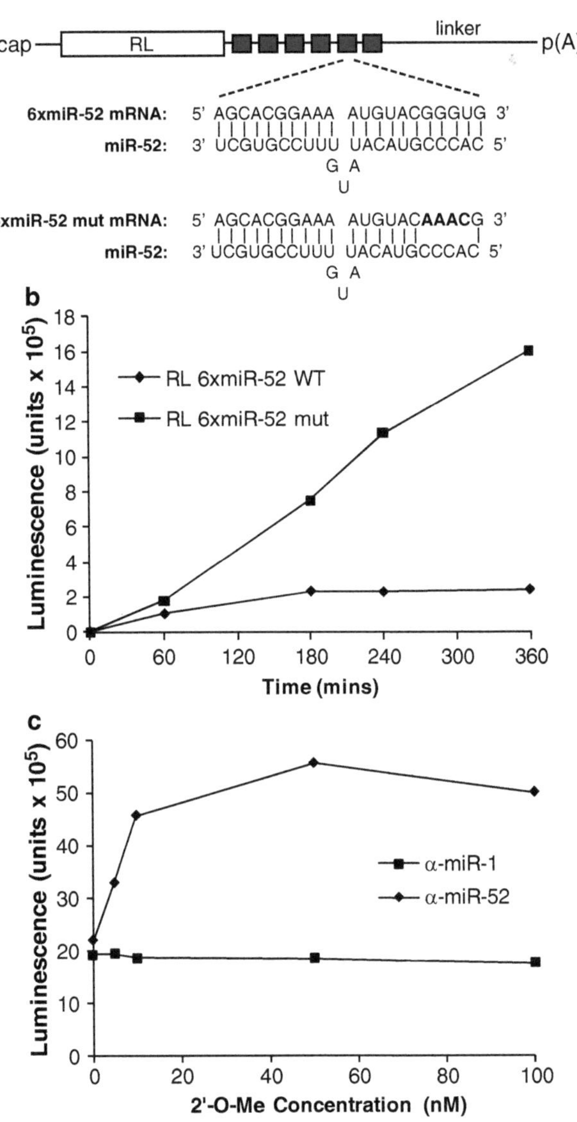

Fig. 2. Translation repression by miRNAs in *C. elegans* embryonic extracts. (**a**) Diagram of the Luciferase miR-52 reporters used to assay for translation repression. (**b**) Translation repression time-course of RL 6xmiR-52 vs. 6xmiR-52 mut reporter mRNAs. (**c**) Dose-response translation de-repression using α-miR-52 (specific) and α-miR-1 (negative control) 2′-*O*-Me.

4. Mix each reaction by tapping the tubes gently. Avoid frothing.
5. Incubate the reactions at 17°C for 0–6 h in a water bath (see Note 15).
6. Once translation reactions are complete (at each time-point), place tubes on ice and withdraw 2 μL from each reaction tube to measure the luciferase activity using the Dual-Luciferase Reporter Assay (see Note 16).

Using this method, translation of RL 6xmiR-52 reaches a plateau and is usually fully repressed between the 1 and the 3-h time-points, while RL 6xmiR-52 mut mRNA remains unrepressed during the entire time-course (Fig. 2b). Since the translation of RL 6xmiR-52 is arrested, and translation of 6xmiR-52 mut mRNA persists, the extent of the "repression" detected using this method depends on the time of incubation. Typically for miR-52, a 3-h time-course leads to an approximately threefold difference in overall translation.

3.5. Alternative Method: miRNA-Mediated Translation Repression as Revealed with 2′-O-Me Inhibitors

In vitro transcription efficiency and the quality of the resulting mRNA are very sensitive to the quality of the DNA template, its linearization, and concentration. Accordingly, the result of the translation assay will vary with each RNA preparation in a manner that depends on parameters that are not only due to the effect of the miRNA. For this reason, it is crucial to prepare the 6xmiR-52 and the 6xmiR-52 mut in parallel, and using the very same conditions. To circumvent the problem of batch-to-batch variation, we propose an alternative approach that relies on a single mRNA reporter. For this, we use 2′-*O*-methylated oligonucleotides (2′-*O*-Me) as miRNA inhibitors to specifically prevent the repression of the RL 6xmiR-52 reporter. 2′-*O*-Me inhibitors encode a sequence that is complementary to the miRNA of interest. Their inclusion in the reaction results in irreversible hybridization with the miRNA and hence, prevents the repression of the target mRNA. Translation repression is revealed when comparing with a non-related 2′-*O*-Me (here α-miR-1), used at the same concentration. As an alternative to 2′-*O*-Me inhibitors, Locked nucleic acids (LNA) may also be employed (12, 13).

1. Thaw the frozen extract, and assemble the master mix as in Table 1b, *with* 2′-*O*-Me.
2. Prior to mRNA addition, the extract is incubated with 1 μL of either α-miR-52 (specific) or α-miR-1 (control) 2′-*O*-Me, which sets a final concentration of 50 nM of 2′-*O*-Me (Note 17). Mix by tapping the tubes gently while avoiding frothing. Preincubate for 30 min at 17°C in a water bath (Note 18).
3. After the 30 min of preincubation, add 1 μL of the RL 6xmiR-52 mRNA target (1 nM final) to each reaction tube, mix with 11.5 μL of the 2′-*O*-Me preincubated mastermix and allow the reaction to proceed at 17°C.
4. Incubate the reactions at 17°C for 3 h in a water bath (Note 19).
5. Place tubes on ice, and withdraw 2 μL from each reaction tube to measure the luciferase activity using the Dual-Luciferase Reporter Assay.

Including α-miR-52 leads to a threefold derepression when monitored at the 3-h time-point, while the addition of the non-related control α-miR-1 does not significantly affect translation at concentrations up to 100 nM (Fig. 2c). Results with this method are usually similar to the 6xmiR-52 mut reporter comparison method, but are less sensitive to mRNA reporter prep-to-prep quality variations.

4. Notes

1. The 6× target site is partially complementary to the guide strand of the miRNA, leading to a "bulge" in the seed-complementary region and hence imperfect base pairing between the miRNA and the mRNA.
2. Note that cloned poly(A) tail-encoding sequences are inherently unstable in bacteria, and should be resequenced every time a preparation is made. Sequencing of *midi* or *maxi* scale preparations is recommended to ensure that batches with predetermined poly(A) tails remain available. Plasmids encoding a poly(A) tail no less than 80 A residues are used.
3. We also successfully used tRNA isolated from cultured cells such as Krebs-2 ascites and *C. elegans* embryos.
4. The optimal concentration for the supplementation with K^+ and Mg^{2+} may vary from batch to batch, especially when the experimenter prepares the extract for the first few times. We have established an optimal range of 1.5–3 mM for Mg^{2+} and 60–75 mM for K^+. Optimally, the salt concentrations should be adjusted for each batch of extract that is prepared. In typical batches, we set the final salt concentrations in translation reactions at 2 mM for $Mg(OAc)_2$ and 65 mM for KOAc.
5. Lumat LB 9507 (Berthold Technologies GmbH & Co. KG) can also be used.
6. Diluting the extract too much can dramatically reduce the translation activity of the embryonic extract.
7. Make sure to keep the pestle in contact or close to the embryo suspension while homogenizing, i.e., no more than 1 cm above the slurry of embryos. Lifting the pestle too high will result in a reduction of yield of the embryonic extract. The slurry is viscous, and will remain on the walls of the homogenizer, making it difficult to recover after homogenization.
8. The protein concentration of the lysates prior to filtration typically ranges from 20 to 60 μg/μl.
9. The step of fractionation on Sephadex™ G-25 Superfine beads *is absolutely required* to obtain translation activity. Centrifugation-based and gravity-based chromatography may

both be used, but the gravity-based method yields more consistent results.

10. To follow the activity, we count the elution fractions that have passed the matrix dead volume. We do this by following the brown-yellowish tint of the extract. Alternatively, you may wish to simply follow the protein concentration by mixing 1 μl of each fraction with a Bradford assay, as you recover them.

11. Fractions 1–4 (when counting *after* the beads dead volume) typically have highest protein concentration, with fractions 2 and 3 usually being the most concentrated. Concentration within these two fractions is only slightly lower than the concentration prior to filtration. The protein concentrations for fractions 1–4 can range from 5 to 35 μg/μl, with batch-to-batch variations.

12. Fractions 1–4 typically yield the highest translation activity. Like protein concentration, the elution profile for translation activity is also typically bell-curved. We usually combine the fractions yielding similar translation activity to prepare the supplemented extract.

13. Fractions can also be frozen prior to supplementation. However, supplementing on the day of the preparation of the extract leads to a better consistency for subsequent experiments.

14. Our system is highly dependent on the presence of both 5′ cap and 3′ poly(A) tail. Translation of Luciferase reporters bearing either regular or ARCA-capped analogs yields translation activity, although translation of the ARCA-capped mRNA is most efficient.

15. Translation is active over temperatures ranging from 10 to 25°C, but the optimal temperature for *in vitro* translation in our *C. elegans* embryonic extract is 17°C.

16. Ensure that the luciferase reagents (substrates mix) are at room temperature prior to mixing and the measurement of luminescence, as this will greatly affect the read out for luciferase activity.

17. When using a different miRNA, a pilot translation experiment with varying concentrations of 2′-*O*-Me should be performed to select for the optimal concentration at which translation is efficiently de-repressed. This is particularly essential when assaying for miRNAs of unknown concentration in the extract.

18. This preincubation, prior to translation, allows for the annealing of the 2′-*O*-Me with the endogenous miRNA (14).

19. In the alternative method, we use a single 3-h time-point. A time-course (as in Subheading 3.4) may also be conducted. The time-course design is often more suitable, as it is more informative.

Acknowledgments

We thank Ahilya Sawh and Mathieu Flamand for their comments on the manuscript. This work was supported by Canadian Institute of Health Research (CIHR), the Canada Foundation for Innovation (CFI), and the Fonds de la Rercherche en Santé du Québec (FRSQ) (Chercheur-Boursier Salary Award J.1) to T.F.D.

References

1. Olsen, P. H., and Ambros, V. (1999) The lin-4 regulatory RNA controls developmental timing in Caenorhabditis elegans by blocking LIN-14 protein synthesis after the initiation of translation, *Dev Biol 216*, 671–680.
2. Filipowicz, W., Bhattacharyya, S. N., and Sonenberg, N. (2008) Mechanisms of post-transcriptional regulation by microRNAs: are the answers in sight?, *Nat Rev Genet 9*, 102–114.
3. Bartel, D. P. (2009) MicroRNAs: target recognition and regulatory functions, *Cell 136*, 215–233.
4. Gebauer, F., and Hentze, M. W. (2007) Studying translational control in Drosophila cell-free systems, *Methods Enzymol 429*, 23–33.
5. Fabian, M. R., Mathonnet, G., Sundermeier, T., Mathys, H., Zipprich, J. T., Svitkin, Y. V., Rivas, F., Jinek, M., Wohlschlegel, J., Doudna, J. A., Chen, C. Y., Shyu, A. B., Yates, J. R., 3rd, Hannon, G. J., Filipowicz, W., Duchaine, T. F., and Sonenberg, N. (2009) Mammalian miRNA RISC recruits CAF1 and PABP to affect PABP-dependent deadenylation, *Mol Cell 35*, 868–880.
6. Wang, B., Love, T. M., Call, M. E., Doench, J. G., and Novina, C. D. (2006) Recapitulation of short RNA-directed translational gene silencing in vitro, *Mol Cell 22*, 553–560.
7. Wu, E., Thivierge, C., Flamand, M., Mathonnet, G., Vashisht, A. A., Wohlschlegel, J., Fabian, M. R., Sonenberg, N., and Duchaine, T. F. (2010) Pervasive and cooperative deadenylation of 3'UTRs by embryonic microRNA families. Mol Cell 24, 40(4):558–70.
8. Hope, I. A., (Ed.) (1999) *C. elegans: a practical approach*, Vol. 1, 1 ed., Oxford University Press, Oxford.
9. Timmons, L., Court, D. L., and Fire, A. (2001) Ingestion of bacterially expressed dsRNAs can produce specific and potent genetic interference in Caenorhabditis elegans, *Gene 263*, 103–112.
10. Scott, M. P., Storti, R. V., Pardue, M. L., and Rich, A. (1979) Cell-free protein synthesis in lysates of Drosophila melanogaster cells, *Biochemistry 18*, 1588–1594.
11. Doench, J. G., and Sharp, P. A. (2004) Specificity of microRNA target selection in translational repression, *Genes & development 18*, 504–511.
12. Chan, J. A., Krichevsky, A. M., and Kosik, K. S. (2005) MicroRNA-21 is an antiapoptotic factor in human glioblastoma cells, *Cancer research 65*, 6029–6033.
13. Orom, U. A., Kauppinen, S., and Lund, A. H. (2006) LNA-modified oligonucleotides mediate specific inhibition of microRNA function, *Gene 372*, 137–141.
14. Mathonnet, G., Fabian, M. R., Svitkin, Y. V., Parsyan, A., Huck, L., Murata, T., Biffo, S., Merrick, W. C., Darzynkiewicz, E., Pillai, R. S., Filipowicz, W., Duchaine, T. F., and Sonenberg, N. (2007) MicroRNA inhibition of translation initiation in vitro by targeting the cap-binding complex eIF4F, *Science (New York, NY 317*, 1764–1767.

Chapter 16

Argonaute Pull-Down and RISC Analysis Using 2′-*O*-Methylated Oligonucleotides Affinity Matrices

Guillaume Jannot, Alejandro Vasquez-Rifo, and Martin J. Simard

Abstract

During the last decade, several novel small non-coding RNA pathways have been unveiled, which reach out to many biological processes. Common to all these pathways is the binding of a small RNA molecule to a protein member of the Argonaute family, which forms a minimal core complex called the RNA-induced silencing complex or RISC. The RISC targets mRNAs in a sequence-specific manner, either to induce mRNA cleavage through the intrinsic activity of the Argonaute protein or to abrogate protein synthesis by a mechanism that is still under investigation. We describe here, in details, a method for the affinity chromatography of the *let-7* RISC starting from extracts of the nematode *Caenorhabditis elegans*. Our method exploits the sequence specificity of the RISC and makes use of biotinylated and 2′-*O*-methylated oligonucleotides to trap and pull-down small RNAs and their associated proteins. Importantly, this technique may easily be adapted to target other small RNAs expressed in different cell types or model organisms. This method provides a useful strategy to identify the proteins associated with the RISC, and hence gain insight in the functions of small RNAs.

Key words: 2′-*O*- methyl oligonucleotides, Argonaute, *Let-7* microRNA, Affinity chromatography, TaqMan chemistry

1. Introduction

RNA-mediated gene silencing pathways can be triggered either by the introduction of exogenous double-stranded RNA molecules (RNA *interference* or RNAi) or by the endogenous expression of microRNAs. They constitute two distinct pathways that lead to posttranscriptional down regulation of targeted messenger RNA (mRNA) through either degradation of the mRNAs or inhibition of protein synthesis. RNAi is commonly used in research laboratory as a simple, but powerful technique to knockdown the expression of specific genes. The microRNA pathway can be described as an endogenous process, which is conserved across many species and is

Tom C. Hobman and Thomas F. Duchaine (eds.), *Argonaute Proteins: Methods and Protocols*,
Methods in Molecular Biology, vol. 725, DOI 10.1007/978-1-61779-046-1_16, © Springer Science+Business Media, LLC 2011

used to control cell homeostasis. It is estimated that 60% of human genes could be regulated by microRNA molecules and it is now becoming clear that the mis-expression of these small RNA molecules contributes to the development of a large spectrum of human diseases. Moreover, the recent findings involving small RNA in the control of spermatogenesis and in the silencing of transposable elements in the germline highlight their role in an important diversity of developmental processes in animals (reviewed in ref. (1). Essential to all small RNA-mediated gene regulation pathways is a ribonucleoprotein complex called RNA-induced silencing complex (RISC) composed of, at least, one 21–23 nucleotides long, single-stranded RNA and a member of the Argonaute protein family (2). The activity of the RISC on its target depends on the base-pairing between the small RNA and the mRNA. When pairing is perfect, the RISC mediates mRNA cleavage through the endonuclease activity of the Argonaute protein. When the pairing is imperfect, the RISC prevents protein synthesis by a mechanism that is still unclear.

In 2004, Hutvágner *et al.* reported that modified antisense oligonucleotides can block RISC-mediated gene regulation in vivo (3) through base pairing with the small guide RNAs in this complex. The replacement of the hydroxyl group by a methyl group, as shown in Fig. 1, makes the oligonucleotide resistant to cellular ribonucleases (4) and allows the sequence-specific purification of the RISC. A biotinylated 2′-*O*-methylated oligonucleotide is bound to the RISC, and pull-down is performed using streptavidin-coupled magnetic beads. This approach is useful to identify the proteins associated with small RNAs in vivo, as well as to characterize their molecular functions (for examples of utilization, see (5–8)).

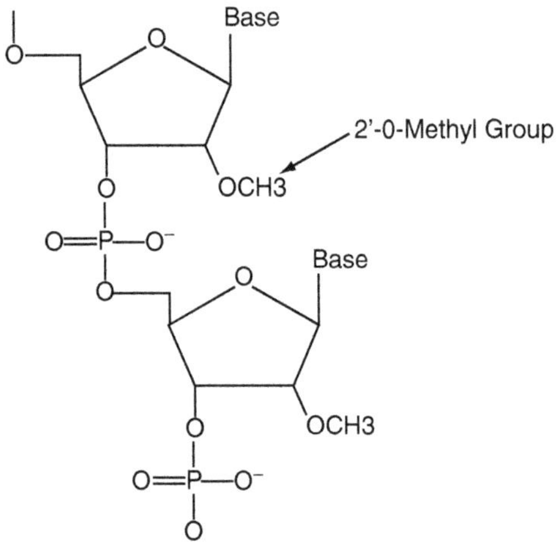

Fig. 1. Representative backbone of a single-stranded 2′-*O*-methylated oligonucleotide is shown. Each hydrogen from 2′ hydroxyl groups was replaced by a CH_3 (methyl) group. This modification confers resistance to most RNase activities.

2. Materials

2.1. Caenorhabditis elegans Culture and Extract Preparation

1. The wild-type *C. elegans* Bristol, N2 strain, was obtained from the *Caenorhabditis* Genetics Center (CGC, University of Minnesota, USA). The transgenic *C. elegans* strain expressing ALG-1 protein tagged at the N-terminus with the Green Fluorescent Protein (GFP) was generated in the laboratory as previously described in (3).
2. *Escherichia coli* (OP50) in TB liquid culture, as a food source for *C. elegans.*
3. TB medium (1 L): 12 g bacto-peptone, 24 g bacto-yeast extract, 4 mL glycerol, in 900 mL of water. Sterilize by autoclaving and cool down to <60°C. Add 100 mL of sterile 10× TB phosphates and store at room temperature (RT).
4. 10× TB phosphate (1 L): 23.1 g KH_2PO_4, 125.4 g K_2HPO_4. Sterilize by autoclaving. Store at RT.
5. Nematode growth medium (NGM) (1 L): 20 g agarose, 3 g NaCl, 2.5 g bacto-peptone. Sterilize by autoclaving and cool down to <60°C. Add, 1 mL 1 M $MgSO_4$, 1 mL 1 M $CaCl_2$, 1 mL cholesterol stock solution (5 mg/mL in ethanol), and 25 mL PPB stock solution.
6. PPB stock solution (1 L): 98 g KH_2PO_4 and 48 g K_2HPO_4. Dissolve the salts in deionized water (final volume 1 L). Sterilize by autoclaving. Store at RT.
7. M9 Buffer (1 L): Dissolve 3 g KH_2PO_4, 6 g Na_2HPO_4, 5 g NaCl, 1 mL 1 M $MgSO_4$ in deionized water (final volume 1 L). Sterilize by autoclaving. Store at RT.
8. Bleaching solution (20 mL): 2 mL of 5 M KOH, 3 mL of sodium hypochlorite, 15 mL of water (see Note 1).
9. Lysis buffer stock solution (1 L): 1 mL of 1 M KAc, 300 μL of 1 M HEPES–KOH pH 7.5, 20 μL of 1 M $Mg(Ac)_2$. Sterilize by filtration. Store at 4°C.
10. Complete lysis buffer (10 mL): 9.40 mL lysis buffer stock solutions, 10 μL 1 M DTT, 50 μL triton X-100, 1 tablet of protease inhibitors (Complete Mini EDTA-free, Roche). Add 2% [v/v] of RNase inhibitor (SUPERaseIn, Ambion) when indicated.
11. Homogenizer: Wheaton dounce tissue grinders 1 mL (Fisher).

2.2. Immobilized 2′-O-Methylated Oligonucleotide Matrices

1. 2′-*O*-methylated oligonucleotides were synthesized by Integrated DNA Technologies (IDT). The modification is represented in Fig. 2.

   Unrelated oligo: 5′-Bio-$_mC_mA_mU_mC_mA_mC_mG_mU_mA_mC_mG_mC_mG_mG_mA_mA_mU_mA_mC_mU_mU_mC_mG_mA_mA_mA_mU_mG_mU_mC$-3′;

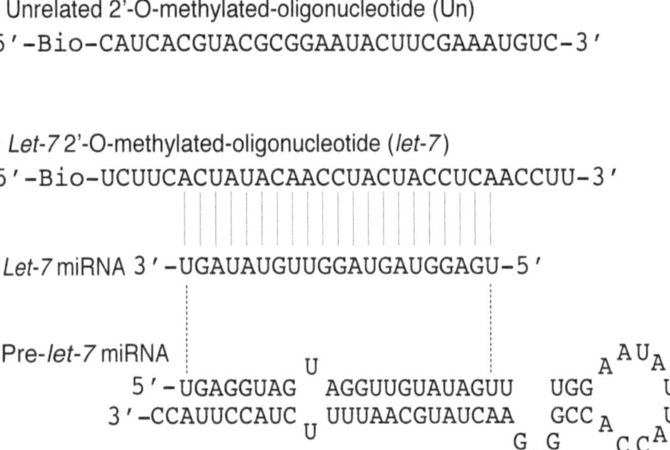

Fig. 2. *Let-7* precursor microRNA is a substrate for the cleavage activity of Dicer to form mature *let-7* microRNA, which is fully complementary to 2′-*O*-methylated oligonucleotides used to pull-down *let-7* RISC complex.

and *let-7* complementary microRNA oligo: 5′-Bio-$_{m}U_{m}C_{m}U_{m}U_{m}C_{m}A_{m}C_{m}U_{m}A_{m}U_{m}A_{m}C_{m}A_{m}A_{m}C_{m}C_{m}U_{m}A_{m}C_{m}U_{m}A_{m}C_{m}C_{m}U_{m}C_{m}A_{m}A_{m}C_{m}C_{m}U_{m}U$-3′ (see Notes 2–5).

2. Dynabeads M-280 streptavidin (Invitrogen): To pull-down the RISC, the modified oligonucleotide is biotinylated at the 5′ end and then coupled to magnetic beads through streptavidin monolayer attached to the surface of beads (see Notes 6 and 7).
3. Dynal MPC-S (Invitrogen): We used a magnetic particle concentrator (referred as magnet) compatible with microcentrifuge tube (1.5 mL tube). Dynabeads are separated from the solution when the tube is inserted into the magnet and the magnetic slide is inserted. Bead-bound material is then attracted to the side of the tube.

2.3. SDS–Polyacrylamide Gel Electrophoresis

1. Resolving buffer 4× (1 L): 182 g Tris (hydroxymethyl) aminomethane, 4 g Sodium Dodecyl Sulfate (SDS) in deionized water. Adjust pH to 8.8 with concentrated HCl. Sterilize by filtration. Store at RT.
2. Stacking buffer 4× (1 L): 60 g Tris (hydroxymethyl) aminomethane, 4 g SDS in deionized water. Adjust pH to 6.8 with concentrated HCl. Sterilize by filtration. Store at RT.
3. Running buffer 10× (1 L): 30.3 g Tris (hydroxymethyl) aminomethane, 144.1 g glycine, 10 g SDS in deionized water. Store at RT (see Note 8).
4. 29:1, 40% Acrylamide/bis solution (J.T Baker). (This is a neurotoxic reagent when unpolymerized, handle with care).

5. *N,N,N,N′*-Tetramethyl-ethylenediamine (TEMED) (this is a corrosive and irritant reagent: handle with care).
6. Ammonium persulfate (APS) 10% (10 mL): 10 g $(NH_4)_2S_2O_8$ in cold deionized water and freeze as 1 mL aliquots at –20°C.
7. Prestained molecular weight markers: Kaleidoscope marker (Bio-Rad).
8. Laemmli buffer 4× (20 mL): 4 mL 1 M Tris (hydroxymethyl) aminomethane–HCl pH 6.8, 0.04 g bromophenol blue, 8 mL glycerol, 1.6 g SDS, and 8 mL 1 M Dithiothreitol (DTT). Store at –20°C as 1 mL aliquots.
9. Ethanol 95%.

2.4. Western Blotting for Argonautes Associated to the RISC

1. Transfer buffer (1.6 L): 5.86 g glycine, 11.64 g Tris (hydroxymethyl) aminomethane, 0.75 g SDS. Store at RT.
2. Methanol.
3. Nitrocellulose Hybond-ECL membrane (Amersham) and Blotting pad (VWR).
4. TBS-T (1 L): 6.05 g Tris (hydroxymethyl) aminomethane, 8.76 g NaCl, 1 mL Tween-20. Adjust pH to 7.5 with concentrated HCl. Store at 4°C.
5. Blocking solution: 5% [w/v] nonfat dry milk in TBS-T.
6. Primary antibody: anti-GFP antibody (Roche) diluted in TBS-T supplemented with 5% [w/v] nonfat dry milk (see Note 9).
7. Secondary antibody: Peroxydase-conjugated Anti-mouse IgG (Roche) diluted in TBS-T supplemented with 5% [w/v] nonfat dry milk.
8. Enhanced chemiluminescence (ECL) reagent (Perkin Elmer) and hyperfilm ECL (Amersham).

2.5. Total RNA Extraction

1. TRI-Reagent solution (Sigma). (The TRI-Reagent solution contains phenol and guanidine thiocyanate, handle with care).
2. Chloroform (chloroform can be fatal if swallowed, inhaled, or absorbed. Manipulate with care in a fume hood).
3. Isopropanol 75% [v/v] and ethanol 75% [v/v].
4. Freshly autoclaved deionized water.
5. 1.5 mL RNase-free microcentrifuge tubes.

2.6. Analysis of Short RNA Integrity by 12% Mini-Polyacrylamide Gel Electrophoresis

1. SequaGel Sequencing System kit (National Diagnostic): (1) SequaGel Concentrate (1 L: 37.5 g of acrylamide, 12.5 g of methylene bis-acrylamide, and 7.5 M urea in a deionized aqueous solution), (2) SequaGel Diluent (7.5 M urea in deionized water), and (3) SequaGel Buffer [0.89 M

Tris–Borate–20 mM EDTA buffer pH 8.3 (10× TBE) and 7.5 M urea] (see Note 10).

2. TEMED.
3. APS 10% (10 mL): 10 g $(NH_4)_2S_2O_8$ in cold water and freeze in 1 mL aliquots at −20°C.
4. Formamide-dye loading buffer: 98% [w/v] deionized formamide, 10 mM EDTA pH 8, 0.025% [w/v] xylene cyanol, and 0.025% [w/v] bromophenol blue.
5. TBE buffer 10× (1 L): 108 g Tris (hydroxymethyl) aminomethane and 40 mL of 0.5 M Na_2EDTA, 55 g boric acid. Adjust to pH 8. Sterilize by autoclaving. Store at RT.
6. Ethidium bromide (EtBr) solution at 5 μg/mL (EtBr is a mutagenic agent, so handle with care).

2.7. Quantitative Real-Time PCR for Let-7 MicroRNA

1. Real-Time PCR machine: 7900HT Fast Real-Time PCR system (Applied Biosystems).
2. MicroAmp Fast optical 96-well reaction plate (Applied Biosystems).
3. Optical adhesive covers (Applied Biosystems).
4. MicroRNAs Primers: We used TaqMan probes-based chemistry. The primers were obtained from Applied Biosystems and correspond to mature *let-7* microRNA (Assay ID: 000377). We used the short nuclear RNA *sn2841* as an endogenous control (Assay ID: 001759).
5. Reverse Transcription Assays: TaqMan microRNA transcription kit (Applied Biosystems).
6. PCR reaction: TaqMan 2× Universal PCR Master Mix, No AmpErase UNG (Applied Biosystems).

3. Methods (see Note 11)

For this protocol, we took *let-7* microRNA-associated with ALG-1 RISC as an example to demonstrate the efficiency of the pull-down using 2′-*O*-methylated oligonucleotides affinity matrices. A transgenic *C. elegans* strain expressing ALG-1 protein tagged at the N terminus with the green fluorescent protein (GFP) was used as described in ref. (3). An extract was generated using staged animals as starting material, a pull-down assay targeting *let-7* microRNA was performed, and the associated GFP::ALG-1 protein was detected by western blot. As a negative control, we used an unrelated 2′-*O*-methylated oligonucleotide that does not share base complementarity with any known microRNA. To confirm *let-7* microRNA depletion, we quantified *let-7* in the

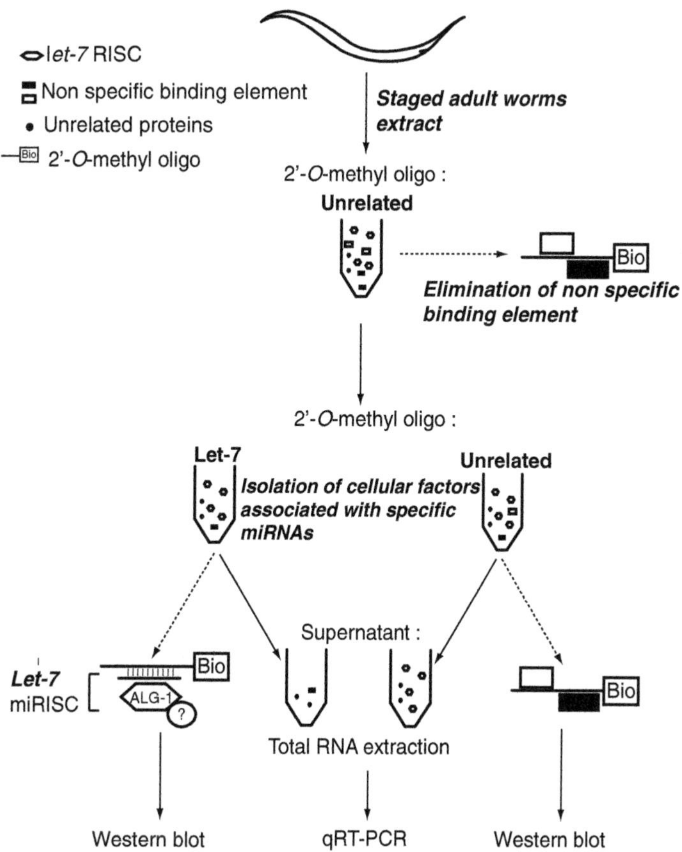

Fig. 3. Experimental procedure. This scheme provides an overview of the technical steps in this method (1) Growing a population of staged *Caenorhabditis elegans*; (2) Extraction of animal proteins; (3) Pull-down of the *let-7* microRNA-associated RISC; (4) Detection of ALG-1 protein by western blot; (5) Analysis of the efficiency of this method by quantification of the level of *let-7* miRNA by qRT-PCR.

recovered unbound fractions by quantitative real-time PCR (qRT-PCR). We also verified the 5.8S and 5S RNA integrity by EtBr staining from total RNA preparations. Figure 3. summarizes the experimental procedure followed.

3.1. Preparation of Agar NGM Plate Seeded with Concentrated OP50

1. Pour 20 mL of NGM medium in 15 cm diameter Petri dishes aseptically, cover with the lids, and let dry overnight.
2. Inoculate one colony of OP50 bacteria into 1 L of TB medium. Incubate 16 h at 37°C under agitation.
3. Split the overnight culture in two sterile centrifuge bottles (700 mL) and spin for 30 min at 3,500 × *g*.
4. Resuspend the pellet with 20 mL of M9 and transfer into 50 mL Falcon tube.

5. Wash three times with sterile M9 and spin for 10 min at 4,500 × *g*.
6. Resuspend the pellet in four volumes of M9. Vortex vigorously to obtain a homogenized OP50 solution.
7. Dispense uniformly 2 mL of concentrated food on each plate, and let them dry in a sterile field with the lid open.
8. Store the seeded plates at 4°C for up to for several days.

3.2. Preparation of a Synchronous *C. elegans* population

All the following centrifugation steps are done at 2,800 × *g* for 30 s. Manipulation must be done in a sterile field (flame).

1. Transfer starved animals from three 35 mm plates to one 150 mm plate. Once population reach adult stage (hermaphrodites bearing fertilized eggs), transfer gravid adult population of *C. elegans* in a 15 mL sterile Falcon tube using the M9 solution.
2. Wash three times by successively transferring 10 mL of M9 and centrifuging to pellet the animals.
3. Remove the supernatants completely and add 10 mL of freshly prepared bleaching solution.
4. Shake vigorously for 5 min and monitor the lysis of worm cuticles by observing the content of the tube under a stereomicroscope.
5. Once half of the worm cuticles are destroyed, centrifuge, remove the supernatant, and add 10 mL of fresh bleaching solution.
6. Shake vigorously. After an additional 1–2 min, most of the animals should be dissolved, leaving only the embryos (eggs) (see Note 12).
7. Centrifuge, remove the supernatant, and wash the eggs three times with 10 mL of M9.
8. Let the eggs hatch in 10 mL of M9 at 20°C under gentle rotation overnight.
9. On the next day, estimate the number of newly hatched larvae (L1) by counting under a stereomicroscope the amount of animals in 1 μL aliquot of homogenous solution (triplicate). Centrifuge the L1 animals, wash, and resuspend them to obtain approximately 300 animals/μL in M9. Dispense 500 μL (around 150,000 animals total) on a large OP50-seeded plate, and grow the animals at 20°C up to the young adult stage (approximately 50 h).
10. Wash the young adults from the plate with sterile M9 and transfer the animals to a 15 mL Falcon tube.
11. Centrifuge and wash three times with M9. Resuspend in 10 mL of M9, and let the suspension rotate gently for 1 h to

completely eliminate the bacteria from the gut of the animals.

12. Centrifuge, eliminate the supernatant, and freeze the pellet of young adults at –80°C until use.

3.3. Preparation of the Crude C. elegans Extract

1. Thaw a pellet of 150,000 young adults at room temperature.
2. Wash the pellet three times with three volumes of complete lysis solution.
3. Add ½ volume of complete lysis solution and transfer into a Wheaton Dounce Tissue grinder.
4. Grind for around 7 min on ice and monitor the lysis efficiency under the microscope by transferring 2 μL of extract on a glass slide. Continue until the worm cuticles are destroyed.
5. Transfer the lysate in a 1.5 mL tube and spin at 17,000 × *g* for 15 min at 4°C.
6. Transfer the supernatant to a fresh 1.5 mL tube on ice. This is the crude *C. elegans* extract.
7. Determine the protein concentration of the crude *C. elegans* extract using one of the standard colorimetric assays, such as Lowry, Biuret, or Bradford (about 10 mg of total protein should be obtained from 150,000 young adults animals).

3.4. Let-7 MicroRNA Pull-Down and RISC Analysis Using 2′-O-Methylated Oligonucleotide Affinity Matrices (see Note 13)

1. Resuspend the magnetic beads from the stock tube by pipetting up and down 20 times.
2. Transfer 60 μL of homogenized bead suspension in a 1.5 mL tube, place the tube into the magnet for 1 min, and discard the supernatant (see Note 14). The beads are washed twice with four volumes of lysis solution containing 2% [v/v] RNase inhibitor.
3. Add 120 μL of the biotinylated unrelated 2 -*O*-methyl oligonucleotide at 1 μM concentration. Incubate for 45 min at RT with gentle rotation (see Note 15). Place the tube in a magnet for 1 min and discard the supernatant.
4. Wash the coated beads twice in stock lysis solution with 2% [v/v] RNase inhibitor.
5. Place the tube in the magnet for 1 min and discard the supernatant.
6. Dilute the quantified crude *C. elegans* extract to a working solution of 13.3 mg/mL.
7. Add 4 mg of crude *C. elegans* extract (300 μL of 13.3 mg/mL suspension) in a 1.5 mL tube containing the unrelated (control) 2′-*O*-methylated oligonucleotide bound to the beads.

8. Incubate at RT with gentle rotation for 45 min. During this incubation, prepare the *let-7*-complementary oligo, and the unrelated matrices (see Note 16).
9. Transfer 30 μL of each bead suspensions into two separate 1.5 mL tubes. Wash the beads and coat with 60 μL of either the unrelated or the *let-7* complementary 2′-*O*-methylated oligonucleotides for 45 min at RT, as described in steps 2–5. Keep the coated beads on ice until used. Discard the buffer just before starting step 10.
10. Once the incubation time of the extract is complete (step 8), place the tube on the magnet for 1–2 min and transfer half of the cleared supernatant either to the unrelated or to the *let-7* coated beads (step 9). Incubate at room temperature with gentle rotation for 45 min (see Note 17).
11. Place the tubes on the magnet for 1–2 min and then transfer the supernatants to new 1.5 mL tubes. Keep the supernatants on ice until total RNA extraction and qRT-PCR analysis.
12. Wash the beads twice with three volumes of stock lysis solution containing 2% [v/v] of RNase inhibitor.
13. Resuspend the beads in Laemmli 2× loading buffer and boil them 10 min before loading on 8% SDS–Polyacrylamide gel (SDS–PAGE) electrophoresis (see Subheading 3.5). As input control, load 100 μg of total proteins from the crude *C. elegans* extract (resuspended and boiled in Laemmli 2× loading buffer).

3.5. SDS–PAGE (8%) Electrophoresis

1. The SDS–PAGE electrophoresis is performed with the "Mini-PROTEAN Tetra Cell" system (Bio-Rad). Ensure that the glass plates are clean. Before use, it is recommended to wash them with detergent first, then with ethanol 95%, and finally to let them dry completely.
2. Prepare a 1.5 mm thick, 8% gel by mixing 2.5 mL of 4× resolving buffer with 2 mL of 40% acrylamide/bis-acrylamide solution, 5 mL of water, 100 μL of APS 10%, and 10 μL of TEMED. Pour the gel while leaving space for the stacking gel (around 1–2 cm) and overlay with 95% ethanol. Let the gel polymerize for 30 min.
3. Discard the layer of ethanol 95% on the gel.
4. Prepare a 6% stacking gel by mixing 1.25 mL of 4× stacking buffer with 750 μL of 40% acrylamide/bis-acrylamide solution, 2.9 mL of water, 50 μL of APS 10%, and 5 μL of TEMED. Pour the stacking gel, and insert the comb. Let the stacking gel polymerize for 30 min.
5. Prepare the running buffer by dissolving 100 mL of 10× stock solution with 900 mL of water.

6. Once the stacking gel is polymerized, transfer the gel assembly in the tank and fill with running buffer.
7. Remove the comb carefully and wash the wells with running buffer using a 10 mL syringe fitted with a 22-gauge needle.
8. Boil your samples for 1 min, spin 30 s at 17,000 × *g*, and load the total sample in the wells. Load one well with a prestained molecular weight marker (see Note 18).
9. Connect the gel chamber to a power supply and run at 160 V for 1.5 h, or until the blue dye front runs off the gel.

3.6. Western Blotting for ALG-1 Protein

1. The samples that are resolved by SDS–PAGE are transferred using a semidry transfer apparatus (Trans-Blot SD semidry Transfer cell, Bio-Rad).
2. Prepare working transfer buffer solution (setup buffer) by mixing 200 mL of transfer buffer with 50 mL of methanol.
3. Carefully remove the gel from the glass plate assembly and put the gel in a clean dish containing 50 mL of setup buffer.
4. Cut a sheet of nitrocellulose membrane and two blotting pads of slightly larger dimensions than the size of the separating gel, and soak in the setup buffer.
5. Prepare a sandwich assembly with one wet blotting pad covered with the nitrocellulose membrane. Carefully transfer the resolving gel on the top of the membrane and cover with a second soaked blotting pad. Ensure that no bubbles are trapped in the resulting sandwich by gently rolling a 15 mL Falcon tube on the top of the assembly.
6. Connect to a power supply and run at 15 V for 50 min.
7. Once the transfer is complete, disassemble the sandwich and wash the nitrocellulose membrane with TBS-T.
8. Incubate the membrane in 25 mL of TBS-T containing 5% (w/v) dry milk for 1 h at RT with gentle agitation.
9. Discard the blocking buffer, and incubate the membrane with a 1:3,000 dilution of the anti-GFP antibody in TBS-T containing 5% [w/v] dry milk overnight at 4°C on a rocking platform.
10. Remove the primary antibody solution and wash the membrane three times with TBS-T for 10 min each.
11. The secondary antibody, an HRP-coupled anti-mouse, is diluted at 1:10,000 in TBS-T and added to the membrane for 45 min at room temperature on a rocking platform.
12. Remove the secondary antibody, and wash the membrane three times for 10 min each with TBS-T.
13. Mix equal volumes of ECL reagent (2 mL each) and keep the mixture at room temperature until used.

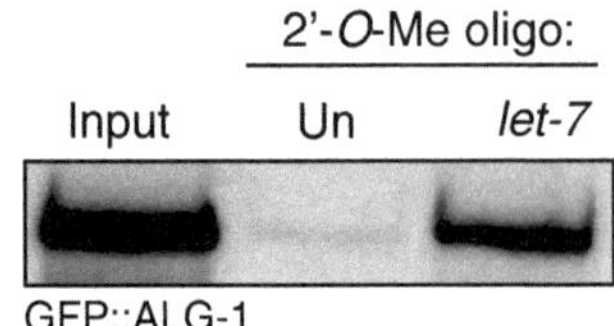

Fig. 4. Western blot analysis of the GFP-tagged ALG-1 protein associated with *let-7* microRNA. The input represents 100 μg of total protein. Each matrix [unrelated (Un) or *let-7* biotinylated 2′-*O*-methylated oligonucleotides] was mixed with 2 mg of protein and pull-down was conducted using Dynabeads coupled to streptavidin. Notice that ALG-1 is detected only in the *let-7* complementary 2′-*O*-methylated oligonucleotide pulldowns.

14. Cut out two clean acetate sheets with dimensions just larger than the size of the nitrocellulose membrane.
15. Fix a laboratory film (Parafilm) on the working bench and place the 4 mL of mixed ECL reagent on the top.
16. Transfer the membrane protein up side down on the ECL drop and incubate for 1 min.
17. Put the membrane side up on the clean acetate sheet and place the other acetate sheet over membrane and smooth out air bubbles.
18. Place the membrane between acetate sheets protector (as in a sandwich), into an autoradiography cassette and head to the dark room with X-ray films for developing.
19. In the dark room, place the film in an autoradiography cassette for a suitable exposure time (1–2 min for GFP::ALG-1 in our experiment). An example of the result is shown in Fig. 4.

3.7. Total RNA Extraction

1. Add three volumes of TRI-Reagent in 1.5 mL tubes containing either 2 mg of protein from the input crude *C. elegans* extract or the supernatant (unbound) samples recovered from the unrelated and the *let-7* matrices. Vortex for 30 s.
2. Add one volume of chloroform, vortex for 10 s and allow to sit at RT for 3 min.
3. Centrifuge the tubes at 13,000 × *g* for 15 min at 4°C.
4. Remove the top aqueous layer and transfer to a fresh 1.5 mL RNase-free tube (do not disturb or retrieve the interphase).
5. Add an equal volume of isopropanol, mix well, and incubate at RT for 10 min.
6. Centrifuge at 13,000 × *g* for 15 min at 4°C.
7. Discard the supernatant and wash the RNA pellet carefully with 300 μL of 75% ethanol.
8. Spin at 7,500 × *g* for 5 min at 4°C.

9. Discard the supernatant, air dry the RNA pellet with the open cap at RT for 5–10 min.
10. Dissolve the pellet in deionized, autoclaved water (around 100–200 μL). To dissolve the pellet, heat at 65°C for 10 min and homogenize the RNA solution by pipetting up and down.
11. The concentration of the purified RNA is determined by measuring the absorbance at 260 nm in a spectrophotometer.
12. Monitor the short RNA quality by running 15 μg of total RNA on 12% mini-polyacrylamide gel electrophoresis (see Subheading 3.8).

3.8. Preparation of 12% Polyacrylamide Gel Electrophoresis

1. The 12% mini-PAGE is performed with the same clean material used for SDS–PAGE assembly.
2. Prepare a 1.5 mm thick, 12% mini-PAGE gel by mixing 4.8 mL of SequaGel concentrate, 4.2 mL of SequaGel Diluent, 1 mL of SequaGel buffer, 80 μL of APS 10%, and 4 μL of TEMED. Pour the gel and insert the comb. Let the gel polymerize for 1 h.
3. Carefully remove the comb and pre-run the gel for 30 min at 150 V in 0.5× TBE before loading the samples.
4. While pre-running, prepare the RNA samples. Add ½ volume of formamide-dye loading buffer with 15 μg of total RNA.
5. Heat the RNA samples at 65°C for 15 min.
6. Once the pre-run is complete, wash the wells with 0.5× TBE buffer using a 10-mL syringe fitted with a 22-gauge needle.
7. Load the warm samples and run at 150 V until the bromophenol blue dye is at the bottom.
8. After completion of the run, allow the plates to cool 10–15 min before separation.
9. Carefully remove the gel from the glass plate assembly and put it in clean dish containing 50 mL of 5 μg/μL EtBr diluted in 0.5× TBE. Incubate for 15 min with gentle shaking.
10. Discard the EtBr solution and wash the membrane with 0.5× TBE.
11. Take a picture of the stained gel with a UV transilluminator to record EtBr-stained RNA fluorescence. An example of a typical experimental result is shown in Fig. 5.

3.9. Quantitative Real-Time PCR for Let-7 MicroRNA

1. Dilute the total RNA samples in water to obtain a 2 ng/μL solution.
2. Perform reverse transcription reaction to convert specific microRNAs to complementary DNA with the TaqMan MicroRNA Reverse Transcription (RT) kit, using either *let-7*

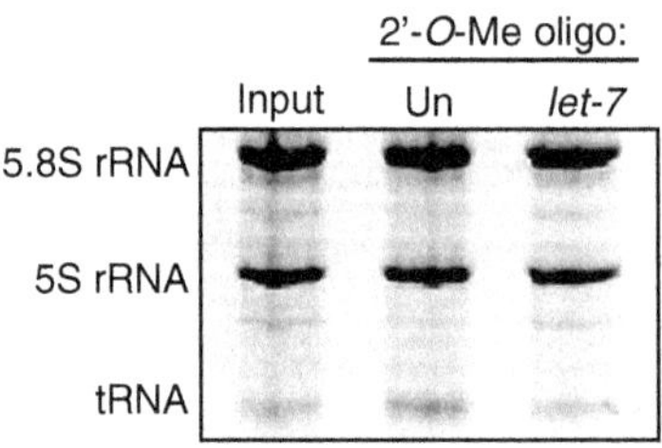

Fig. 5. Assessment of the integrity of the 5.8S and 5S rRNAs in the unbound fractions recovered from either the unrelated (Un) or *let-7* complementary matrices. *Discrete bands* are visible, demonstrating the integrity of the short RNA preparations.

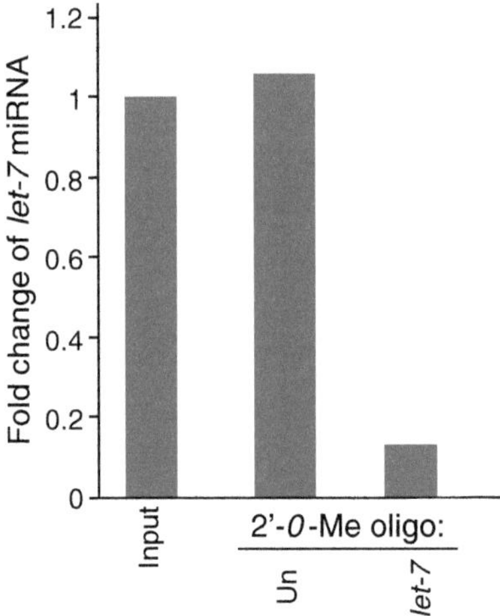

Fig. 6. qRT-PCR analysis of the *let-7* miRNA remaining in the unbound fractions recovered from either the unrelated (Un) or the *let-7* complementary matrices. When using the *let-7* complementary matrix, almost 90% of the mature *let-7* miRNA is depleted from the starting material. This demonstrates the efficiency and the specificity of the affinity chromatography.

microRNA-specific primer or the *sn2841* short RNA primer as an endogenous control. We use 10 ng of total RNA per reaction.

3. Perform Quantitative RT-PCR analyses of *let-7*, and *sn284* with TaqMan microRNA Assays kit following manufacturer procedures.
4. Data analysis was performed using the comparative C_t method using the endogenous control to normalize the level of microRNA.
5. A typical experimental result is shown in Fig. 6.

4. Notes

1. Be aware that sodium hypochlorite solution becomes less effective with time. Volume can be changed from 2 to 5 mL to obtain a complete lysis of *C. elegans* cuticules with an older solution.
2. Biotinylated 2′-*O*-methylated oligonucleotides can be designed to pull-down any small RNA-specific complex. The nucleotide sequences must be fully complementary to the chosen small RNA, and five extra nucleotides are added on each side of the complementary region to increase the stability of the small RNA complex-oligonucleotide association and the efficiency of the pull-down.
3. Ensure that the RNA molecule has been purified by HPLC techniques.
4. Upon reception of the synthesized modified RNA oligonucleotides, aliquot at 100 μM in deionized, freshly autoclaved water (use RNase-free tubes). Store aliquots at –20°C.
5. 2′-*O*-methylated oligonucleotides are stable in water solution at 4°C for up to 2 weeks, and stable for at least 6 months if stored at –20°C. Dried oligos stored at –20°C in a nuclease-free environment should be stable for several years.
6. It is very important to store the vial containing beads upright to keep them in liquid suspension. The performance will decrease if the beads air dry. Do not freeze.
7. Dynabeads are not supplied in RNase-free solutions, it is very important to use buffer containing nuclease inhibitors during the binding and washing steps.
8. Running buffer can be used for up to two runs, if it is cooled down at room temperature between successive runs.
9. The same primary antibody diluted in TBS-T and supplemented with dry milk can be reused two or three times. This solution can be stored at 4°C for several days or at –20°C for 2–3 weeks.
10. As urea may precipitate if these solutions are refrigerated, the solution should be stored in a tightly capped container and in a dark area at room temperature. However, the urea can be rapidly redissolved when warmed to room temperature.
11. For this protocol manuscript, we used the nematode *C. elegans* as a biological sample. The RISC pull-down assay using 2′-*O*-methylated oligonucleotides as affinity matrices can easily be applied to any other biological samples. The binding of RNA molecules to the beads should be optimized for a lysis buffer adequate for the specific biological sample used.

12. Do not allow the reaction to proceed for longer than 3 min as this will result in death of the embryos.
13. For better results and to avoid the microRNA/RISC/protein degradation, the pull-down assay should be performed (if possible) directly after the crude *C. elegans* extract is prepared.
14. We used 30 μL of suspension solution containing 10 mg/mL of beads for 2 mg of protein extract.
15. The binding capacities of beads are around to 200 pmol of RNA molecule for 1 mg of beads. Thirty microliter of suspension solution corresponds to 300 μg of beads, so 60 pmol of 2′-*O*-methylated oligonucleotide molecules are necessary for the binding. This is why the oligonucleotide solution is diluted to 1 pmol/μL (1 μM final).
16. Some proteins can interact more strongly than others with the matrix. To limit nonspecific binding, you may increase the quantity of beads bound to the unrelated oligo in the first step (clearing step) or repeat this step two or three times.
17. If the cell extract is too viscous, some beads may not be separated from the suspension by the magnet. To circumvent this problem, centrifugation may be used to pellet the beads.
18. Sometimes, a small amount of beads are loaded on the gel, along with the samples. This does not prevent the proper running process.

Acknowledgments

We are grateful to the members of our laboratory for comments on this methods manuscript. Nematode strain, N2 was provided by the *Caenorhabditis* Genetics Center, which is funded by the NIH National Center for Research Resources (NCRR). Our research is funded work by the Canadian Institutes of Health Research (CIHR), the Natural Sciences and Engineering Research Council of Canada, and the Cancer Research Society. M.J.S. is a CIHR New Investigator.

References

1. Stefani, G., and Slack, F. J. (2008) Small non-coding RNAs in animal development, Nat Rev Mol Cell Biol **9**, 219–230.
2. Hutvagner, G., and Simard, M. J. (2008) Argonaute proteins: key players in RNA silencing, Nat Rev Mol Cell Biol **9**, 22–32.
3. Hutvágner, G., Simard, M. J., Mello, C. C., and Zamore, P. D. (2004) Sequence-specific inhibition of small RNA function, PLoS Biol **2**, E98.
4. Inoue, H., Hayase, Y., Imura, A., Iwai, S., Miura, K., and Ohtsuka, E. (1987) Synthesis and hybridization studies on two

complementary nona(2′-O-methyl)ribonucleotides, Nucleic Acids Res **15**, 6131–6148.

5. Fabian, M. R., Mathonnet, G., Sundermeier, T., Mathys, H., Zipprich, J. T., Svitkin, Y. V., Rivas, F., Jinek, M., Wohlschlegel, J., Doudna, J. A., Chen, C. Y., Shyu, A. B., Yates, J. R., 3rd, Hannon, G. J., Filipowicz, W., Duchaine, T. F., and Sonenberg, N. (2009) Mammalian miRNA RISC recruits CAF1 and PABP to affect PABP-dependent deadenylation, Mol Cell **35**, 868–880.
6. Aoki, K., Moriguchi, H., Yoshioka, T., Okawa, K., and Tabara, H. (2007) In vitro analyses of the production and activity of secondary small interfering RNAs in *C. elegans*, EMBO J **26**, 5007–5019.
7. Yigit, E., Batista, P. J., Bei, Y., Pang, K. M., Chen, C. C., Tolia, N. H., Joshua-Tor, L., Mitani, S., Simard, M. J., and Mello, C. C. (2006) Analysis of the *C. elegans* Argonaute family reveals that distinct Argonautes act sequentially during RNAi, Cell **127**, 747–757.
8. Mayr, C., Hemann, M. T., and Bartel, D. P. (2007) Disrupting the pairing between let-7 and Hmga2 enhances oncogenic transformation, Science **315**, 1576–1579.

Chapter 17

Cloning Argonaute-Associated Small RNAs from *Caenorhabditis elegans*

Weifeng Gu*, Julie M. Claycomb*, Pedro J. Batista, Craig C. Mello, and Darryl Conte

Abstract

Small RNA pathways fulfill a plethora of gene-regulatory functions in a variety of organisms. In the nematode worm, *Caenorhabditis elegans*, a number of endogenous small RNA pathways have been described, including the microRNA pathway, the 21U/piRNA pathway, the 26G-RNA pathways, and the 22G-RNA pathways. Argonaute proteins are key effector molecules of each pathway that, together with their small RNA cofactors regulate various processes including developmental timing, fertility, transposon silencing, and chromosome segregation. Although several of the 26 Argonautes in the worm have been studied to date, a number have yet to be fully characterized or their small RNA binding complement defined. The identification of small RNAs that copurify with an Argonaute family member is central to understanding the targets and assessing the function of that Argonaute. Here we discuss the rationale for generating reagents to immunoprecipitate Argonaute complexes and provide a cohesive protocol for the cloning and Illumina deep-sequencing of Argonaute-associated small RNAs in *C. elegans*.

Key words: Argonaute, *C. elegans*, IP cloning, small RNA, Illumina deep-sequencing, Genomics, RNAi, Silencing

1. Introduction

Argonaute proteins interact with small RNA guide sequences to direct a variety of cellular processes (1). The identification of small RNAs that copurify with individual Argonaute proteins has been instrumental to understanding the mechanisms and functions of Argonaute family members. In many cases, the expression of a particular subset of small RNAs is abolished when the Argonaute to which they bind is mutated (2–7). In some cases, however,

* These authors contributed equally to this work.

Tom C. Hobman and Thomas F. Duchaine (eds.), *Argonaute Proteins: Methods and Protocols*, Methods in Molecular Biology, vol. 725, DOI 10.1007/978-1-61779-046-1_17, © Springer Science+Business Media, LLC 2011

particular small RNAs are not fully abolished in Argonaute mutants, thus making it difficult to assess which subset of small RNAs are dependent on the Argonaute (4, 8). This difficulty can be overcome by identifying the small RNAs that copurify with an Argonaute. For example, despite a dramatic developmental defect, *csr-1(tm892)* mutants displayed a relatively mild small RNA defect (8). By cloning and deep sequencing the small RNAs that copurify with CSR-1, we were able to identify the endogenous small RNA cofactors of CSR-1. The identity of the loci targeted by these small RNAs, together with the phenotypic analysis of *csr-1(tm892)* mutants, led to a model for the function of the CSR-1 small RNA pathway.

There are at least four distinct classes of small RNAs that have been identified in *Caenorhabditis elegans*, including miRNAs, 21U-RNAs, 22G-RNAs, and 26G-RNAs (4, 8–14). Each class of small RNA exhibits distinct biochemical features that directly affect their ability to be cloned. The de novo products of RNA-dependent RNA polymerase, the 22G-RNAs are 22 nt small RNAs with a 5′ guanosine that is triphosphorylated. In order to efficiently clone the 22G-RNAs, small RNA samples must be enzymatically treated to generate a 5′ end that is compatible with the 5′ ligation reaction. The ability to efficiently clone 22G-RNAs revealed that they are the major class of small RNA in *C. elegans* and that they fall into distinct pathways, defined by the Argonautes with which they interact (4, 8). This chapter discusses rationale for generating reagents for immunoprecipitation of Argonautes (Subheading 3.1) as well as detailed protocols for harvesting *C. elegans* samples (Subheading 3.2), enrichment of small RNA from the control/input sample (Subheading 3.3), immunoprecipitation of Argonaute complexes and RNA extraction (Subheading 3.4), gel purification of 18–40 nt small RNAs (Subheading 3.5), and details for cloning small RNAs for deep-sequencing using the Illumina platform (Subheadings 3.6–3.11). The small RNA cloning protocol is broken into individual sections, including the treatment of 5′ triphosphorylated small RNAs with pyrophosphatase (Subheading 3.6), 3′-adapter ligation (Subheading 3.7), 5′-adapter ligation (Subheading 3.8), cDNA synthesis (Subheading 3.9), determination of library amplification parameters (Subheading 3.10) and final amplification, purification and quantification of small RNA amplicons (Subheading 3.11). Methods of data processing and analysis are not described, but some discussion is provided in Subheading 3.12. A suggested timeline for the procedure is provided in Fig. 1.

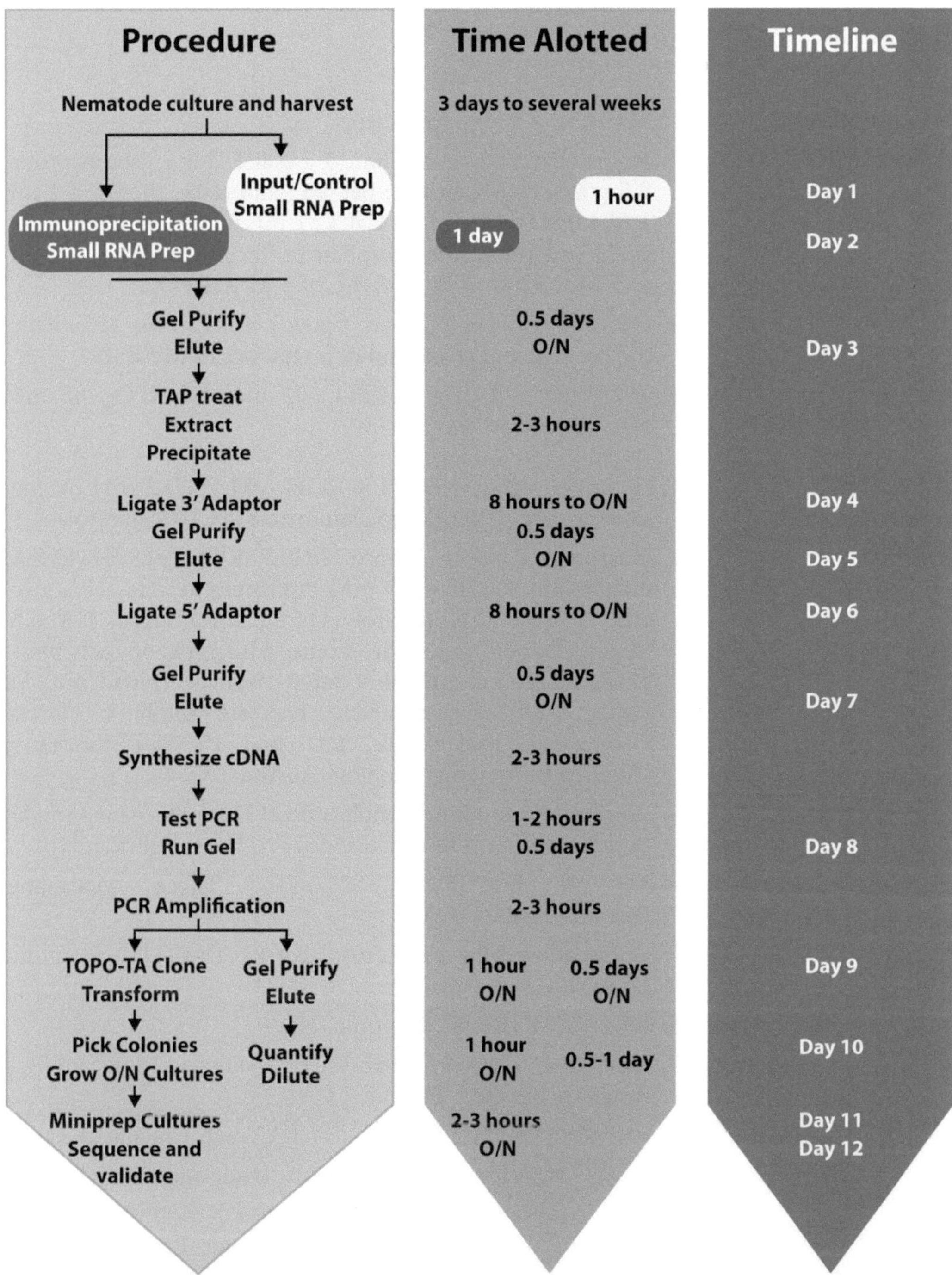

Fig. 1. Flowchart of IP/small RNA isolation and cloning procedure. The basic steps in the protocol are shown, along with the amount of time that is required to perform individual steps. In parallel, a day-by-day outline is presented, to provide a sense for the overall time-frame of the protocol.

2. Materials

2.1. Caenorhabditis elegans Culture

1. 150 mm × 15 mm Petri Plates.
2. Nematode growth medium: 3 g NaCl, 2.5 g bactopeptone, 8.5 g agar, 8.5 g agarose per 1 L; autoclave, then add 1 mL 1 M $MgSO_4$, 1 mL 1 M $CaCl_2$, 1 mL 5 mg/mL cholesterol, and 25 mL potassium phosphate buffer (PPB). PPB is 132 mL of 1 M K_2HPO_4 with 868 mL of 1 M KH_2PO_4.
3. OP50 *Escherichia coli* food: Grow 1 L OP50 for 24 h. Pellet the bacteria and resuspend in 5 volumes of M9 Buffer.
4. M9 Buffer: 22 mM KH_2PO_4, 22 mM Na_2HPO_4, 85 mM NaCl, 1 mM $MgSO_4$.

2.2. Input RNA Preparation and IP Reagents

1. IP Buffer: 30 mM HEPES–KOH (pH 7.4), 2 mM magnesium acetate, 100 mM potassium acetate, 10% glycerol.
2. Complete IP buffer: 30 mM HEPES–KOH (pH 7.4), 2 mM magnesium acetate, 100 mM potassium acetate, 10% glycerol, 2 mM dithiothreitol (DTT), 0.1% Igepal CA 630 (Sigma-Aldrich Corp., St. Louis, MO), 1% of each phosphatase inhibitor cocktails 1 and 2 (Sigma-Aldrich Corp., St. Louis, MO), 4× complete protease inhibitor (Roche Diagnostics, Indianapolis, IN), and 1% SUPERase•In™ (Applied Biosystems/Ambion, Austin, TX).
3. Wheaton Dura-Grind Stainless Steel Dounce Tissue Grinder (7 mL capacity, Wheaton Science Products, Millville, NJ).
4. 1.5 mL Low-Binding/Siliconized Tubes (Medsupply Partners, Atlanta, GA).
5. Gel Loading Tips and Aerosol Barrier Tips (USA Scientific, Ocala, FL).
6. RNase ZAP (Biohit, Neptune, NJ).
7. Bio-Rad D_C Protein Assay Kit or Quickstart Bradford Dye Reagent (Bio-Rad, Hercules, CA).
8. TRI Reagent (Molecular Research Center, Cincinnati, OH).
9. 5 mL capacity Potter-Elvehjem tissue grinder (Kimble-Kontes, Vineland, NJ).
10. Bromopropyl chloride (BCP; Molecular Research Center, Cincinnati, OH).
11. Phase Lock Gel Columns, heavy (5 Prime, Inc., Gaithersburg, MD).
12. Protein A/G PLUS Agarose Beads (Santa Cruz Biotechnology, Inc., Santa Cruz, CA).
13. mirVana™ miRNA Isolation Kit (Applied Biosystems/Ambion, Austin, TX).

2.3. Cloning Linkers, PCR Oligonucleotides, RNA Modification, and Ligation

All Oligonucleotides were purchased from IDT (Coralville, IA) and are DNA unless noted otherwise:

1. Size markers – custom 18, 24, 26, and 40 nt RNA oligonucleotides were resuspended at 100 μM in water. Dilute to 10 μM in water to make a working stock. Use 1 μL (1 pg) (see Note 16).
2. 3′-Adapter Oligo (MicroRNA cloning linker #1):
   AppCTGTAGGCACCATCAAT/ddC/
3. 5′-Adapter Oligo (DNA/RNA hybrid oligo – Ribonucleotides in *bold* and preceded by "r"):

   TCTAC**rArGrUrCrCrGrArCrGrArUrC** (+ optional 4 nt RNA barcode)
4. First round PCR oligos:

   PL1 GTTCTACAGTCCGACGATC

   PR1 ATTGATGGTGCCTACAG
5. Second round PCR oligos (*underlined* sequences correspond to adapter attachment sequences for hybridization and cluster building; *bold underlined* sequence corresponds to the Illumina sequencing primer):

   PL2 AATGATACGGCGACCACCGACAGGTTCAG AGTT**CTACAGTCCGACGATC**

   PR2 CAAGCAGAAGACGGCATACGAATTGATG GTGCCTACAG
6. Tobacco Acid Pyrophosphatase (TAP; Epicentre, Madison, WI).
7. SUPERase•In™ (Applied Biosystems/Ambion, Austin, TX).
8. 40 U/μL T4 RNA ligase (with 1 mg/mL BSA, Takara Bio, Otsu, Shiga, Japan).
9. 30% Acrylamide/bis solution, 29:1 (Bio-Rad, Hercules, CA) and *N,N,N,N′*-tetramethylethylenediamine (TEMED; Sigma-Aldrich Corp., St. Louis, MO).
10. 5× TBE running buffer: 0.45 M Tris–borate, 1 mM ethylenediamine tetraacetic acid (EDTA). Dissolve 54 g Tris base, 27.5 g boric acid, and 20 mL EDTA (pH 8.0) in 800 mL of water and adjust the volume to 1 L. Pass the solution through a 0.22 μm filter and store at room temperature. Dilute to 1× with water prior to use.
11. 15% Denaturing gel solution: 15% Acrylamide/bis and 7 M urea in 1× TBE. Prepare stock by mixing 100 mL of 30% Acrylamide/bis solution, 84.1 g urea (Sigma-Aldrich Corp., St. Louis, MO), 40 mL of 5× TBE, and 50 mL of water. Bring the volume to 200 mL with water.
12. Ammonium persulfate (APS): Prepare 10 mL of a 10% (w/v) solution in water and store at 4°C for up to 2 weeks.

13. Electrophoresis system capable of running 16 cm polyacrylamide gels, 0.8 mm thick. We use the Model V16 Polyacrylamide Gel Electrophoresis System (Whatman, Inc., Florham Park, NJ). This electrophoresis system comes with glass plates 19.7 cm × 16 cm and 19.7 cm × 18.5 cm. The 0.8 mm spacers, a 20-well comb (5 mm wells) and large binder clamps are sold separately.
14. Aluminum plate, 19 cm × 10 cm × 0.3 cm, custom made.
15. Formamide loading buffer: Gel Loading Buffer II (Applied Biosystems/Ambion, Austin, TX).
16. Ethidium bromide, 10 mg/mL stock, or Sybr Gold nucleic acid stain, 10,000× stock (Invitrogen, Carlsbad, CA).
17. UV trans-illuminator with wavelength 365 nm.
18. RNA/DNA Elution Buffer: 10 mM Tris–Cl (pH 7.5), 1 mM EDTA (pH 8.0), 3 M NaCl.
19. 0.45 μm Cellulose Acetate spin filters (Corning, Corning, NY or Pall Life Sciences, Ann Arbor, MI).
20. 5 mg/mL Glycogen (Applied Biosystems/Ambion, Austin, TX).
21. Phenol::chloroform::isoamyl alcohol (25:24:1, pH 6.7) (Fisher BioReagents, Fairlawn, NJ).

2.4. RT, PCR, and Topo Cloning Components

1. SuperScript III First Strand cDNA Synthesis Kit (Invitrogen, Carlsbad, CA).
2. Ex Taq DNA Polymerase Kit (Takara Bio, Otsu, Shiga, Japan).
3. 0.2 mL PCR 8-tube strips (USA Scientific, Ocala, FL).
4. Bio-Rad Criterion gel electrophoresis system and empty cassettes, 18-well and 26-well (Bio-Rad, Hercules, CA).
5. 6× DNA loading buffer: Bromophenol blue (0.25%), xylene cyanol (0.25%), sucrose (40%).
6. 10 bp Ladder (Invitrogen, Carlsbad, CA).
7. 100 bp Ladder (New England Biolabs, Ipswich, MA).
8. Microcon YM-10 Columns (Millipore, Billerica, MA).
9. TOPO-TA Cloning Kit (Invitrogen, Carlsbad, CA).

3. Methods

The method described here assumes some experience with molecular cloning as well as familiarity with the growth and manipulation of *C. elegans*. We will not go into details for cloning fusions but will instead provide rationale for choosing epitopes against which Argonaute-specific antibodies were raised, and for choosing sites for insertion of ectopic epitopes. Although our focus is on *C. elegans* in this chapter, many of the methods are of general use and can be applied to any system. For organisms that do not express Argonaute-associated triphosphorylated small RNAs, the cloning method can be easily modified to directly clone monophosphorylated small RNAs by omitting Subheading 3.6, sample treatment with tobacco acid pyrophosphatase (TAP).

3.1. Argonaute Antibody Generation and Epitope Tagging

1. Choice of epitope for generating antibodies.
   The epitopes for polyclonal peptide and antibodies generated against CSR-1, PRG-1, and ERGO-1 were chosen based on antigenicity and hydrophilicity predictions, which theoretically approximate the surface characteristics of a protein (4, 8, 15). Admittedly, this is an inexact science but one that has yielded reasonable success. There are several tools available for determining these characteristics. The Protein Analysis Toolbox function of the MacVector® software (Accelrys, Inc.) provides a suite of hydrophilicity and antigenicity algorithms to choose from as well as a "Surface Probability" algorithm (Fig. 2). Other molecular biology software and online tools (e.g. http://www.cbs.dtu.dk/services/NetSurfP/) with similar capabilities can be used to predict antigenicity. Alternatively, many companies that generate custom antibodies also offer peptide antigenicity prediction and synthesis services (e.g. http://www.pbcpeptide.com/Feedback.htm).

   In general, the N-terminal portion of the Argonautes is a region of greater hydrophilicity and is less highly conserved among the Argonautes, making it a useful region from which to identify peptide sequences that are both antigenic and unique to a particular Argonaute. In principle, threading of the primary amino acid sequence onto known Argonaute crystal structures could also provide clues as to exposed surfaces or flexible regions of the protein that may be useful targets for peptide antibodies (see Note 1).
2. Choice of epitope tagging site for generating transgenic *C. elegans*.
   An alternative to generating costly antibodies against the Argonautes is to generate epitope-tagged transgenic animals, although the time necessary to generate transgenic strains and validate that they are appropriately expressed and functional, for example by rescuing the mutant phenotype, may parallel that of generating antibodies (in the range of 1–3 months from cloning to rescue). Epitopes that have been successfully used include the green fluorescent protein (GFP), hemagglutinin A (HA) and Flag, for which antibodies are commercially available. An obvious potential benefit of GFP

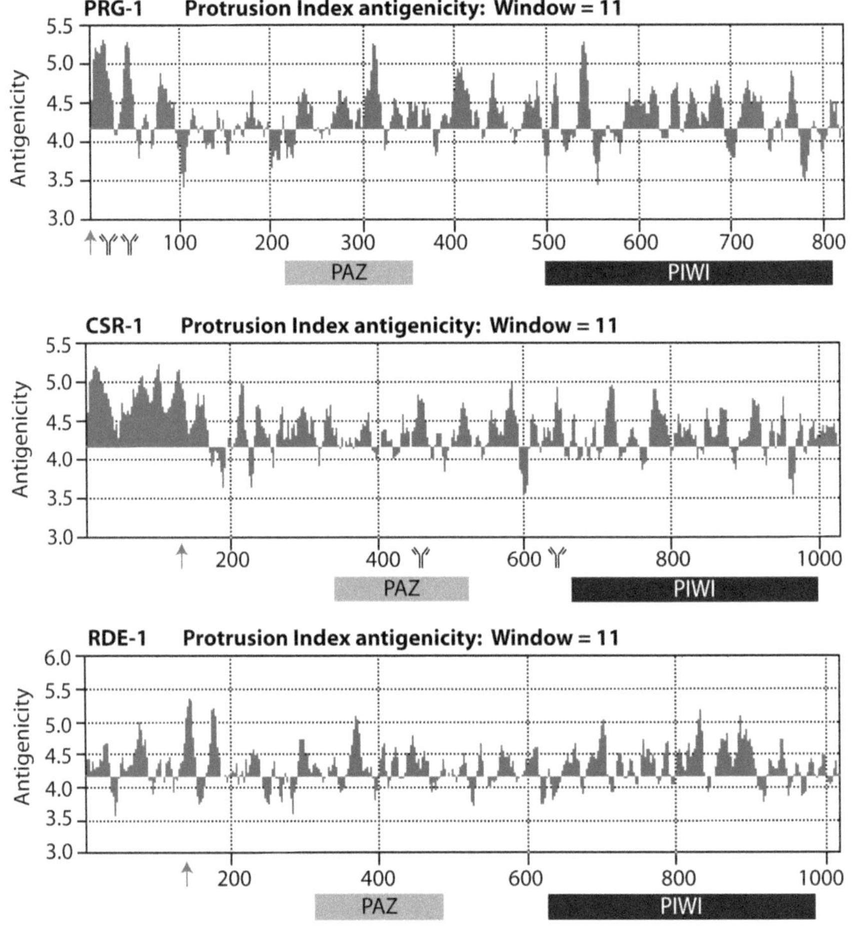

Fig. 2. Protrusion Index Antigenicity for the Argonautes PRG-1 (*top*), CSR-1 (*middle*) and RDE-1 (*bottom*). Protein sequences were analyzed using the "Protein Analysis Toolbox" function of MacVector, to determine which regions of the proteins would be useful for generating antibodies. Regions with higher antigenicity indices are likely to be exposed on the surface of the protein and useful for generating antibodies. *Arrows* indicate positions where epitope tags have successfully been used. The locations of peptides used for immunization are indicated by Y-shaped antibody cartoons. *Gray boxes* show the approximate positions of the PAZ domains. *Black boxes* show the approximate positions of the PIWI domains.

translational fusions is that localization of the Argonaute can be examined in the live animal (3, 4, 6, 8, 15–17).

Terminal tagging: Inserting an epitope at the N terminus of an Argonaute is the simplest strategy and has proven successful in many cases (3, 4, 6, 16, 17). It is essential to verify that the full-length fusion protein is expressed and functional. For example, in the case of RDE-1, an N-terminal GFP epitope appeared to be cleaved from RDE-1 in vivo (H. Tabara, personal communication). In this case, it was necessary to resort to inserting an epitope at an internal site within the protein (see below).

C-terminal epitope tagging of Argonautes is not recommended. A number of mis-sense mutations that render RDE-1 nonfunctional cluster at extreme C terminus of RDE-1 ((18); C.C. Mello, unpublished observation), suggesting that the C-terminal residues of Argonaute proteins have an important function. Indeed, the first crystal structure of an Argonaute protein revealed that the penultimate residue is buried within the PIWI domain and plays an important role in binding to the small RNA (19).

Internal tagging: Considerations for the choice of an internal site for epitope fusions include the degree of conservation, as well as the hydrophilicity or antigenicity profile. Regions of an Argonaute protein that are less conserved between family members are better targets for epitope tagging, because the most highly conserved regions correspond to the functional domains. Therefore, less conserved regions of the protein that were predicted to be hydrophilic or exposed surfaces were targeted for inserting epitope tags. This strategy has worked for small epitopes, 2XHA into RDE-1 (20) and 3XFlag into CSR-1 (8), as well as the larger GFP epitope into CSR-1 (see Note 2) (8).

3. Generation of transgenic strains: Three methods of transgenesis are widely used in *C. elegans* and described in detail elsewhere, including: traditional microinjection of plasmid based constructs to create simple (high-copy) or complex (low-copy) extrachromosomal arrays that can subsequently be integrated into the genome (21, 22); biolistic transformation to create low copy-number integrated transgenes (23); and transposon-mediated, single-copy insertion of a transgene (24). Some knowledge of the expression pattern of the Argonaute is desirable and will influence the method chosen for transgenesis (see Note 3). In the case of germline-expressed Argonautes, low- or single-copy transgene insertions are essential to achieve sufficient expression levels and to avoid robust germline silencing that is directed against highly repetitive elements such as extrachromosomal arrays.

3.2. *Caenorhabditis elegans* Growth and Harvesting

1. The protocol assumes prior knowledge of the developmental expression pattern of the Argonaute of interest (see Note 3), as well as basic knowledge of the culturing and handling of *C. elegans*. The amount of material required for an IP/cloning experiment depends in part on the stage and level of expression of the particular Argonaute and the immunoprecipitation efficiency of the antibody. These parameters must be determined empirically.
2. Assuming that the Argonaute of interest is expressed in the adult and/or embryo, cultivate synchronous populations of

worms on ten 150 mm × 15 mm plates at a density of 100,000 per plate at 15° or 20°C (see Note 4).

3. When 5–10 embryos are visible within most of the adult animals, harvest the worms by washing them from the plates with M9 buffer and transfer to 15-mL conical tubes (2–3 plates of worms per tube). Pellet the worms by centrifugation in a clinical centrifuge at 800 × *g* for 20 s. Remove the supernatant by aspiration, using a 5″ Pasteur pipet attached to a vacuum flask, being careful to avoid the worm pellet. Wash each worm pellet three or more times with 10 mL of M9 buffer. The supernatant should be clear of bacteria and most debris (see Note 5).
4. Resuspend the worm pellets in 7 mL of fresh M9 and incubate for 30 min on a rocking platform at room temperature. This step allows the worms to digest any food remaining in the gut, which could contribute to increased levels of non-worm sequences in the small RNA cloning.
5. Pellet the animals and wash twice with 10 mL of 4°C water (see Note 6).
6. After the second water wash, remove most, but not all, of the water so that the pellets are slightly loose. Transfer approximately 200 μL of pellet (about 25,000–50,000 worms or half of one plate) to a new 15-mL conical tube. This tube will be used to prepare the input/control RNA sample (see Subheading 3.3). Centrifuge the control/input tube at 800 × *g* for 20 s to pellet the worms, and remove as much of the residual water as possible without disturbing the 200 μL pellet. Place the 200 μL pellet of input/control worms on ice and add 5 volumes (1 mL) of TRI Reagent.
7. Combine the remaining three-to-five worm pellets into two tubes (2–3 mL pellet per tube). Wash each pellet with10 mL of cold IP buffer. Pellet the animals at 800 × *g* for 20 s. Remove as much buffer as possible without disturbing the pellet.
8. Snap-freeze the two 2–3 mL worm pellets in a dry-ice/ethanol bath. These pellets may be stored at –80°C until you are ready to perform the IP.
9. Immediately proceed to Subheading 3.3.

3.3. Input/Control RNA Preparation

1. Transfer the input/control TRI Reagent/worm slurry (~1.2 mL) to a glass Potter-Elvehjem tissue grinder (a Wheaton Dura-Grind Stainless Steel Dounce Tissue Grinder, 7 mL capacity, may also be used) (see Note 7). Dounce until most of the worms are broken (see Note 8).
2. Transfer the lysate to a fresh 1.5-mL microcentrifuge tube and vortex at >800 rpm for 5–10 min at room temperature (see Note 9).

3. Pellet the insoluble worm debris by centrifugation at 10,000 × *g* for 5 min at room temperature. Transfer the lysate to a new siliconized 1.5-mL tube. Note the volume of the transferred lysate.
4. Add 0.1 volume of the phase separation reagent BCP to the RNA sample, and shake for five more minutes.
5. Transfer the mixture into a prepacked Phase Lock Gel tube (see Note 10). Centrifuge at 16,000 × *g* for 4 min at 4°C to separate the aqueous and organic phases.
6. Transfer the aqueous phase (top) to a fresh siliconized tube, and add an equal volume of isopropanol. Mix the sample and centrifuge at 12,000 × *g* for 8 min at 4°C. Wash the pellet with 500 μL of ice cold 70% ethanol. Store the pellet in 70% ethanol at −20°C.
7. When ready to proceed, pellet the RNA by centrifugation at 20,000 × *g* for 10 min at 4°C. Resuspend the final pellet in 50–100 μL of 10 mM Tris (pH 7.5). Dilute 1 μL in 9 μL of water and determine the concentration using a Nanodrop spectrophotometer. Yield is generally in the range of 2–5 μg/μL.
8. Next enrich the small RNA fraction from the total RNA sample using a modification of Ambion's mirVana™ small RNA isolation protocol. Place 200 μg of total RNA in a new siliconized 1.5-mL microcentrifuge tube and adjust the volume to 80 μL using 10 mM Tris (pH 7.5) (see Note 11).
9. To 80 μL of sample, add 400 μL (5 volumes) of mirVana™ 5× Lysis/Binding Buffer and mix.
10. Add 48 μL (0.1 volume) of miRNA Homogenate Additive solution, mix and incubate the sample at room temperature for 5 min.
11. Add 176 μL (1/3 volume) of 100% ethanol, mix and centrifuge the sample at 2,500 × *g* for 4 min to pellet large RNAs (>200 nt).
12. Transfer the supernatant to a fresh 1.5-mL microcentrifuge tube and fill with isopropanol. Place the sample at −20°C for 30 min or −80°C until frozen.
13. Pellet the small RNA (<200 nt) at 20,000 × *g* for 10 min and wash once with 70% ethanol. If convenient, pellets may be stored in 70% ethanol at −20°C.
14. Remove the 70% ethanol, dry the small RNA pellet and resuspend in 25 μL of 10 mM Tris, pH 7.5. Determine the yield by quantifying 1.5 μL on a Nanodrop spectrophotometer.
15. When ready to gel purify 18–40 nt RNAs (Subheading 3.5), prepare sample for electrophoresis by mixing up to 40 μg of enriched small RNA with 20 μL of formamide loading buffer.

Evaporate the sample to 20 μL using a SpeedVac at the lowest heat setting or with the heat turned off. The sample can be split between two wells (10 μL per well) of the gel for fractionation and purification of 18–40 nt small RNAs (see Note 12).

3.4. Argonaute Immuno-precipitation and RNA Purification

1. To proceed with immunoprecipitation, add an equal volume (~5 mL) of Complete IP buffer to frozen worm pellet (Subheading 3.2, step 7) and thaw the pellet on ice.
2. Transfer the slurry to a Wheaton Dura-Grind Stainless Steel Dounce Tissue Grinder (7 mL capacity) and dounce (2–3 mL of worm slurry at a time) until the worms are mostly destroyed and only portions of the cuticle remain. After douncing, pool the lysates and distribute equally into siliconized 1.5-mL microcentrifuge tubes.
3. Centrifuge the lysates at 13,000 × g for 10 min at 4°C to pellet worm debris. Transfer the supernatant to a new tube, avoiding the pellet and viscous layer of lipids just above the pellet. Determine the concentration of protein in the lysate using a Lowry or Bradford protein quantification assay. Generally, the concentration of the lysates are between 15 and 20 mg/mL.
4. Split the lysate into approximately twenty 5 mg aliquots. Dilute each aliquot of lysate with buffer to a total volume of 1 mL (see Note 13).
5. Add 25 μL of a 50% slurry of equilibrated protein A/G agarose beads to each aliquot and clear the lysates for 1 h at 4°C.
6. Pellet the protein A/G beads by centrifugation at 800 × g for 10 s. Transfer the supernatants to fresh tubes.
7. Add an appropriate, predetermined amount of antibody to each aliquot of lysate and incubate for 1–2 h at 4°C on a rocking platform or tumbler (see Note 14).
8. Add 50 μL of a 50% slurry of protein A/G agarose beads equilibrated with IP buffer to each IP reaction and incubate the beads with the lysate for 1 h at 4°C.
9. Pellet the beads by centrifugation at 800 × g for 10 s. Allow the beads to settle further by placing the tubes on ice for 1 min. Remove the supernatants and freeze an aliquot for analysis of the unbound fraction, if desired.
10. Wash the beads 6 times with 1 mL of IP buffer for 7 min per wash. Two additional washes should be performed with IP buffer lacking detergent, for 7 min each. Pellet the protein A/G immune complexes after each wash, as above.
11. After the final wash, combine the beads from all tubes. Rinse the tubes with 200 μL IP buffer several times to recover as many beads as possible. Pellet the pooled beads and remove the supernatant. Add 1 volume of fresh IP buffer to the beads.

12. Add 1 volume (equal to the bead + buffer volume) of TRI Reagent to the beads and vortex or shake the beads at >1,000 rpm on an orbital shaker for 15 min at room temperature.
13. Add 0.1 volume of BCP to the sample and shake for an additional 5 min.
14. Transfer the sample to prepacked phase lock gel tubes and spin at 16,000 × *g* for 4 min at 4°C. Transfer the aqueous phase to a new siliconized 1.5-mL tube. Add 20 μg of glycogen and 1.5 volumes of isopropanol to the tube. Invert to mix and precipitate at −20°C for 30 min or at −80°C until frozen. Pellet the Ago-associated RNA by centrifugation at 20,000 × *g* for 20 min and wash with ice cold 70% ethanol. If convenient, the samples can be stored in 70% ethanol at −20°C.
15. Just prior to running the samples on the size fractionation gel, centrifuge the sample at 16,000 × *g* for 1 min at room temperature, remove the 70% ethanol, and dry the small RNA pellet. Resuspend the pellet in 10 μL of formamide loading buffer. The sample is ready for gel purification of 18–40 nt RNAs.

3.5. Gel Purification of 18–40 nt RNA

1. Prepare a 0.8-mm thick, 15% polyacrylamide/7 M urea gel by mixing 25 mL of 15% denaturing gel solution, 250 μL of 10% APS, and 25 μL of TEMED. Pour the gel and insert the 20-well comb (5 mm wells). Allow gel to polymerize for 30 min (see Note 15).
2. Prepare samples and 18 and 40 nt RNA oligonucleotide size markers for loading by adding 10 μL of formamide loading buffer and concentrate in a SpeedVac until only 10 μL is remaining (see Note 16). Between 1 and 3 pg of each oligonucleotide size marker should be used per lane. In addition to the 18 and 40 nt markers, 24 and 26 nt markers may be used.
3. Once the gel has polymerized, assemble the gel, aluminum plate, and electrophoresis unit using four large binder clamps. Align the long edge of the aluminum plate with the bottom of the wells and use two clamps to attach the aluminum plate and gel to the upper reservoir of the apparatus. Use the two remaining clamps to attach the very top of the gel to the upper reservoir.
4. Fill the upper and lower reservoirs with 1× TBE running buffer and ensure that the upper reservoir is not leaking. Remove the comb and excess polymerized gel that may interfere with loading. Rinse the urea from the wells using a syringe loaded with 1× TBE and carefully load the samples and size markers into the bottom of the appropriate wells using a gel-loading tip (see Notes 17–19).
5. Run the gel at 15 W until the Bromophenol blue (leading dye) migrates ~10 cm into the gel (see Note 20).

6. At the end of the run, disassemble the gel and stain in 1× TBE containing 50 μg/mL of ethidium bromide for 5 min (see Note 21). Destain the gel in water or 1× TBE for 5 min to enhance visualization by trans-illumination. While the gel is destaining, clean the trans-illuminator with RNase ZAP. Photograph the gel at wavelength 365 nm.
7. Using the size markers as a guide, excise the small RNAs from just below the 18 nt to just below the 40 nt marker for each IP and input sample at UV 365 nm (see Notes 22 and 23).
8. Place each gel slice into a 1.5-mL siliconized microcentrifuge tube, crush the gel using a 1-mL micropipette tip (see Note 24) and add 750 μL (at least two gel volumes) of RNA/DNA Elution Buffer to the crushed gel slice. Close the tube and wrap the top with parafilm. Elute the samples overnight at room temperature with agitation (see Notes 25 and 26).
9. Briefly pellet the gel slice in a microcentrifuge and transfer the eluate containing the small RNA sample to a 0.4 μm cellulose acetate spin filter (see Note 27). Minimize transfer of gel fragments to avoid clogging the spin filter. Filter the sample through the spin column by centrifugation at 10,000 × *g* for 1 min or as long as necessary and transfer the eluate to a new 1.5-mL siliconized microcentrifuge tube.
10. Add 10 μg of glycogen and 750 μL of isopropanol to each tube containing ~700 μL of filtered eluate (see Note 28). Mix the sample and store at –20°C for at least 30 min or –80°C until frozen, to precipitate the small RNAs.
11. Pellet the small RNA precipitate by centrifugation at 16,000 × *g* for 10 min at 4°C. Carefully remove most of the isopropanol using a 1-mL pipet tip. Add 500 μL of ice cold 70% ethanol and invert the tube several times to wash the pellet (see Note 29). Centrifuge at 16,000 × *g* for 1 min at room temperature and aspirate all but 10–20 μL of the 70% ethanol using a 1-mL pipet tip. Briefly spin the sample again and remove the remaining 70% ethanol using a fine micropipet tip (see Note 30). The pellet will dry quickly.

3.6. Sample Treatment

1. A large fraction of endogenous small RNAs in *C. elegans* are triphosphorylated on the 5′ residue (4, 9, 13). It is necessary to treat the small RNAs enzymatically to obtain 5′ monophosphate small RNAs, to which a 5′ linker can be ligated. This can be done by either sequential dephosphorylation and phosphorylation steps (CIP-PNK method) or by treatment with tobacco acid pyrophosphatase (TAP method). For IP-cloning procedures, we prefer the TAP method over the CIP-PNK method because much fewer degradation products of tRNA, rRNA, and mRNA are cloned using the TAP method and the TAP method requires one less step than the CIP-PNK method (see Notes 31 and 32).

2. Dry the pellet completely before proceeding with TAP reaction.
3. Resuspend the small RNA pellet with the following mixture:

Water	8.5 μL
10× TAP Buffer	1 μL
20 U/μL SUPERase•In™	0.5 μL
10 U/μL TAP	0.1–0.5 μL

4. Mix and incubate reactions at 37°C for 1 h.
5. Dilute the reaction to 100 μL with TE (pH 7.5) and transfer to a phase-lock column. Add 100 μL of phenol::chloroform:: isoamyl alcohol, mix by inversion and centrifuge at 16,000 × *g* for 5 min. Repeat the extraction in the same tube with an additional 100 μL of phenol::chloroform::isoamyl alcohol.
6. Transfer the aqueous (top) phase to a siliconized 1.5-mL microcentrifuge tube. Add 10 μL of 3 M sodium acetate (pH 5.2) and 440 μL of 100% ethanol and precipitate the TAP treated small RNA at −20°C for 30 min (see Note 33). Pellet the small RNA precipitate by centrifugation at 16,000 × *g* for 20 min at 4°C. Carefully aspirate the ethanol supernatant and wash the pellet with 500 μL of cold 70% ethanol.

3.7. Ligation of 3′ Adapter

1. Prepare 50 μL of 10× 3′ Ligation Buffer (store at −20°C in 10 μL aliquots):

1 M Tris–Cl (pH 7.5)	25 μL	(0.5 M final)
1 M $MgCl_2$	5 μL	(0.1 M final)
1 M DTT	5 μL	(0.1 M final)
Water	15 μL	

2. Prepare 10 μL of fresh reaction mix for each *n* + 1 sample (see Note 34).

10× 3′ Ligation Buffer	1 μL
20 U/μL SUPERase•In™	0.5 μL
1 mg/mL BSA	1 μL
100 μM 3′-adapter Oligo	0.5 μL
100 μM DMSO	1 μL
40 U/μL T4 RNA Ligase	0.5 μL
Water	5.5 μL

3. Centrifuge the TAP treated RNA pellet, remove the 70% ethanol and dry the pellet completely. Resuspend the RNA pellets in 10 μL of freshly prepared reaction mix and transfer to 0.2-mL

PCR tubes. Incubate at 15°C for 2 h, followed by 4°C overnight in a thermal cycler (see Note 35).

4. Add 5 μL of formamide loading buffer and fractionate the small RNA::3′-adapter ligation products on a 15% polyacrylamide/7 M urea gel as described in Subheading 3.5. This time, run the gel until the Bromophenol blue reaches the bottom of the gel in order to resolve the ligation products. An example is shown in Fig. 3a (see Note 36).
5. Excise, elute, and precipitate the small RNA::3′-adapter products as described in Subheading 3.5 (see Note 37).
6. Wash the precipitated ligation products with ice cold 70% ethanol as described in Subheading 3.5.
7. Remove the 70% ethanol before proceeding with the 5′ ligation.

3.8. Ligation of 5′ Adapter

1. Prepare 10 μL of fresh reaction mix for each $n+1$ sample:

10× Ligation Buffer	1 μL
20 U/μL SUPERase•In™	0.5 μL
1 mg/mL BSA	1 μL
200 μM 5′-adapter Oligo	1 μL
DMSO	1 μL
40 U/μL T4 RNA Ligase	0.5 μL
Water	5 μL

Fig. 3. Examples of steps required for the production of small RNA libraries. (**a**) PAGE of 3′-adaptor ligation: 26, 24, and 18 nt size markers and two small RNA samples were ligated to the 3′-adapter Oligo (+3′ linker lanes). Oligonucleotide size markers without ligation (26, 24, and 18 nt) are shown for comparison. (**b**) PAGE of 5′ Adaptor ligation: The samples and markers ligated in (**a**) were ligated to the 5′-adapter Oligo (+5′ linker lanes). Note that the 5′ ligation is not as efficient as the 3′ ligation, as a population of unligated molecules persists in each sample (+3′ linker). Free or unligated 5′ linker migrate more quickly and are indicated at the bottom of the gel (5′ linker). (**c**) Determination of appropriate PCR cycle number: Two cDNA libraries were amplified by two rounds of PCR and samples were removed at cycles 2, 4, 6, and 8 of the second round of PCR (cycle number, *below lanes*). Samples were run and analyzed by native PAGE along with a 10-bp marker. Suitable amplicons migrate at about 110 bp (+RT, *white arrowhead*), and the positions of spurious products (short oligo products and heteroduplexes, *boxed*) and primers are also indicated. As a control, amplification of –RT samples is used and results in no product. In determining the appropriate number of PCR cycles for a particular sample, it is important to choose the number of PCR cycles that achieve the maximum amplification of the desired product while minimizing the accumulation of heteroduplex molecules. In this example, the first round of PCR was 6 cycles. For both samples, 4 cycles of PCR in the second round of amplification would be ideal. (**d**) PAGE and quantification of PCR amplified libraries: After scaling-up the PCRs and gel purifying the small RNA amplicons, the concentration of each library is measured by Nanodrop spectrophotometry. The libraries were diluted to 10 nM and increasing amounts (indicated below each lane) were compared to known quantities (1–6 ng) of a 100-bp marker to verify the concentration. In this example, a 90-bp "Control" library of known quantity was compared to Sample I.

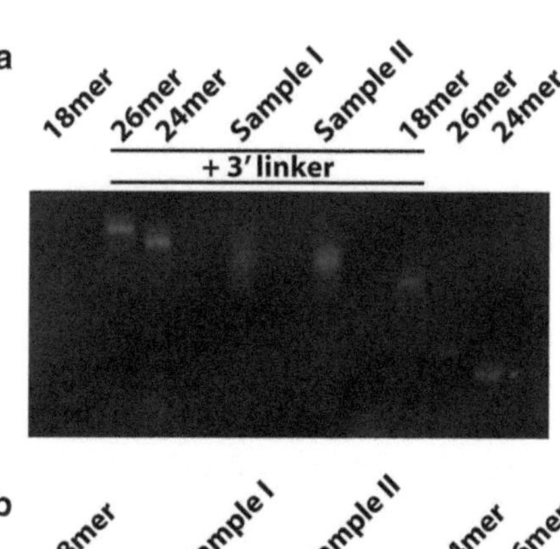
a
18mer
26mer
24mer
Sample I
Sample II
18mer
26mer
24mer
+ 3′ linker

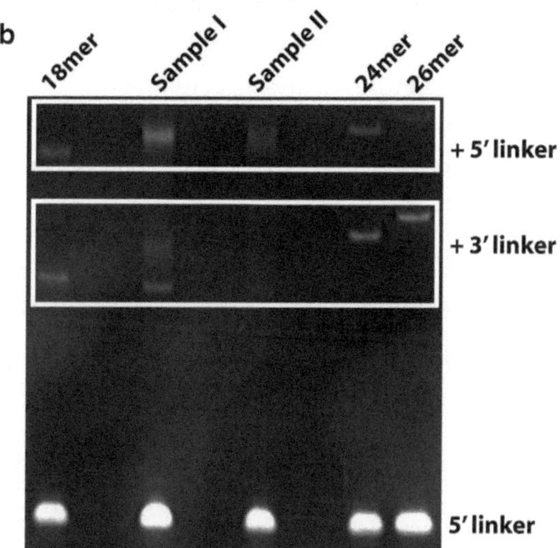
b
18mer
Sample I
Sample II
24mer
26mer
+ 5′ linker
+ 3′ linker
5′ linker

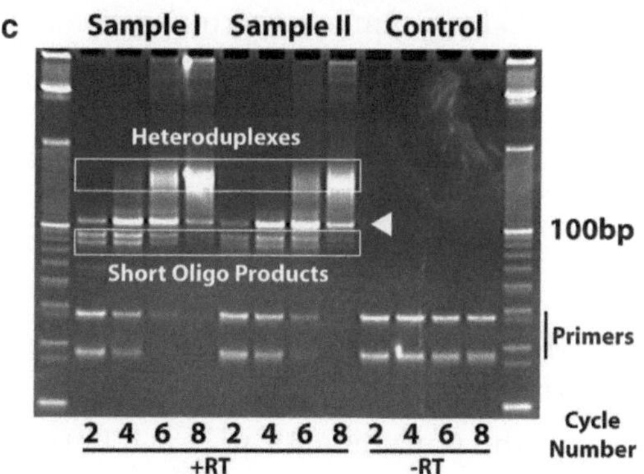
c
Sample I
Sample II
Control
Heteroduplexes
Short Oligo Products
100bp
Primers
2 4 6 8 2 4 6 8 2 4 6 8
+RT
-RT
Cycle Number

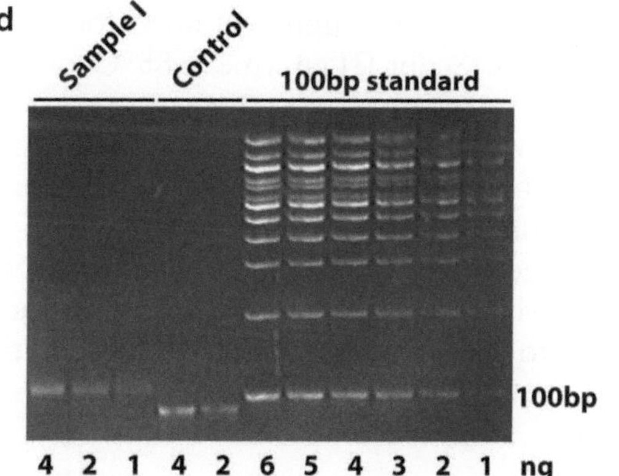
d
Sample I
Control
100bp standard
100bp
4 2 1 4 2 6 5 4 3 2 1 ng

2. Resuspend the RNA pellets in freshly prepared reaction mix and transfer to 0.2-mL PCR tubes. Incubate at 15°C for 6 h, followed by 4°C overnight in a thermal cycler.
3. Fractionate the 5′-adapter::small RNA::3′-adapter ligation products on a 15% polyacrylamide/7 M urea gel as described in Subheading 3.5. However, this time, run the gel until the Bromophenol blue reaches the bottom of the gel in order to resolve the ligation products. An example is shown in Fig. 3b.
4. Excise, elute and precipitate the 5′-adapter::small RNA::3′-adapter ligation products as described in Subheading 3.5.
5. Use the RNA ligation products for cDNA synthesis below (see Note 38).

3.9. cDNA Synthesis

1. Set up 0.2-mL PCR tubes for cDNA synthesis reactions. Thaw the components of the SuperScript III kit and place on ice.
2. To each RNA ligation pellet, add the following:

100 μM PR1	0.25 μL	(1.25 μM)
10 mM dNTP Mix	1.1 μL	(0.5 mM)
Water	11.95 μL	(13.3 μL total)

3. Transfer the mixture to the 0.2-mL PCR tubes, incubate at 65°C for 5 min in a thermal cycler and then on ice for 2 min.
4. To each sample add the following components (see Note 39):

5× SuperScript Buffer	4.4 μL	(1× final)
100 mM DTT	2.2 μL	(10 mM final)
20 U/μL RNaseOUT	1.1 μL	(1 U/μL final)

5. Mix the sample and remove 2 μL to a separate tube labeled with the sample name and "–RT."
6. To the remaining 19 μL ("+RT"), add 1 μL of SuperScript III enzyme.
7. Incubate the samples at 50°C for 1 h and heat inactivate the SuperScript III enzyme at 85°C for 5 min (see Note 40).

3.10. Library Amplification

1. Two short rounds of PCR are used to amplify the small RNA library. Shorter oligos that lack the Illumina adapter sequences for the initial amplification step. The Illumina adapter sequences are added during the subsequent round, during which an aliquot is removed from each sample and analyzed to determine the optimal cycle conditions.

2. Prepare 48 μL of primary amplification mix for *n*+1 samples (including one of the "–RT" samples):

10× Ex Taq Buffer	5 μL
10 μM PL1	0.5 μL
10 μM PR1	0.5 μL
2.5 mM dNTP	5 μL
2.5 U/μL Ex Taq Polymerase	0.5 μL
Water	36.5 μL

3. Add 48 μL of primary reaction mix to 2 μL of cDNA and run the following parameters in a thermal cycler (see Note 41):

1 cycle	94°C	30 s
6 cycles	94°C	20 s
	55°C	20 s
	72°C	20 s
1 cycle	4°C	Hold

4. Prepare 10 μL of secondary reaction mix per *n*+1 samples:

10× Ex Taq Buffer	1 μL
10 μM PL2	2.5 μL
10 μM PR2	2.5 μL
Water	4 μL

5. For every sample, prepare a strip of four tubes, labeled "2," "4," "6," and "8" and the sample name and each containing 1 μL of 6× DNA loading buffer.
6. Add 10 μL of secondary reaction mix to each primary reaction and run the following parameters in a thermal cycler:

1 cycle	94°C	30 s
8 cycles	94°C	20 s
	55°C	20 s
	72°C	20 s
1 cycle	4°C	Hold

7. At the end of cycles 2, 4, 6, and 8, pause the thermal cycler, place the tubes on ice, remove 4 μL from each reaction and mix with 1 μL of 6× DNA loading buffer in the appropriately labeled tube from step 5 above (see Note 42). Place the tubes back into thermal cycler and resume program.

8. Prepare 15 mL of non-denaturing 10% polyacrylamide (29:1) gel solution by mixing 5 mL of 30% acrylamide (Bio-Rad), 3 mL of 5× TBE and 7 mL of water (see Note 43). Immediately prior to pouring the gel, add 150 μL of 10% APS and 15 μL of TEMED. Pour the gel into a Bio-Rad Criterion cassette, insert comb and clamp the cassette to the teeth of the comb with two large binder clamps. Allow the gel to polymerize for 30 min (see Notes 44–46).
9. Once the gel has polymerized, remove the tape at the bottom of the gel cassette and place the cassette into the Criterion tank. Add 1× TBE running buffer to the upper and lower reservoirs and carefully remove the comb from the gel. Load the samples and 10 bp ladder and run the gel at 250 V until the Xylene cyanol dye has migrated half an inch from the bottom of the gel.
10. At the end of the run, disassemble the gel and stain in 1× TBE containing 50 μg/mL of ethidium bromide for 5 min. Destain the gel in water for 5 min to enhance visualization by trans-illumination and photograph the gel.
11. Determine the optimal number of cycles that will be used for final amplification of each library. Remember that shorter oligonucleotide primers that lack the Illumina adapter sequences are present and these will give rise to products that are ~50 nt shorter than the desired amplicon, as indicated in Fig. 3c. The Illumina primers are limiting in the reaction (see Notes 47 and 48). The optimal cycle for scaling-up the amplification of the library will be the cycle with maximal product and minimal heteroduplex formation (see Note 49). For the example shown in Fig. 3c, the optimal cycle parameters for the second round of amplification would be 3–4 cycles for Sample I and 4 cycles for Sample II.
12. Set up three PCRs (as described in Subheading 3.10, steps 2–6) for each sample and amplify the small RNA library based on the cycle parameters determined in the previous step.

3.11. Final Purification Steps, Quality Control, and Quantification

1. Perform the steps in this subheading as quickly as possible without allowing the DNA amplicon to fully dry and without exposing to air for an extended period of time – amplicons may denature and reanneal, thus forming heteroduplexes that make final quantification difficult.
2. Prepare an 18-well, non-denaturing, 10% polyacrylamide gel as described in Subheading 3.10, step 8. Allow the gel to polymerize for 30 min.
3. Pool the 3× 60 μL PCRs, dispense into a prepacked phase lock column and extract once with 250 μL of phenol::chloroform (pH 6.7). Cenrifuge at 15,000 ×g for 5 min at room temperature.

4. Transfer the aqueous (top) phase containing the small RNA library amplicon to a Microcon YM-10 column. Concentrate the aqueous phase to ~30 μL by centrifugation at 13,500 × *g* for 20 min at 4°C.
5. Mix the concentrated amplicon(s) with 6 μL of 6× DNA loading buffer. Load the sample into three lanes of the 10% polyacrylamide gel. Run the gel at 10 W until the Xylene cyanol dye is 1–2 cm from the bottom of the gel. Disassemble and stain the gel with 50 μg/mL ethidium bromide in 1× TBE for 5 min. Destain with water for 5 min. Photograph the gel and excise the ~100–130 bp DNA band from each lane, cutting as close as possible to the amplicon in order to minimize gel volume and maximize elution. Combine the gel slices from each of the three lanes and crush with a 1-mL pipet tip. Immediately add 750 μL of RNA/DNA Elution Buffer and elute the amplicon(s) as described in Subheading 3.5.
6. Add 10 μg of glycogen and at least 1 volume (~800 μL) of isopropanol and precipitate the amplicon at −20°C for 30 min. Pellet the DNA precipitate by centrifugation at 20,000 × *g* for 10 min at 4°C.
7. Working one sample at a time, use a 1-mL pipet tip to remove all but 10 μL of the isopropanol, taking care not to disturb or aspirate the pellet (see Notes 31 and 50).
8. Immediately add 500 μL of ice cold 70% ethanol to the pellet. Vortex briefly and centrifuge at 16,000 × *g* for 1 min at room temperature. Remove all but 10 μL of the 70% ethanol from each sample. Briefly centrifuge samples at 16,000 × *g* at room temperature to get all ethanol to the bottom of the tube.
9. Again working one sample at a time, use a fine pipet tip to remove the last traces of 70% ethanol from the tube and immediately resuspend in 20 μL of 10 mM Tris (pH 8.0) (see Note 50).
10. Quantify 1.5 μL of each sample using a nanodrop spectrophotometer (see Note 51).
11. Prepare a non-denaturing 10% polyacrylamide gel as described in Subheading 3.10, step 8. Dilute 3 ng of small RNA amplicon to a total volume of 5 μL with water or 10 mM Tris (pH 8.0) and add 1 μL of 6× DNA loading buffer. Load six different quantities of the New England Biolabs 100 bp ladder, corresponding to 1–6 ng of the 100 bp fragment, respectively (see Note 52). Run the gel at 10 W until the Xylene cyanol dye is 1–2 cm from the bottom of the gel. Disassemble and stain the gel with 50 μg/mL ethidium bromide in 1× TBE for 5 min. Destain with water for 5 min and photograph the gel. An example of this analysis is shown in Fig. 3d.

12. In addition to inspecting the quantity and quality of libraries by electrophoresis, clone 4 μL of the gel-purified amplicon by TOPO-TA cloning, prepare miniprep DNA and sequence at least 20 clones. This is enough to be sure that the majority of amplicons encode the 5′ linker, the insert and the 3′ linker (see Note 53). Using the method described, we expect most clones from a wild-type input sample to correspond to 22G-RNA or secondary siRNAs, with a few miRNAs and possibly a 21U-RNA.
13. Dilute the samples to a final concentration of 10 nM (or ~0.73 ng/μL of 110 bp product). The sample is ready to submit for Illumina sequencing. If your samples are barcoded, be sure to inform the deep-sequencing service of which nucleotide positions are nonrandom so that they can be omitted from the matrix that is generated for base calling.

3.12. Data Analysis and Validation of Results

Before analysis, the raw sequences and quality scores must be trimmed to remove the barcodes and the 3′ linker sequences. In addition, the reads are grouped according to barcodes (if used) at this time. Next, the reads are aligned to a reference genome. Methods for alignments to a reference genome are a matter of preference and range from custom scripts (available online or developed in house) to software that is available as part of the deep-sequencing pipeline. In addition, multifunctional third party software such as CLC Genomics Workbench (CLC Bio) and GeneSifter (Geospiza) are becoming commonly used tools to analyze deep-sequencing data. Note that some software packages or custom scripts simply do not consider reads matching repetitive genomic loci. Therefore, it is difficult to get a complete profile of repetitive elements, such as telomeres, using such approaches. One critical step prior to further analysis is to identify and remove RNA degradation products, which could be highly enriched in the IP sample depending on the quality of the RNA library. Generally, any sense reads from structural RNAs including rRNAs, tRNAs, and others should be suspected to be degradation products and removed before further analysis. Analysis of the miRNAs should be less stringent for the ends because of RNA modifications, RNA processing (promiscuously cut by Dicer), or annotation mistakes.

To identify a class of Argonaute-associate small RNAs, it is important to determine the enrichment of a group of RNAs in the IP sample over the input sample. One way to do this is to examine the size distribution of reads that start with each nucleotide. For example, immunoprecipitation of PRG-1 enriched for 21 nt reads that start with 5′U (7), while CSR-1 and WAGO-1 IPs enriched for 22 nt reads that start with 5′G (4, 8). Another approach is to tally small RNA reads (that have been normalized to the total read number as well as the number of matches to the genome) that target individual loci in both the IP and input samples and compare the two directly (4, 8). This approach may allow for the identification of a particular class of transcripts that are targeted by the Argonaute. For instance, analysis of WAGO-1-associated 22G-RNAs identified repetitive elements and pseudogenes as targets of this particular Argonaute (4).

After determining which small RNAs are enriched in an IP sample by deep sequence analysis, it is important to validate the results by independent experimentation. For example, Argonaute/small RNA interactions may be tested by (1) 2′-O-methylated oligo pull-down of a particular small RNA species, followed by western blotting for the Argonaute of interest (6) (see Chapter 16); (2) immunoprecipitation of the Argonaute followed by quantitative real-time PCR analysis or northern blot analysis of particular small RNA species (2, 6, 7, 15); (3) small RNA cloning and deep-sequencing analysis to identify a small RNA population that is depleted when the Argonaute is mutated (2, 4, 7, 8, 15, 16). Furthermore, coupling deep-sequencing analyses with phenotypic analyses of Argonaute mutants provides a powerful approach for understanding the endogenous functions of Argonautes in *C. elegans*.

4. Notes

1. When identifying regions of an Argonaute protein to use for generating antibodies, cross-reactivity with homologous Argonautes is an important consideration, especially in *C. elegans*, where some of the 26 Argonautes are highly related. It is imperative to test the specificity of the antisera by western blotting against lysates from both wild-type and Argonaute mutant. Deletion alleles for each of the worm Argonautes are available from the *Caenorhabditis* Genetics Center (6).
2. Interestingly, a 3XFlag epitope inserted 28 amino acids N-terminal of the PIWI domain of CSR-1 did not rescue the *csr-1(tm892)* mutant phenotype, suggesting that the fusion protein is non-functional. Furthermore, the IP efficiency of the rescuing RDE-1 and CSR-1 fusions may be low. These points emphasize the importance of generating multiple transgenic lines as well as differently tagged forms.
3. The developmental time period during which the Argonaute of interest is most highly expressed can be determined by quantitative real-time PCR to detect mRNA levels throughout development, fluorescence-imaging using antibodies to endogenous protein or GFP-tagged transgenic strains, or by western blot analysis of developmentally staged samples.

4. 10 plates of 100,000 worms each will yield about 5 mL of worm pellet, which corresponds to about 100 mg of protein lysate after extraction in Subheading 3.4.
5. It is important to maintain the worms clear of contamination, and to minimize the amount of OP50 *E. coli* retained in the harvested worms, by extensively washing the worms with M9 Buffer.
6. Cold water causes the worms to decrease their movement and form a compact pellet. We use cold water instead of placing the worms on ice to settle the worms, because prolonged exposure to cold invokes a cold shock response.
7. All dounce homogenizers should be thoroughly cleaned with RNase Zap, 10% SDS, or some other RNase-removal cleanser. It is also a good practice to clean benches with an RNase-removal cleanser prior to beginning this protocol.
8. Dounce worms on ice and monitor worm disruption after 10 or 15 strokes by removing 2 μL of sample, dropping it onto a glass slide, and examining under a dissection microscope. Repeat if necessary.
9. Always use aerosol barrier tips to prevent contamination between samples and by RNases. Likewise, always using siliconized tubes helps to minimize the absorption of precious samples to the walls of the microcentrifuge tube.
10. Pack the phase-lock column by centrifugation at ~15,000 × *g* for 1 min at room temperature immediately before adding sample.
11. Up to 1 mg of total RNA can be processed in the modified mirVana™ procedure as described. If the total RNA volume exceeds 80 μL, you may split the sample and perform parallel purifications, ultimately combining the sample in the final steps.
12. Under the gel conditions described here, a maximum of 15–20 μg of RNA is run per lane. If you are working with <20 μg of RNA, mix the sample with 10 μL of formamide loading buffer, evaporate and load sample in a single lane. Formamide loading buffer will not evaporate significantly in the SpeedVac, if the sample is concentrated below the boiling temperature of formamide.
13. Each IP is done in parallel, as pooling the lysates may not achieve the same IP efficiency.
14. The antibody amount and IP time must be determined empirically for each antibody. As a guideline, anti-CSR-1 peptide antibodies were used at 20 μg/mL, anti-PRG-1 was used at 20 μg/mL, for WAGO-1, anti-Flag M2 antibody (Sigma-Aldrich) was used at a concentration of 10 μg/mL.

15. The elution efficiency is determined by the equilibrium of RNA distributed between the solid phase (gel) and the liquid phase (elution buffer). Therefore, a thin gel is preferred over a thick gel particularly when resolving a very small amount of RNA.
16. RNA oligos can be used as size markers for the initial gel purification of small RNA and to follow the ligation procedure. If RNA oligos are used to follow the ligation steps, they must be phosphorylated immediately prior to the 5′ ligation step. After gel purification, elution, and precipitation of the oligo::3′-adapter ligation product, phosphorylate the oligo::3′-adapter product with T4 Polynucleotide Kinase. To the oligo::3′-adapter pellet add 14.2 μL water, 2 μL 10× PNK Buffer, 1 μL SUPERase•In™, 0.8 μL ATP (50 mM), 2 μL PNK (10 U/μL), mix and incubate at 37°C for 1 h. Extract and precipitate the phosphorylated product and proceed with the 5′ ligation.
17. It is not essential to prerun the gel to warm it prior to loading. The formamide denatured small RNAs are run in the presence of urea and the aluminum plate will help warm-up the gel quickly and evenly.
18. It is important to leave 1–2 blank lanes between sample lanes on the gel to minimize the potential for cross-contamination of samples. Cross-contamination can occur during loading or during band excision if samples are loaded too closely together.
19. Do not exceed the loading capacity of each lane. Under the conditions described, we load a maximum of 15–20 μg of mirVana™ purified small RNA. In addition, load small volumes (10 μL or less) to minimize sample "trailing" along the edge of the lane. As much as possible, be careful not to disturb the sample layer while loading or removing the pipette tip. Small volumes can be achieved by adding 10 μL of formamide loading buffer to the sample and evaporating the sample in a SpeedVac with the temperature set below the boiling point of formamide. This is usually the lowest heat setting (~30–40°C). Alternatively, turn off the heat setting, as it is not essential.
20. If your power supply does not have the capacity to run according to Wattage, run the gel at 500–700 V constant, with an current limit set at 0.02 A. The short run is enough to resolve the desired size species of ~18–40 nt without making the gel fragments too large. This will minimize the number of tubes required for elution and increase the elution efficiency.
21. SYBR® Gold nucleic acid gel stain (Invitrogen) can be used in any gel detection steps in place of ethidium bromide. SYBR® Gold is much more sensitive than ethidium bromide, but can be easily saturated. A 1:20,000 or greater dilution of SYBR® Gold is sufficient for staining. Otherwise, use the stain in the same manner as described for ethidium bromide.

22. It may be helpful to use straight edges to identify the upper and lower boundaries across the gel when determining where to excise bands. Cut as close to the lane as possible. In cases where samples are faint or invisible, use the loading Bromophenol blue and Xylene cyanol dyes to identify the lane and proceed as if you could see the small RNA.
23. When cutting samples from the gel, be as quick as possible to minimize the formation of adducts. Be sure to use the appropriate protective equipment – face shield, long sleeves, and gloves to avoid a UV burn.
24. It may be helpful to seal the end of the 1-mL pipet tip shut by melting, to keep any pieces of the gel from being taken up into the tip during crushing of the gel slice.
25. A multi-tube vortex or a tumbler can be used with similar results to agitate samples overnight.
26. A double elution will recover almost all of the small RNA, but is typically not necessary.
27. When preparing to filter the eluate from gel extraction, it may be helpful to transfer the eluate to a new siliconized microcentrifuge tube and pellet any aspirated gel fragments prior to loading the spin column to prevent it from clogging. Nanosep spin filters can be used in place of Corning spin filters. Both spin filters have a maximum capacity of 500 μL. After the initial spin, transfer the filtered eluate to a fresh 1.5-mL siliconized microcentrifuge tube. Reload the spin filter with any remaining eluate and centrifuge at 10,000 × *g* for 1 min. Combine with the initial filtered eluate.
28. There is some loss of elution buffer during the elution process and after filtering.
29. Small RNA samples can be stored in 70% ethanol at –20°C until it is convenient to proceed to the next step.
30. We do not use a vacuum aspirator to remove alcohol from precipitations during the purification of amplicons, because the sample is too precious to risk losing by aspiration. A round gel-loading tip is useful for removing residual ethanol from RNA pellets, because it holds more than 10 μL and its shape minimizes the chance of aspirating the pellet.
31. Certain small RNAs, notably the 21U-RNAs and the 26G-RNAs, are modified at the 3′ end with a 2′-*O*-methyl group and are ligated at a much lower efficiency. However, these species can be enriched after treatment with periodate (10, 15).
32. We recognize that various labs have different preferences for which particular version of T4 RNA ligase they utilize in the ligation reactions of small RNA cloning protocols. We have found that, for our purposes, the Takara T4 RNA ligase is the

best enzyme available, and we do not recommend substitutions for this enzyme in this protocol.

33. When precipitating the TAP-treated RNA, there is still glycogen remaining in the sample from the previous precipitation step after gel elution. There is no need to add more glycogen.
34. The 3′-adapter Oligo is activated with a 5′-App moiety that serves as the rate-limiting ligation intermediate of T4 RNA ligase. Therefore ATP is not added to the 3′ ligation reaction.
35. If necessary, the ligation times can be shortened. We have performed a minimum 8 h ligation successfully.
36. The small RNA::3′-adapter ligation products will be ~40 nt and may be more easily visualized than the small RNA was during the initial purification.
37. Precipitated ligation products can be stored in 70% ethanol at –20°C until it is convenient to proceed to the next step.
38. If your original small RNA sample was very low concentration or not visible during the initial and subsequent gel purifications (Subheadings 3.5, 3.7, and 3.8), use the entire ligation product for cDNA synthesis to maximize sample scale and minimize the number of cycles necessary to amplify the small RNA library.
39. You can prepare a reaction cocktail of RT mix if you have multiple samples. The final volume should be 21 μL at this point.
40. It is optional to treat the +RT and –RT reactions with 1 and 0.1 μL, respectively, of RNase H. We normally omit this step.
41. The number of PCR cycles provided here is a typical example. However, depending on the amount of starting material, you may need 20 or more cycles in the first round. For example, if you start with ~1 pg or 1 fg of 22 mer, use ~20 cycles in the first round of amplification.
42. Instead of pausing the thermal cycler, the second round of PCR can be performed using a 2-cycle program, and running this cycle repeatedly.
43. The native gel solution should be freshly prepared as premixed 10% polyacrylamide is unstable.
44. We are aware that some laboratories use high-resolution agarose gels to separate amplicons, but in our experience, polyacrylamide gels provide superior separation of products.
45. We find the Bio-Rad Criterion cassettes to be a convenient mid-size system that permits the analysis of as many 26 samples. One-time use is an environmentally unfriendly downfall of the system, but they save time and are the appropriate size for our needs. The cycle parameters for four different libraries

can be analyzed on a single Criterion 26-well gel. If cost is a consideration, generic alternatives to Bio-Rad cassettes are available; however we have not tested them.

46. The Criterion cassette combs do not form a tight seal between the inner and outer plates of the cassette, and gel solution leaks and polymerizes between the teeth of the comb and the cassette. This results in wells that can be difficult to load. We use two large binder clamps to clamp the cassette to the comb in order to make uniform, clean wells.
47. The long Illumina oligonucleotides used for the second round of PCR result in more primer-dimers than the short Illumina oligonucleotides used in the first round of PCR. Therefore, the second round of PCR should not exceed 10 cycles.
48. As the primers begin to deplete, the amplicons begin to denature and anneal during each cycle. This results in heteroduplexes with perfectly annealed terminal sequences and 20–30 nt mis-matches in between Fig. 3c. These heteroduplexes accumulate in the later cycles and migrate as a smear with a retarded mobility in the gel. The products within the smear may also contain a significant amount of primer-dimers heteroduplexed with less abundant small RNA clones.
49. The optimal cycle for final amplification must be determined empirically for each library. It may be necessary to increase the number of cycles in the first round of amplification in order to generate enough amplicon for subsequent purification.
50. Do not let the amplified DNA pellet dry completely as it may result in denaturation of the amplicons. This, in turn, will lead to inaccurate quantification and dilution of the final libraries.
51. We do not trust Nanodrop spectrophotometer results at or below ~10 ng/μL. Even at readings of ~20 ng/μL, it is imperative to quantify by electrophoresis. Therefore, the nanodrop results are used as a starting point and samples are then compared to known quantities of a size standard by electrophoresis. This is also a quality control step. For example, you will be able to determine whether the library has denatured and reannealed or the extent to which other background bands such as primer-dimers were isolated during the procedure. If the library has denatured and reannealed, you will observe a smear that migrates slowly in the gel. While these denatured amplicons are ultimately not a problem for deep sequencing, they complicate quantification of the library.
52. The concentration of the 100 bp ladder from New England Biolabs is 500 ng/μL. Dilute 10 μL of the stock into 374 μL of 10 mM Tris (pH 8.0) and add 96 μL of 6× DNA loading buffer. One microliter of this dilution contains ~1 ng of the 100 bp fragment.

53. The TOPO-cloned sequences should have the following structure:

    5′-AATGATACGGCGACCACCGACAGGTTCA GAGTT**CTACAGTCCGACGATC** -(optional 4 nt barcode)-…small_RNA… CTGTAGGCACCATCAATTCGTATGCCGT CTTCTGCTTG-3′

Acknowledgments

The authors thank members of the Mello lab for helpful discussion; James C. Carrington and laboratory for initial help with small RNA cloning protocols and deep sequencing; E. Kittler and the UMass Deep-Sequencing Core for processing Illumina samples. J.M. Claycomb was an HHMI fellow of the LSRF. P.J. Batista was supported by SFRH/BD/11803/2003 from Fundação para Ciência e Tecnologia, Portugal. C.C. Mello is a Howard Hughes Medical Institute Investigator. This work was supported by R01 grant GM58800 (C.C. Mello) and Ruth L. Kirschstein N.R.S.A. GM63348 (D. Conte) from the NIGMS.

References

1. Carthew, R. W., and Sontheimer, E. J. (2009) Origins and Mechanisms of miRNAs and siRNAs. *Cell* **136**, 642–655.
2. Das, P. P., Bagijn, M. P., Goldstein, L. D., Woolford, J. R., Lehrbach, N. J., Sapetschnig, A., Buhecha, H. R., Gilchrist, M. J., Howe, K. L., Stark, R., Matthews, N., Berezikov, E., Ketting, R. F., Tavare, S., and Miska, E. A. (2008) Piwi and piRNAs act upstream of an endogenous siRNA pathway to suppress Tc3 transposon mobility in the *Caenorhabditis elegans* germline. *Molecular Cell* **31**, 79–90.
3. Wang, G., and Reinke, V. (2008) A *C. elegans* Piwi, PRG-1, regulates 21U-RNAs during spermatogenesis. *Curr. Biol.* **18**, 861–867.
4. Gu, W., Shirayama, M., Conte, D., Jr., Vasale, J., Batista, P. J., Claycomb, J. M., Moresco, J. J., Youngman, E. M., Keys, J., Stoltz, M. J., Chen, C. C., Chaves, D. A., Duan, S., Kasschau, K. D., Fahlgren, N., Yates, J. R., 3rd, Mitani, S., Carrington, J. C., and Mello, C. C. (2009) Distinct argonaute-mediated 22G-RNA pathways direct genome surveillance in the *C. elegans* germline. *Molecular Cell* **36**, 231–244.
5. Grishok, A., Pasquinelli, A. E., Conte, D., Li, N., Parrish, S., Ha, I., Baillie, D. L., Fire, A., Ruvkun, G., and Mello, C. C. (2001) Genes and mechanisms related to RNA interference regulate expression of the small temporal RNAs that control *C. elegans* developmental timing. *Cell* **106**, 23–34.
6. Yigit, E., Batista, P. J., Bei, Y., Pang, K. M., Chen, C. C., Tolia, N. H., Joshua-Tor, L., Mitani, S., Simard, M. J., and Mello, C. C. (2006) Analysis of the *C. elegans* Argonaute family reveals that distinct Argonautes act sequentially during RNAi. *Cell* **127**, 747–757.
7. Batista, P. J., Ruby, J. G., Claycomb, J. M., Chiang, R., Fahlgren, N., Kasschau, K. D., Chaves, D. A., Gu, W., Vasale, J. J., Duan, S., Conte, D., Jr., Luo, S., Schroth, G. P., Carrington, J. C., Bartel, D. P., and Mello, C. C. (2008) PRG-1 and 21U-RNAs interact to form the piRNA complex required for fertility in *C. elegans*. *Molecular Cell* **31**, 67–78.
8. Claycomb, J. M., Batista, P. J., Pang, K. M., Gu, W., Vasale, J. J., van Wolfswinkel, J. C., Chaves, D. A., Shirayama, M., Mitani, S., Ketting, R. F., Conte, D., Jr., and Mello, C. C. (2009) The Argonaute CSR-1 and its 22G-RNA cofactors are required for holocentric chromosome segregation. *Cell* **139**, 123–134.
9. Pak, J., and Fire, A. (2007) Distinct populations of primary and secondary effectors during RNAi in *C. elegans*. *Science* **315**, 241–244.

10. Ruby, J. G., Jan, C., Player, C., Axtell, M. J., Lee, W., Nusbaum, C., Ge, H., and Bartel, D. P. (2006) Large-scale sequencing reveals 21U-RNAs and additional microRNAs and endogenous siRNAs in *C. elegans*. *Cell* **127**, 1193–1207.
11. Lee, R. C., and Ambros, V. (2001) An extensive class of small RNAs in *Caenorhabditis elegans*. *Science* **294**, 862–864.
12. Ambros, V., Lee, R. C., Lavanway, A., Williams, P. T., and Jewell, D. (2003) MicroRNAs and other tiny endogenous RNAs in *C. elegans*. *Curr. Biol.* **13**, 807–818.
13. Sijen, T., Steiner, F. A., Thijssen, K. L., and Plasterk, R. H. (2007) Secondary siRNAs result from unprimed RNA synthesis and form a distinct class. *Science* **315**, 244–247.
14. Lim, L. P., Lau, N. C., Weinstein, E. G., Abdelhakim, A., Yekta, S., Rhoades, M. W., Burge, C. B., and Bartel, D. P. (2003) The microRNAs of *Caenorhabditis elegans*. *Genes & Dev.* **17**, 991–1008.
15. Vasale, J. J., Gu, W., Thivierge, C., Batista, P. J., Claycomb, J. M., Youngman, E. M., Duchaine, T. F., Mello, C. C., and Conte, D., Jr. (2010) Sequential rounds of RNA-dependent RNA transcription drive endogenous small-RNA biogenesis in the ERGO-1/Argonaute pathway. *Proc. Natl. Acad. Sci. USA* **107**, 3582–3587.
16. Conine, C. C., Batista, P. J., Gu, W., Claycomb, J. M., Chaves, D. A., Shirayama, M., and Mello, C. C. (2010) Argonautes ALG-3 and ALG-4 are required for spermatogenesis-specific 26G-RNAs and thermotolerant sperm in *Caenorhabditis elegans*. *Proc. Natl. Acad. Sci. USA* **107**, 3588–3593.
17. Guang, S., Bochner, A. F., Pavelec, D. M., Burkhart, K. B., Harding, S., Lachowiec, J., and Kennedy, S. (2008) An Argonaute transports siRNAs from the cytoplasm to the nucleus. *Science* **321**, 537–541.
18. Tabara, H., Sarkissian, M., Kelly, W. G., Fleenor, J., Grishok, A., Timmons, L., Fire, A., and Mello, C. C. (1999) The rde-1 gene, RNA interference, and transposon silencing in *C. elegans*. *Cell* **99**, 123–132.
19. Song, J. J., Smith, S. K., Hannon, G. J., and Joshua-Tor, L. (2004) Crystal structure of Argonaute and its implications for RISC slicer activity. *Science* **305**, 1434–1437.
20. Tabara, H., Yigit, E., Siomi, H., and Mello, C. C. (2002) The dsRNA binding protein RDE-4 interacts with RDE-1, DCR-1, and a DExH-box helicase to direct RNAi in *C. elegans*. *Cell* **109**, 861–871.
21. Mello, C. C., Kramer, J. M., Stinchcomb, D., and Ambros, V. (1991) Efficient gene transfer in C.elegans: extrachromosomal maintenance and integration of transforming sequences. *The EMBO Journal* **10**, 3959–3970.
22. Kelly, W. G., Xu, S., Montgomery, M. K., and Fire, A. (1997) Distinct requirements for somatic and germline expression of a generally expressed *Caenorhabditis elegans* gene. *Genetics* **146**, 227–238.
23. Praitis, V., Casey, E., Collar, D., and Austin, J. (2001) Creation of low-copy integrated transgenic lines in *Caenorhabditis elegans*. *Genetics* **157**, 1217–1226.
24. Frokjaer-Jensen, C., Davis, M. W., Hopkins, C. E., Newman, B. J., Thummel, J. M., Olesen, S. P., Grunnet, M., and Jorgensen, E. M. (2008) Single-copy insertion of transgenes in *Caenorhabditis elegans*. *Nature Gen.* **40**, 1375–1383.

Chapter 18

Immunoprecipitation of piRNPs and Directional, Next Generation Sequencing of piRNAs

Yohei Kirino, Anastassios Vourekas, Eugene Khandros, and Zissimos Mourelatos

Abstract

Piwi interacting RNAs (piRNAs) are small (~25 to ~30 nucleotide) and are expressed in the germline. piRNAs bind to the Piwi subclade of Argonaute proteins and form the core ribonucleoproteins (RNPs) of piRNPs. We describe a method for the massive identification of piRNAs from immunopurified piRNPs. This strategy may also be used for immunopurification and directional sequencing of RNAs from other RNPs that contain small RNAs.

Key words: piRNA, piRNP, Argonaute, Piwi, Xili, Xiwi, Mili, Miwi, Miwi2, Aub, Ago3, Y12, Next gen sequencing, Illumina, cDNA, Immunoprecipitation, RNA-Immunoprecipitation, RNA-IP, T4 RNA ligase, Reverse Transcriptase, Polymerase Chain reaction, PCR, RT-PCR, Posttranscriptional RNA processing, Gene silencing

1. Introduction

Piwi interacting RNAs (piRNAs) comprise a class of ~25 to ~30 nucleotide (nt) RNAs, which that are expressed in germline cells and bind to the Piwi subclade of Argonaute proteins. Mice express three Piwi proteins termed Mili (1), Miwi (2), and Miwi2 (3). *Drosophila melanogaster* expresses three Piwi proteins termed Aubergine (Aub) (4), Piwi (5), and Ago3 (6–8). *Xenopus tropicalis* and *X. laevis* express three Piwi proteins termed Xili, Xiwi, and Xiwi2 (9, 10). The sequence diversity of piRNAs is tremendous and hundreds of thousands of unique piRNAs have been described in diverse species (11). Many piRNAs are derived from transposable or repetitive elements and also target transposons by antisense complementarity (12). Many genic and intergenic regions

Tom C. Hobman and Thomas F. Duchaine (eds.), *Argonaute Proteins: Methods and Protocols*,
Methods in Molecular Biology, vol. 725, DOI 10.1007/978-1-61779-046-1_18, © Springer Science+Business Media, LLC 2011

also give rise to piRNAs but the significance of these piRNAs remains to be determined.

Here, we describe a method for the immunoprecipitation of piRNPs and the isolation and directional sequencing of associated small RNAs. This approach combines techniques developed for the immunoprecipitation of RNPs (13, 14) with techniques developed for the directional adapter ligation to small RNAs (15) and their identification by Illumina next generation sequencing. In principle, this methodology may be used for the immunopurification and directional sequencing of RNAs from any RNP that contains small RNAs.

2. Materials

1. Germline tissue (such as mouse testis; *X. laevis* oocytes or testis; *D. melanogaster* ovaries or testis).
2. PBS (Fisher).
3. Lysis buffer: 20 mM Tris–HCl, pH 7.4, 200 mM sodium chloride, 2.5 mM magnesium chloride, 0.5% NP-40, 0.1% Triton X-100, one tablet of Complete Protease Inhibitor EDTA-free (Roche) per 50 ml of lysis buffer.
4. RNasin (Promega).
5. Recombinant Protein G Agarose beads (Invitrogen).
6. Sonicator (Sonics Vibra-Cell or equivalent).
7. dNTPs (Roche).
8. ATP (Roche).
9. Millipore (MilliQ) water.
10. T4 RNA Ligase (New England Biolabs-NEB).
11. pBR322 DNA-Msp I Digest (DNA markers; NEB).
12. DNA polymerase I, large (Klenow) fragment (NEB).
13. T4 polynucleotide kinase (T4 PNK; NEB).
14. Calf intestinal alkaline phosphatase (CIP; NEB).
15. Complete protease inhibitor EDTA-free (Roche).
16. Nonimmune mouse serum or mouse IgG.
17. Anti-piRNP antibodies. For example, Y12 monoclonal antibody (Abcam, ab3138); see Note 1.
18. RNA loading buffer 1–2× (95% formamide, 18 mM EDTA, Xylene cyanol, bromophenol blue; Ambion).
19. Glycogen (at 5 mg/ml from Ambion).
20. Ethanol, 100%.

21. 3 M sodium acetate, pH 5.2.
22. 10× TBE (Ambion).
23. Acrylamide/Bis 19:1 40% (w/v) solution (Ambion).
24. Urea (Ambion).
25. 10% (w/v) ammonium persulfate (APS; dissolved in water).
26. TEMED (Biorad).
27. 10% Urea/PAGE solution (1 l): combine in a glass beaker: 480 g urea, 250 ml of 40% acrylamide/Bis (19:1), 100 ml 10× TBE, and water up to 1 l. Stir until completely dissolved, filter-sterilize, and store up to a year at RT in an aluminum-covered bottle (to protect from light).
28. 20% Urea/PAGE (1 l): combine in a glass beaker: 480 g urea, 500 ml of 40% acrylamide/Bis (19:1), 100 ml 10× TBE, and water up to 1 l. Stir until completely dissolved, filter-sterilize, and store as described above.
29. SE 400 Sturdier Gel electrophoresis apparatus with 18 × 24 cm glass plates (Amersham).
30. [γ-^{32}P] ATP at 3,000 Ci/mmol, 10 mCi/ml (NEN).
31. [α-^{32}P] dCTP, at 3,000 Ci/mmol, 10 mCi/ml (NEN).
32. Phenol/chloroform/isoamyl alcohol (25:24:1); pH 7.9 (Fisher).
33. Elution buffer (0.1% SDS, 0.3 M NaOAC, 100 μM EDTA).
34. Ethidium bromide (10 mg/ml).
35. Superscript II reverse transcriptase (RT; Invitrogen).
36. Trizol (Invitrogen).
37. Isopropanol.
38. Chloroform.
39. OR2− buffer (82.5 mM NaCl, 2.5 mM KCl, 1.0 mM Na_2HPO_4, 5.0 mM HEPES (pH 7.8)) (1 l). Mix 20.6 ml of 4 M NaCl, 1.25 ml of 2 M KCl, 10 ml of 100 mM Na_2HPO_4, 10 ml of HEPES, 500 mM, pH 8.3, and MilliQ water to 1,000 ml.
40. OR2+ buffer (82.5 mM NaCl, 2.5 mM KCl, 1.0 mM Na_2HPO_4, 5.0 mM HEPES (pH 7.8), 1.0 mM $MgCl_2$, 1.0 mM $CaCl_2$) (1 l). Mix 20.6 ml of 4 M NaCl, 1.25 ml of 2 M KCl, 10 ml of 100 mM Na_2HPO_4, 10 ml of HEPES, 500 mM, pH 8.3 1 ml $MgCl_2$, 10 ml of 100 mM $CaCl_2$, and MilliQ water to 1,000 ml.
41. BSA (Sigma A9418).
42. Soybean trypsin inhibitor type II-S (Sigma T9128).
43. Collagenase type IA (Sigma C9891).

44. Collagenase stock solution (0.25% BSA, 0.25% soybean trypsin inhibitor type II-S, 5% collagenase type IA in OR2–) (20 ml): Dissolve 1 g of collagenase powder, 200 mg BSA powder, 200 mg soybean trypsin inhibitor type II-S powder in 20 ml OR2–. Mix well and store at –20°C in 1.2 ml aliquots.
45. Collagenase working solution: To 1.2 ml of collagenase stock add 1.8 ml of OR2–; total volume is 3 ml. Final concentration of collagenase in working solution is 2 mg/ml.
46. Small RNA Sample Prep Kit, version 1.5 (Illumina, catalog # FC-102-1009; v1.5). This kit contains the 5′- and 3′-adapters, polymerase chain reaction (PCR) primers (sequences are indicated below), the enzymes, and other consumables. The oligonucleotide sequences are protected by copyright, which is owned by Illumina. Oligonucleotide sequences © 2006 and 2008 Illumina, Inc.

5′-RNA adapter (SRA 5): 5′ GUUCAGAGUUCUACAGUC CGACGAUC.

3′-RNA adapter (SRA 3): 5′ P-UCGUAUGCCGUCUUCU GCUUGU.

RT Primer: 5′ CAAGCAGAAGACGGCATACGA.

Small RNA PCR Primer 1 (GX1): 5′ CAAGCAGAAGACG GCATACGA.

Small RNA PCR Primer 2 (GX2): 5′ AATGATACGGCGA CCACCGACAGGTTCAGAGTTCTACAGTCCGA.

Sequencing Primer: 5′ CGACAGGTTCAGAGTTCTACAGT CCGACGATC.

3. Methods

The outline of the procedures and representative gels are shown in Fig. 1. All procedures and centrifugations are performed on ice or at 4°C unless otherwise indicated. Use RNase-free solutions, tubes, and pipettes (see Note 2).

3.1. piRNP Immunoprecipitation

3.1.1. Binding of Antibodies to Protein-G Agarose Beads

1. Bind the Y12 monoclonal antibody or nonimmune mouse serum (IgG; serves as negative control) on protein G agarose beads. For this, use 30 μl of bed volume of protein G agarose beads per immunoprecipitate. Resuspend, and wash the beads three times with 1 ml of lysis buffer (see Note 1).
2. Aspirate the last wash, while taking care not to dry the beads. Add 1 ml of lysis buffer and 5 μl of Y12 ascites (or 5 μl nonimmune mouse serum for the control immunoprecipitate).

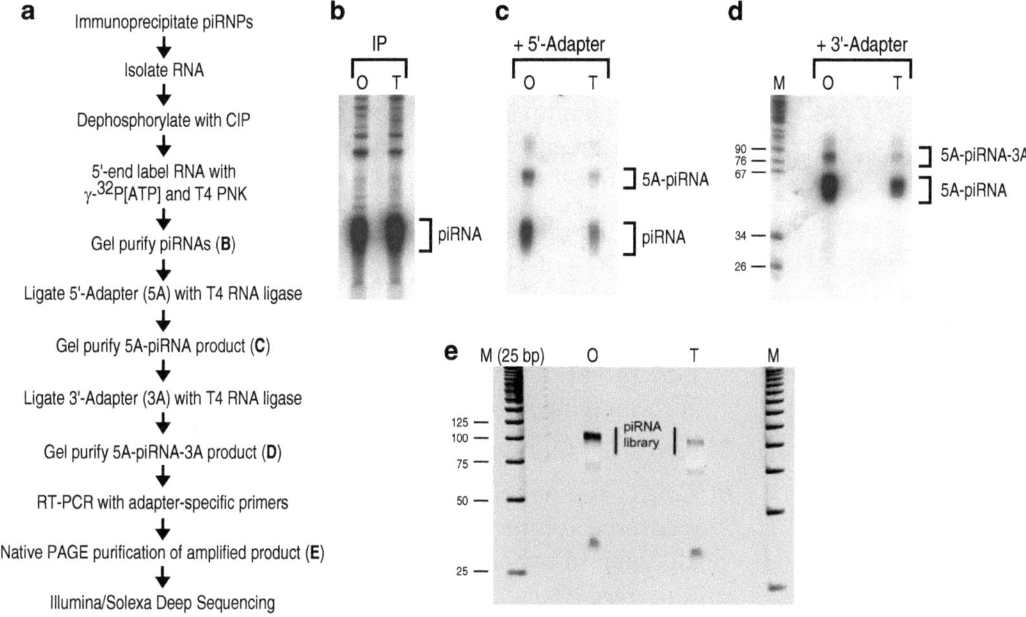

Fig. 1. piRNP immunoprecipitation, directional adapter ligation, and piRNA library preparation. (**a**) Outline of the experimental steps. (**b**) Immunoprecipitations (IP) were performed from Xenopus laevis defolliculated oocytes (O) or testis (T) cells with the Y12 monoclonal antibody that IPs X. laevis Piwi proteins. RNA was isolated, dephosphorylated with CIP, 5′-end radiolabeled with [γ-^{32}P] ATP and T4 PNK and analyzed on a 15% urea/PAGE. (**c**) A 5′-adapter (SRA-5; 5A) was ligated to gel-purified piRNAs with the T4 RNA ligase, and the ligation reaction was resolved on 15% urea/PAGE. 5A-piRNA ligated products and unligated piRNAs are indicated. (**d**) A 3′-adapter (SRA-3; 3A) was ligated to gel-purified 5A-piRNA with T4 RNA ligase and the ligation reaction was resolved on 15% urea/PAGE. The final 5A-piRNA-3A ligation products and the unligated 5A-piRNA are indicated. The marker (M) is a radiolabeled pBR322 DNA-Msp I digest. (**e**) The gel-purified 5A-piRNA–3A RNA product was used for RT-PCR and the resulting piRNA library was analyzed by 8% native PAGE and visualized by ethidium bromide staining. The DNA band corresponding to the piRNA library (indicated) was gel purified and used for Illumina sequencing.

3. Rotate for 45 min.
4. Wash the antibody-beads three times with 1 ml lysis buffer by centrifuging at 2,300×*g* for 5 s and by aspirating.

3.1.2. Preparation of the Tissue Lysate

1. The tissue lysate may be prepared while the antibodies are binding to the protein G agarose beads. Wash the tissue with PBS. Resuspend the tissue (such as defolliculated *X. laevis* oocytes, or minced *X. laevis*, or mouse testis) in 1 ml of lysis buffer (see Notes 3 and 4). Add RNasin to a final concentration of 0.1 U/μl.
2. Sonicate briefly (three times, 8–10 s each) using 40% output (on a Sonics Vibra-Cell sonicator or the equivalent).
3. Centrifuge the cell lysate at 20,000×*g* (16,000×*g* in an Eppendorf microcentrifuge) for 20 min.
4. Collect the supernatant and save an aliquot as a reference for total lysate proteins and discard the pellet.

3.1.3. Immunoprecipitation and RNA Isolation

1. Use 1/2 of the supernatant with your specific antibody-beads (i.e., Y12-beads from Subheading 3.1.1) and use the other 1/2 with the negative control antibody. Perform the immunoprecipitation in microcentrifuge tubes. If the volume of the immunoprecipitate is less than 1 ml, complete to 1 ml with lysis buffer. Rotate in the cold room for 1 h.
2. Wash the beads five times with 1 ml lysis buffer.
3. Add 500 μl of Trizol reagent to the washed beads. Vortex for 30 s to 1 min. Let the tube sit at room temperature (RT) for 5 min.
4. Add 150 μl of chloroform and vortex briefly. Let the tube sit at room temperature (RT) for 2 min.
5. Spin at 20,000 × *g* for 20 min at RT.
6. Collect the upper aqueous phase (avoid the interphase; the recovered volume will be approximately 300 μl). Add 3 μl of glycogen (5 mg/ml) and vortex briefly.
7. Add 350 μl of isopropanol and vortex briefly. Place the tube at −20°C for 20 min.
8. Spin at 20,000 × *g* for 30 min at 4°C.
9. Carefully aspirate the supernatant and let the pellet air dry (see Note 5).
10. Resuspend the pellet in 21.5 μl of MilliQ water. Proceed with piRNA isolation or store the RNA at −80°C (see Note 6).

3.2. piRNA Isolation

3.2.1. Dephosphorylation

1. Combine the following (total reaction volume 25 μl):

RNA	21.5 μl
10× NEB3 buffer	2.5 μl
CIP alkaline phosphatase	1 μl

2. Incubate at 37°C for 30 min.
3. Add 175 μl of water. Add an equal volume (200 μl) of phenol/chloroform/isoamyl alcohol and vortex for 30 s to 1 min.
4. Spin at 20,000 × *g* for 2 min at RT. Collect the upper (aqueous) phase and extract again by adding an equal volume of phenol/chloroform/isoamyl alcohol. Vortex for 30 s to 1 min and spin at 20,000 × *g* for 2 min at RT.
5. Collect the upper (aqueous) phase, add 2 μl glycogen (5 mg/ml), and 20 μl of 3 M sodium acetate (NaOAC), pH 5.2. Add 550 μl of ice-cold 100% ethanol, mix, and place at −80°C for 30 min.
6. Spin at 20,000 × *g* for 30 min at 4°C.

7. Carefully aspirate the supernatant and wash the pellet with 500 μl of ice-cold 70% ethanol. Spin at 20,000 × *g* for 10 min at 4°C.
8. Carefully aspirate the supernatant and air-dry the pellet.
9. Resuspend the pellet in 12 μl of MilliQ water.

3.2.2. 5′-End Labeling of RNA

1. For each labeling reaction combine (for a total reaction volume 15 μl):

RNA (CIP-treated)	10.5 μl
10× T4 PNK buffer	1.5 μl
[γ-^{32}P] ATP	2 μl
T4 PNK	1 μl

2. Incubate at 37°C for 1 h. Add 15 μl of RNA loading buffer.

3.2.3. piRNA Gel Purification

1. To prepare 15% urea/PAGE simply make a 1:1 mix of 10 and 20% urea/PAGE solutions.
2. Cast the gel apparatus using 0.75-mm combs and 18 × 24 cm glass plates. Dispense 40 ml of 15% urea/PAGE solution in a 50-ml Falcon tube. To polymerize, add 200 μl of 10% APS and 30 μl of TEMED, mix well, and immediately pour the gel.
3. Load most, or all of the labeled RNA. Also load radio-labeled DNA or RNA markers (see below for the preparation of radiolabeled DNA markers) and the RNA from the negative control antibody. There is no need to heat the RNA prior to loading. Run the gel until the bromophenol blue dye is at the bottom of the gel.
4. Disassemble the glass plates and lift the gel on a piece of old, exposed film. Cover with Saran wrap and expose the wet gel to a film by placing it in a cassette, between two intensifying screens, at −80°C. Use a radioactive pen or other means (e.g., preflashing) to align the gel with the film. Exposure time varies with the amount of RNA precipitated and loaded on gel. Usually ~5 h is sufficient.
5. Excise the gel piece corresponding to labeled piRNAs with a clean razor blade and place it in a microcentrifuge tube.
6. Add 400 μl of elution buffer and incubate at 37°C for 12–16 h.
7. Collect the elution buffer and place in the Illumina Spin-X cellulose acetate column. Centrifuge for 2 min at maximum speed in a table-top centrifuge at RT.
8. Add 4 μl of glycogen and 1 ml of ice-cold 100% ethanol. Place at −80°C for 30 min.

9. Spin at 20,000 × *g* for 30 min at 4°C.
10. Carefully aspirate the supernatant and wash the pellet with 500 μl of ice-cold 70% ethanol. Spin at 20,000 × *g* for 10 min at 4°C.
11. Carefully aspirate the supernatant and air-dry the pellet (do not over-dry).
12. Resuspend the pellet in 6 μl MilliQ of water.

3.3. Ligation of Adapters to piRNAs

The following steps (subheadings 3.3 and 3.4) are performed with adapters and reagents provided by the "Small RNA Sample Prep Kit" from Illumina. The 5′-adapter (5′-SRA) is 26 nucleotides and the 3′-adapter (3′-SRA) is 22 nucleotides. The sequence of the adapters and primers is shown in Subheading 2.

3.3.1. 5′ Adapter Ligation and Purification

1. Set up the following 10 μl reaction using the reagents provided with the Illumina kit:

RNA (gel-purified piRNA)	5.7 μl
SRA 5¢ adapter	1.3 μl
10× T4 RNA ligase buffer	1.0 μl
RNAse OUT	1.0 μl
T4 RNA ligase	1.0 μl

2. Incubate at 4°C for 16 h.
3. Add 10 μl of the SRA gel-loading dye to the ligation reaction.
4. Gel-purify the 5′-ligated piRNA product as above – the band should be approximately 56 nt long. After the gel purification, resuspend the eluted RNA in 7 μl of MilliQ water.

3.3.2. 3′ Adapter Ligation and Purification

1. Set up the following 10 μl reaction using the reagents provided with the Illumina kit:

RNA (5′-SRA-piRNA)	6.4 μl
SRA 3′ adapter	0.6 μl
10× T4 RNA ligase buffer	1.0 μl
RNAse OUT	1.0 μl
T4 RNA ligase	1.0 μl

2. Incubate at 4°C for 16 h.
3. Add 10 μl of SRA gel-loading dye to the ligation reaction.
4. Gel-purify the 5′, 3′-ligated piRNA product as above – the band should be approximately 78 nt long. After the gel purification, resuspend the eluted RNA in 5 μl of MilliQ water.

3.4. Amplification and Purification of Adapter-Ligated piRNAs

3.4.1. Reverse Transcription

1. Combine the following:

RNA	4.5 μl
SRA RT primer	0.5 μl

2. Heat the above mix at 65°C for 10 min. Place on ice and spin.
3. Add to the above, while still on ice:

5× First strand buffer	2 μl
DTT (0.1 M)	1 μl
RNase OUT	1 μl
dNTPs (12.5 mM)	0.5 μl

4. Heat the above mix at 48°C for 3 min.
5. Add 1 μl of SuperScript II reverse transcriptase and incubate at 44°C for 1 h.
6. Inactivate the enzyme by incubating at 65°C for 20 min. Chill on ice, spin down briefly, and store at −20°C or proceed with the PCR.

3.4.2. PCR

1. For a 50-μl reaction, assemble in a thin-walled PCR tube the following:

Water	28 μl
5× Phusion HB buffer	10 μl
dNTPs (25 mM)	0.5 μl
Primer GX1	0.5 μl
Primer GX2	0.5 μl
Phusion DNA polymerase	0.5 μl
Template (RT reaction)	10 μl

2. Perform the following PCR program:
   (a) 98°C for 30 s
   (b) 15 cycles of:
       98°C for 10 s
       60°C for 30 s
       72°C for 15 s
   (c) 72°C for 10 min
   (d) 4°C hold
3. Following the PCR, add 10 μl of 6× DNA loading buffer. The total volume is now 60 μl.

3.4.3. Native PAGE Purification of the PCR-Amplified piRNA Library

1. Prepare an 8% native PAGE gel as follows:

MilliQ water	17.5 ml
10× TBE	2.5 ml
40% of 19:1 polyacrylamide:bis	5 ml
20% APS	20 μl
TEMED	20 μl

2. Run the entire piRNA library sample (60 μl) on gel using the provided 25 bp marker as a reference.
3. Disassemble the gel and stain in 50 ml of 1× TBE with 5 μl of ethidium bromide (10 mg/ml) for 3 min.
4. Cut out the gel band corresponding to approximately 100–105 nucleotides (this is the piRNA library).
5. Elute the DNA from the gel fragment using 100 μl of Illumina kit 1× gel elution buffer, shaking at 30°C.
6. Collect the elution buffer and spin on provided Spin-X Illumina column for 2 min at 20,000 × *g*.
7. To the eluate, add 1 μl of glycogen, 10 μl of 3 M NaOAC, pH 5.2, 325 μl of −20°C 100% ethanol, and mix.
8. Spin immediately at 20,000 × *g* at RT for 20 min.
9. Remove the supernatant and wash the pellet with 500 μl of 70% ethanol at room temperature.
10. Air-dry the pellet and resuspend in 10 μl of Illumina resuspension buffer.

3.5. Illumina Next Generation Sequencing

Proceed with next generation sequencing on Illumina GAII analyzer as per the manufacturer's instructions (see Note 7). The directional adapter ligation allows the easy identification of the strand polarity of the sequenced piRNAs as their 5′-ends always follows the SRA 5′ adapter.

3.6. Preparation of Radiolabeled pBR322 DNA-Msp I Digest Markers

1. Combine the following (total reaction volume 20 μl):

pBR322/Msp I Digest (NEB)	1 μl (=1 μg)
10× EcoPolI (Klenow) buffer	2 μl
[α-^{32}P] dCTP	5 μl
DNA polymerase I (Klenow)	1 μl
Water	11 μl

2. Incubate at 30°C for 15 min.
3. Add 200 μl of water and 200 μl of phenol/chloroform/isoamyl alcohol, and vortex. Extract and ethanol precipitate the labeled marker as described above. After the final wash with

70% ethanol, dry the pellet and resuspend in 100 μl of RNA loading buffer.

4. Heat the labeled marker at 95°C for 3 min. Cool on ice, and spin briefly. This is the stock (very highly radioactive) marker. Make a 1:100 dilution of an aliquot from the stock in the RNA loading buffer to make working dilution of marker. Store the stock and the dilution at −20°C (see Note 8).

4. Notes

1. Piwi proteins from diverse species (such as mouse, *D. melanogaster*, *X. laevis*) contain symmetrical dimethylarginines (sDMAs) and these sDMAs are recognized by the Y12 monoclonal antibody (9, 16). Other antibodies directed against specific Piwi proteins (e.g., anti-Mili (9); anti-Miwi (17); anti-Aub (6), etc.) may also be used to immunoprecipitate the piRNAs bound to specific Piwi proteins.
2. The strategy outlined above may be adopted for immunopurification and the cloning of RNAs from any RNP that contains small RNAs (e.g., microRNAs), provided that these RNAs contain 5′ and 3′ ends that are amenable to ligation.
3. Defolliculation of *X. laevis* oocytes is performed by collagenase treatment of *X. laevis* as follows. Briefly, tease apart ovary into small clumps using forceps, and place the ovary in OR2− buffer. Wash with two changes of 100 ml OR2− buffer. Place ovary clumps in small glass Petri dish and add 3 ml of collagenase working solution. Shake gently on platform shaker for ~1 h at room temperature, checking frequently for the dissociation of the oocytes from follicles. When most of the oocytes are dissociated, stop and discard the few undissociated clumps. Rinse four times with 100 ml of OR2− buffer and wash at least five times with 100 ml of OR2+ buffer. Place the oocytes into 1.5 ml tubes, discard the buffer, and store at −80°C. To prepare oocyte lysate for immunoprecipitation, add 5 packed oocyte volume of lysis buffer into the tube, break the oocytes by pipetting, and proceed with the sonication protocol described in Subheading 3.
4. Before homogenizing the mouse testis, detunicate by gently grasping and peeling off the tunica albuginea using finely tipped forceps.
5. Do not over-dry the RNA pellets, as they can become very hard to resuspend. Resuspension is best if the pellet is damp and appears glassy.
6. The purified RNA may also be used for other applications, such as northern.

7. Quality control and quantification of the piRNA library may be performed by electrophoresis of a small aliquot of the purified PCR on an Agilent 2100 Bioanalyzer by following the manufacturer's protocol. The bionalyzer is typically an integral component of core facilities that perform Illumina sequencing.
8. The marker is good for at least 3 months, but adjustments need to be made on how much to load depending on the decay.

Acknowledgment

Supported by NIH grants GM072777 and NS056070 to ZM.

References

1. Kuramochi-Miyagawa, S., Kimura, T., Ijiri, T. W., Isobe, T., Asada, N., Fujita, Y., Ikawa, M., Iwai, N., Okabe, M., Deng, W., Lin, H., Matsuda, Y., and Nakano, T. (2004) Mili, a mammalian member of piwi family gene, is essential for spermatogenesis, *Development* ***131***, 839–849.
2. Deng, W., and Lin, H. (2002) miwi, a murine homolog of piwi, encodes a cytoplasmic protein essential for spermatogenesis, *Dev Cell* ***2***, 819–830.
3. Carmell, M. A., Girard, A., van de Kant, H. J., Bourc'his, D., Bestor, T. H., de Rooij, D. G., and Hannon, G. J. (2007) MIWI2 is essential for spermatogenesis and repression of transposons in the mouse male germline, *Dev Cell* ***12***, 503–514.
4. Harris, A. N., and Macdonald, P. M. (2001) Aubergine encodes a Drosophila polar granule component required for pole cell formation and related to eIF2C, *Development* ***128***, 2823–2832.
5. Cox, D. N., Chao, A., Baker, J., Chang, L., Qiao, D., and Lin, H. (1998) A novel class of evolutionarily conserved genes defined by piwi are essential for stem cell self-renewal, *Genes Dev* ***12***, 3715–3727.
6. Gunawardane, L. S., Saito, K., Nishida, K. M., Miyoshi, K., Kawamura, Y., Nagami, T., Siomi, H., and Siomi, M. C. (2007) A slicer-mediated mechanism for repeat-associated siRNA 5′ end formation in Drosophila, *Science* ***315***, 1587–1590.
7. Brennecke, J., Aravin, A. A., Stark, A., Dus, M., Kellis, M., Sachidanandam, R., and Hannon, G. J. (2007) Discrete small RNA-generating loci as master regulators of transposon activity in Drosophila, *Cell* ***128***, 1089–1103.
8. Li, C., Vagin, V. V., Lee, S., Xu, J., Ma, S., Xi, H., Seitz, H., Horwich, M. D., Syrzycka, M., Honda, B. M., Kittler, E. L., Zapp, M. L., Klattenhoff, C., Schulz, N., Theurkauf, W. E., Weng, Z., and Zamore, P. D. (2009) Collapse of germline piRNAs in the absence of Argonaute3 reveals somatic piRNAs in flies, *Cell* ***137***, 509–521.
9. Kirino, Y., Kim, N., de Planell-Saguer, M., Khandros, E., Chiorean, S., Klein, P. S., Rigoutsos, I., Jongens, T. A., and Mourelatos, Z. (2009) Arginine methylation of Piwi proteins catalysed by dPRMT5 is required for Ago3 and Aub stability, *Nat Cell Biol* ***11***, 652–658.
10. Lau, N. C., Ohsumi, T., Borowsky, M., Kingston, R. E., and Blower, M. D. (2009) Systematic and single cell analysis of Xenopus Piwi-interacting RNAs and Xiwi, *EMBO J* ***28***, 2945–2958.
11. Thomson, T., and Lin, H. (2009) The biogenesis and function of PIWI proteins and piRNAs: progress and prospect, *Annu Rev Cell Dev Biol* ***25***, 355–376.
12. Malone, C. D., and Hannon, G. J. (2009) Small RNAs as guardians of the genome, *Cell* ***136***, 656–668.
13. Lerner, M. R., and Steitz, J. A. (1979) Antibodies to small nuclear RNAs complexed with proteins are produced by patients with systemic lupus erythematosus, *Proc Natl Acad Sci USA* ***76***, 5495–5499.

14. Mourelatos, Z., Dostie, J., Paushkin, S., Sharma, A., Charroux, B., Abel, L., Rappsilber, J., Mann, M., and Dreyfuss, G. (2002) miRNPs: a novel class of ribonucleoproteins containing numerous microRNAs, *Genes Dev* ***16***, 720–728.

15. Elbashir, S. M., Lendeckel, W., and Tuschl, T. (2001) RNA interference is mediated by 21- and 22-nucleotide RNAs, *Genes Dev* ***15***, 188–200.

16. Heo, I., and Kim, V. N. (2009) Regulating the regulators: posttranslational modifications of RNA silencing factors, *Cell* ***139***, 28–31.

17. Kirino, Y., Vourekas, A., Kim, N., de Lima Alves, F., Rappsilber, J., Klein, P. S., Jongens, T. A., and Mourelatos, Z (2010). Arginine methylation of vasa protein is conserved across phyla, *J Biol Chem,* ***285,*** 8148–8154.

Chapter 19

Generation of an Inducible Mouse ES Cell Lines Deficient for Argonaute Proteins

Hong Su and Xiaozhong Wang

Abstract

Argonautes (Agos) are core effectors of RNA silencing. In several nonmammalian organisms, multiple Agos are known to exhibit specialized functions for distinct RNA silencing pathway. Mammals have four closely related Agos. To examine the functions of mammalian Agos in the microRNA silencing pathway, we generated mouse embryonic stem (ES) cells that are nullizygous for all Agos. This chapter describes a variety of techniques including BAC recombineering, gene targeting, and inducible Cre-loxP recombination, used to generate inducible Ago knock-out ES cells. The Ago-deficient ES cells provide an important tool for the study of mammalian RNA silencing.

Key words: Mouse embryonic stem cells, BAC-mediated recombination, Cre-loxP, Gap repair, Argonaute

1. Introduction

RNA silencing modulates gene expression and regulates diverse biological processes. Argonaute proteins are the core effector proteins that directly interact with small RNAs to mediate the silencing effect in both siRNA and miRNA pathways. There are four Ago proteins in human and mice. Based on studies in nonmammalian systems, it has been suggested that different Argonaute proteins have distinct functions in RNA silencing. For instance, in *D. melanogaster*, Ago1 is required for mature miRNA production, while Ago2 is required for siRNA-triggered mRNA degradation (1, 2). In mammals, only Ago2 is required for the cleavage activity in the siRNA pathway, while all Ago1–Ago4 appear to be associated with miRNAs (3, 4). The species-specific differences underscore the need to further understand the RNAi pathway in a mammalian system.

Tom C. Hobman and Thomas F. Duchaine (eds.), *Argonaute Proteins: Methods and Protocols*, Methods in Molecular Biology, vol. 725, DOI 10.1007/978-1-61779-046-1_19, © Springer Science+Business Media, LLC 2011

Mouse ES cells are powerful tools for analyzing gene function because endogenous genes can be manipulated through homologous recombination in these cells. Thus, we used a genetic approach in ES cells to examine the function of mouse Ago subfamily members (Ago1–4) in miRNA silencing and the role of individual Ago proteins in this process.

In mice, Ago1, 3, 4 are tandemly located on chromosome 4, spanning more than 150 kb, while Ago2 is located on chromosome 15 (Fig. 1). To generate Ago1–4 deficient cells, we first deleted Ago1, 3, and 4 using chromosomal engineering technology (5) and then removed both allele of endogenous Ago2 using a conventional gene targeting method. Due to potentially deleterious effect associated with miRNA silencing defect, we introduced a hAgo2 transgene, which can be excised by a 4-hydroxytamoxifen (4OH-T) induced Cre expression. An outline of the procedures used to generate Ago1, 2, 3, $4^{-/-}$ ES cells is shown in Fig. 1. Each targeting event was selected using different

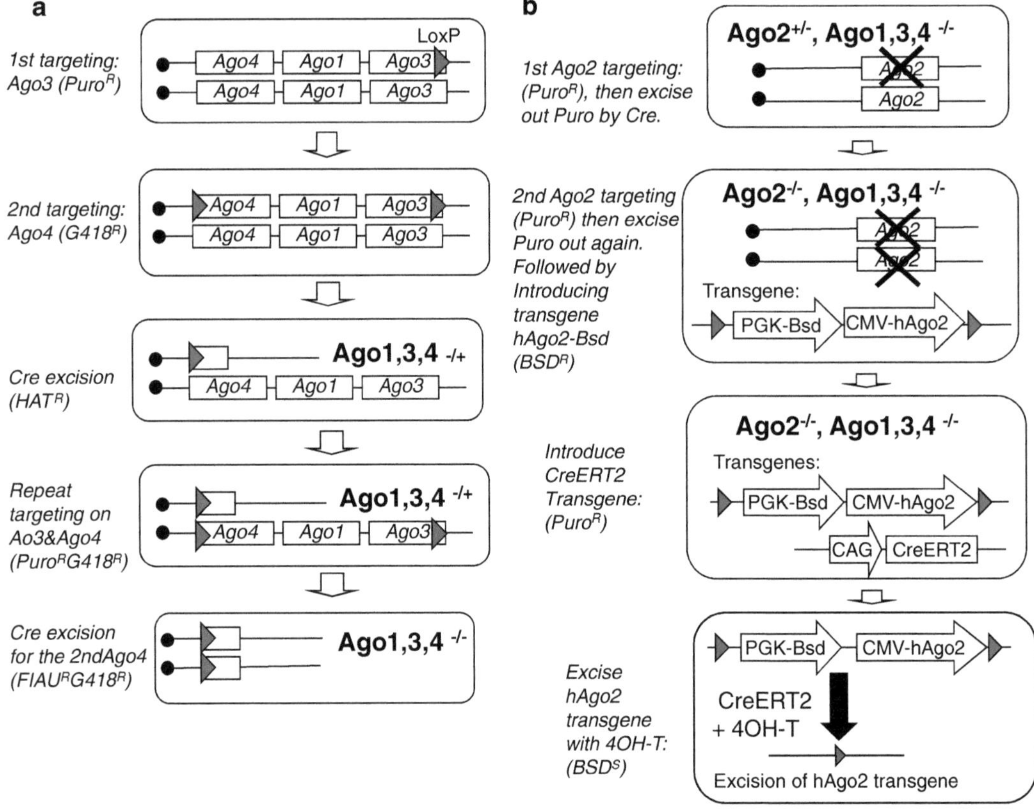

Fig. 1. A genetic strategy to generate inducible ES cells deficient for Ago1–4. (**a**) Multiple steps of gene targeting and Cre-mediated excision were employed to first generate Ago1, Ago3, $Ago4^{-/-}$ ES cells. (**b**) mAgo2 endogenous alleles were deleted by targeting in the presence of floxed hAgo2 transgene. A CreERT2 transgene was introduced to generate the inducible $Ago2^{-/-}$; Ago1, 3, $4^{-/-}$ line. Parts reprinted with permission from ref. 6.

drug resistance markers. Some of drug selection markers used in the first around of gene targeting were recycled after Cre-mediated excision to remove the drug selection markers that were flanked by *lox*P sites.

The Ago-deficient ES cells generated in this study provide valuable reagents for dissecting RNAi pathway (6). This chapter describes the strategy and methods used to generate homozygous Ago mutant ES cells and focuses on how to construct targeting vectors using bacterial artificial chromosome (BAC)-mediated recombination.

2. Materials

2.1. Culture Condition for ES Cells and STO Feeder Cells

1. NM5 ES cell line and Mitomycin C-treated feeder cells (see Note 1).
2. Culture medium for ES cells (M15): Knockout Dulbecco's Modified Eagle's Medium (Knockout-DMEM, Invitrogen) supplemented with 15% fetal bovine serum (FBS, defined grade, Thermo Scientific Hyclone), 2 mM l-glutamine, 100 U/ml penicillin and 0.1 mg/ml streptomycin (100× stock, Invitrogen), and 100 μM of beta-mercaptoethanol. One-month shelf life at 4°C.
3. Culture medium (M10) for feeder cells: Knockout-DMEM supplemented with 10% FBS (Premium Select grade, Atlanta Biologicals), 100 U/ml penicillin, and 0.1 mg/ml Streptomycin. One-month shelf life at 4°C.
4. 0.25% Trypsin (Invitrogen).
5. 0.1% porcine gelatin (Sigma) in ddH_2O. Stored at room temperature.
6. Mitomycin C (Sigma) used for preparing mitotically inactive feeder cells, final concentration is 1 mg/100 ml M10.
7. Phosphate-buffered saline (PBS, Invitrogen).
8. Light paraffin oil (EM Science): Sterilize by filtration.
9. Sterile transfer pipettes (Samco).

2.2. Electroporation, Drug Selection, and Screening of ES Cells

1. Drugs and their working concentration: puromycin (Puro, 2 μg/ml, InvivoGen), geneticin (G418, 350 μg/ml, Invitrogen), FIAU (1-(2′-deoxy-2′-fluoro-β-D-arabino-furanosy)-5-iodouracil) (0.2 mM), hypoxanthine aminopterin thymidine (HAT Supplement (1×), Invitrogen), 50× HT supplement (1×, Invitrogen), and Blasticidin (BSD 5 μg/ml, Invitrogen).
2. Mitomycin-treated feeder cells that are resistant to drugs applied for selection ($G418^R$: iSNL, $G418^R Puro^R$: iSNLP, $G418^R Puro^R BSD^R$: iSNLPB).

3. Gene Pulser Xcell™ with a capacitance extender (Bio-Rad) and electroporation cuvettes (4 mm gap, BTX).
4. 12-Channel pipettor and 8-channel aspirator system (ThermoFisher).
5. Flat-bottom and U-bottom 96-well tissue culture plates.
6. Lysis buffer for 96-well DNA extraction: 10 mM Tris-HCl pH = 7.5, 10 mM EDTA pH = 8.0, 10 mM NaCl, 0.5% sarcosyl, 1.5 mg/ml protease Type XIV (Sigma P5147, freshly added).

2.3. Molecular Cloning and BAC Recombination

1. Plasmids: pSC101-BAD-gbaA (Gene Bridges), BACs containing the gene of interest, plasmids with appropriated drug cassettes and poly-cloning sites.
2. DH10B electro-competent cells.
3. MicroPulser (Bio-Rad) and electroporation cuvettes (1 mm gap, BTX).
4. Incubator and horizontal shakers at 30 and 37°C.
5. 10% L-Arabinose (Sigma) in water, sterilized by filtration.
6. Drugs:

Drug	Stock	Working concentration
Chloramphenicol	34 mg/ml in ethanol	12.5 μg/ml for low copy plasmids, 25 μg/ml for multicopy plasmids
Ampicillin	100 mg/ml stock in 50% EtOH	50 μg/ml for low copy plasmids, 100 μg/ml for multicopy plasmids
Tetracycline	5 mg/ml stock in 75% EtOH	5 μg/ml
Zeocin	100 mg/ml stock in HEPES buffer, pH 7.25	50 μg/ml

7. Various restriction endonucleases with matching buffers.
8. Reagents for polymerase chain reaction (PCR) and long-range PCR kit (Roche).
9. Ice-cold double-distilled water (ddH_2O), keep at 4°C for at least 4 h (best if overnight).

3. Methods

3.1. Principle and Design in Chromosomal Engineering

Chromosomal engineering is based on Cre-loxP system originated from bacterial phage P1. The principle of chromosome engineering is illustrated in Fig. 2. Basically, two *lox*P-containing targeting vectors are consecutively delivered into desired

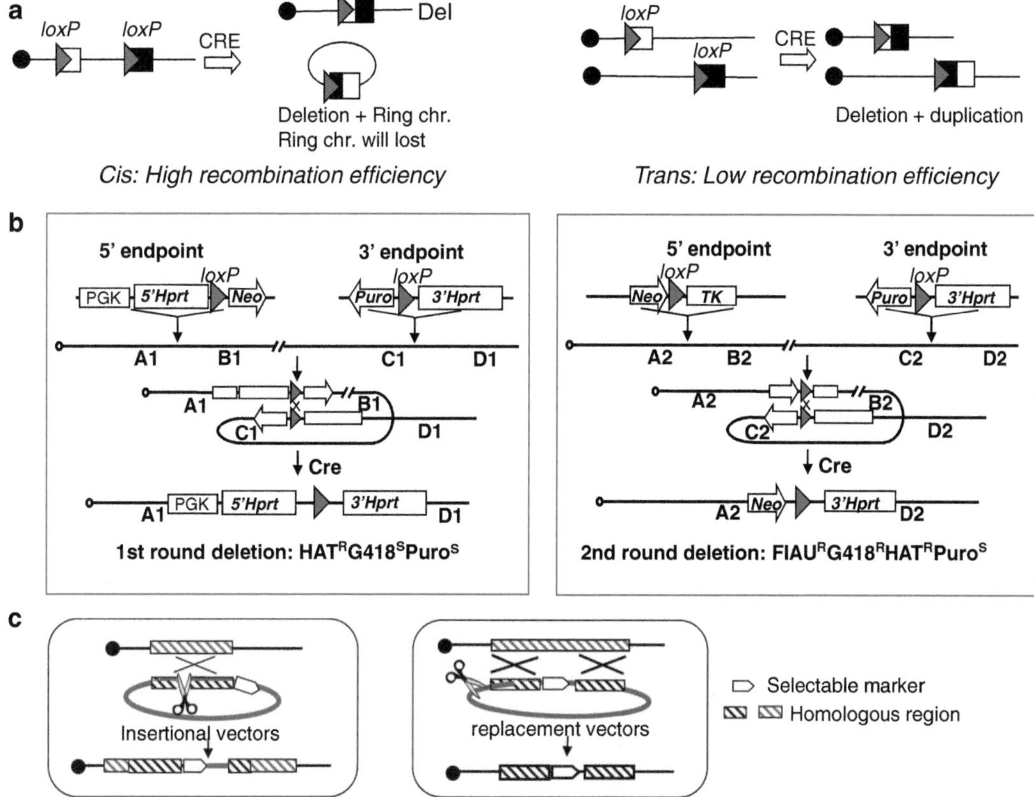

Fig. 2. Principle of chromosome engineering. (**a**) Recombination products and efficiencies of Cre-loxP-mediated recombination depending upon whether *lox*P sites are in *cis* or *trans* configuration. *Cis*: 2 *lox*P sites are on the same chromosome; *trans*: 2 *lox*P sites are on different chromosomes. (**b**) Schematic of the strategy designed to achieve two rounds of Cre-loxP-mediated recombination used for generating Ago1, 3, $4^{-/-}$ cells. (**c**) Two types of targeting vectors: Insertion vectors and replacement vectors. *Scissors mark* the positions of double strand breaks generated by restriction enzymes.

endpoints through gene targeting. Expression of Cre results in chromosomal rearrangements (e.g., deletion) that can be identified based on drug selection. For the first round of deletion (Fig. 2b, left), Cre-mediated recombination leads to the reconstitution of two nonfunctional 5′HPRT and 3′HPRT cassettes into a functional HPRT minigene, thus serving as a positive selection for recombinants. Neomycin (Neo, $G418^R$) and puromycin (Puro) serves as positive drug markers for gene targeting. The two targeting vectors are designed in such a way that both 5′HPRT and 3′HPRT cassettes are outside of the *lox*P-flanked region, while neomycin and puromycin are inside. Upon the expression of Cre, ES cells heterozygous for Ago1, 3, and 4 are $HAT^RG418^SPuro^S$. For the second round deletion (Fig. 2b, right), thymidine kinase (TK) and puromycin are placed between 2 *lox*P sites and used to identify the deletions, while neomycin and puromycin are used as markers for gene targeting steps (see Note 2). The deletion events can be identified by the loss of TK

(FIAU^R). Because Cre-mediated recombination between two *lox*P sites that are located on the same chromosome (*cis*) has 1,000 times higher efficiency than that on a different chromosome (*trans*, Fig. 2a), chromosomal translocation events resulting from *trans*-configuration are rare compared with the deletions. Moreover, deletions and translocations exhibit different drug resistant phenotypes. After Cre-mediated deletion, the Ago1, 3, 4$^{-/-}$ cells should exhibit Puro^{S}G418^{R}FIAU^{R}HAT^R. The deletion events were also further confirmed by Southern analysis.

Gene targeting plays an important role in chromosomal engineering. All the deletion endpoints are introduced by gene targeting. A gene targeting vector consists of two key parts: genomic sequences used for homologous recombination and drug markers used for selection. Based on the position of a double strand break (DSB), targeting vectors can be divided into insertion vectors and replacement vectors (7) (Fig. 2c). Insertion vectors contain a DSB in the region homologous to the chromosomal counterpart, and usually recombine with host chromosomes through a single reciprocal crossover (O-form). Replacement vectors contain a DSB at the edge of homologous sequences and combine with the host genomic sequences though double-reciprocal crossover, which lead to a gene conversion. Because insertion vectors contain continuous genomic fragments thus needing only one round of BAC recombination, we choose to introduce the deletion endpoints (Ago3 and Ago4 targeting vectors) with insertion vectors. For the deletion of Ago2, we designed the Ago2 targeting construct as a replacement vector, which was generated through two rounds of BAC recombination. The next two sections describe how we constructed both insertion vectors and replacement vectors using BAC recombination.

3.2. Construction of Insertion Targeting Vectors Through BAC Recombination

We take advantage of the end sequence-mapped BAC library and well-established Red/ET system (8, 9) to generate gene targeting vectors. By expression of phage proteins RecE/RecT, DNA fragments can be precisely modified through homologous recombination in bacterial hosts. Briefly, each round of recombination-based cloning requires four major steps: (1) construction of a mini-retrieving vector; (2) preparation of recombination competent bacterial host; (3) recombination; (4) confirmation. Here, we use insertion vectors as an example to describe the general steps in detail (Fig. 3).

3.2.1. Construction of Mini-retrieving Vector

The homologous arms used in targeting vectors are subcloned using PCR-amplified fragments from a BAC clone. Mini-retrieving vectors are an intermediate plasmid used in the step of homologous recombination in BACs. It contains three important parts, 5′ homologous arm, 3′ homologous arm, and a drug selection cassette, which later serves as a selectable marker in homologous recombination in ES cells (Fig. 3a).

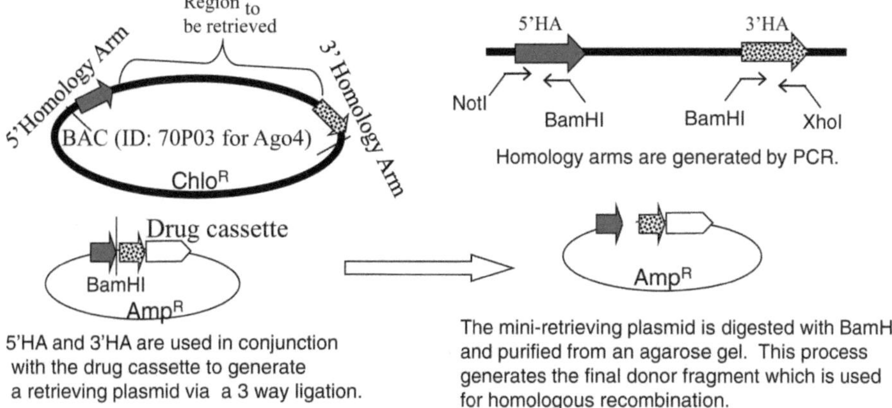

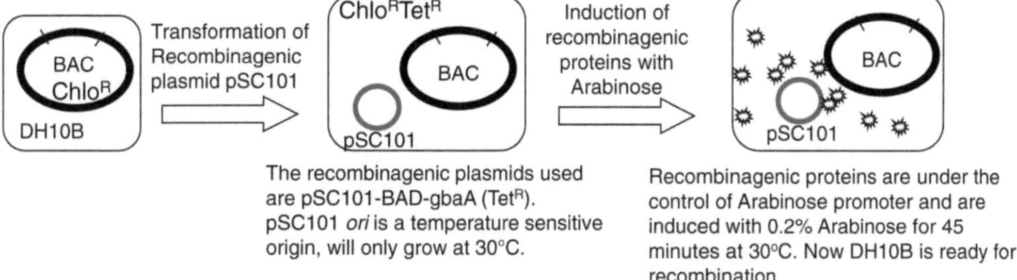

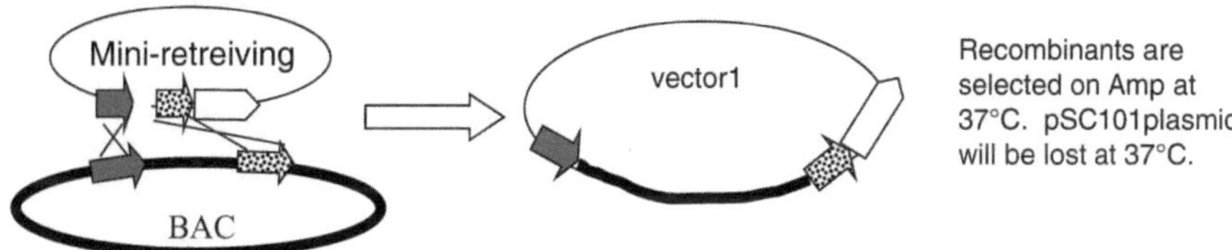

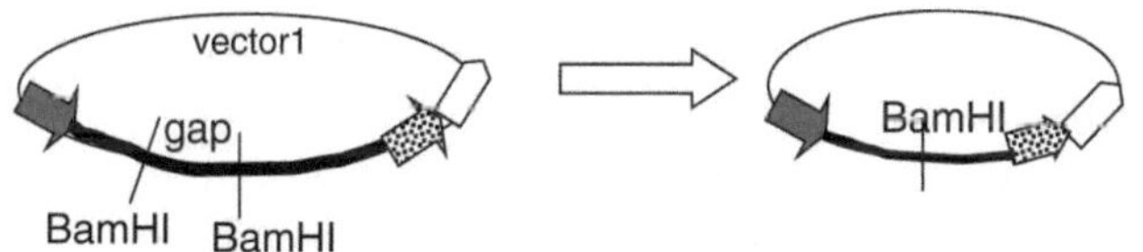

Fig. 3. Schematic illustration of the four steps involved in making an insertion vector. For details see Subheading 3.2.

1. Using BACs (see Notes 3 and 4) as a template, both 5′ and 3′ homologous arms are amplified using primers with appropriated restriction enzyme sites (see Note 5). The 5′ and 3′ homologous arms are short DNA sequences of 100–250 bp, flanking the desired region to be retrieved from BAC (see Note 6). The mouse genome sequence database allows

streamlined design of PCR primers. The PCR products should be confirmed using various restriction enzymes or sequencing.

2. The PCR products and drug cassettes are inserted into a plasmid backbone with β-lactamase gene (Ampicillin^{R}) through a three-way ligation.
3. The mini-retrieving vector is linearized with a restriction enzyme and the digested fragments are double gel-purified using a Qiagen kit (see Note 7).

3.2.2. Preparation of Bacteria for Homologous Recombination

Here we use a DH10B clone carrying a BAC as an example to explain how to transform pSC101-BAD-gbaA into a bacterial host (Fig. 3b).

1. Streak a BAC clone on an LB agar plate with 12.5 μg/ml of chloramphenicol at 37°C for overnight.
2. Inoculate a 2.0 ml LB liquid culture with 12.5 μg/ml of chloramphenicol with a single BAC colony and culture overnight at 37°C. Remember to *pre-cool* ddH_2O at 4°C.
3. Next day, centrifuge 1 ml culture in a table-top centrifuge at 11,000 rpm for 30 s (remember to perform the following procedures on ice).
4. Discard supernatant and suspend the cell pellet with 1 ml of *pre-cooled* ddH_2O.
5. Spin down the cells for 30 s at 11,000 rpm in a table-top centrifuge.
6. Keep the tube on ice, discard the supernatant and re-suspend the cells with 1 ml of *pre-cooled* ddH_2O.
7. Spin down again for 30 s at 11,000 rpm in a table-top centrifuge and then discard supernatant while leaving 20–30 μl in the tube.
8. Add 1 μl of plasmid pSC101-BAD-gbaA (about 2 ng).
9. Mix well and transfer into an electroporation cuvette (1 mm gap).
10. Electroporate at 1,250 V ("EC1" for Bio-Rad MicroPulser). The time constant should be between 4 and 6 ms.
11. Add 1 ml of LB medium to the cuvettes and transfer the LB into an eppendorf tube.
12. Incubate at 30°C for 60 min with shaking.
13. Plate bacteria on LB agar plates containing tetracycline 5 μg/ml (for pSC101-BAD-gbaA) and 12.5 μg/ml of chloramphenicol (for BAC).
14. Incubate the plate at 30°C overnight. Now the cells are ready for homologous recombination (see Note 8).

3.2.3. Generation of an Insertion Targeting Vector Using BAC Recombination

1. Inoculate a single colony of pSC101-BAD-gbaA transformed BAC into a 3 ml LB medium with tetracycline (5 μg/ml) and chloramphenicol (12.5 μg/ml). Culture the bacteria with constant shaking at 30°C overnight.
2. The next day, dilute 0.5 ml of overnight culture into 25 ml low salt LB in a flask containing tetracycline (5 μg/ml) and chloramphenicol (12.5 μg/ml), the final OD_{600} is between 0.02 and 0.04. Shake the culture at 30°C for 2–3 h till OD_{600} reaches 0.2–0.3.
3. For each BAC, split the culture into two flasks (about 10 ml each), one induced with l-Arabinose and one without (negative control).
4. Add L-Arabinose to a final concentration of 0.1–0.2% (dilute from 10% stock, 100×) to induce the expression of recombinase. Shake at 30°C for 30 min until OD_{600} reaches ~0.4.
5. Pre-cool some 15 ml conical tubes and the centrifuge that you are about to use to 4°C.
6. Transfer the cultures to a 15 ml conical tubes to a table-top centrifuge and spin down the bacteria at 4,000 rpm for 8 min at 4°C.
7. Discard the supernatant and place the tubes on ice. Wash the bacteria with 1 ml ice-cold ddH_2O twice in pre-cooled microfuge tubes.
8. After the second wash, remove and discard the supernatant using a 1 ml pipette while leaving about 30–60 μl of solution to resuspend the bacteria.
9. Add ~50 ng (about 1–3 μl, see Note 9) of mini-retrieving vectors (linearized and double gel-purified) and place mixture into an ice-cooled electroporation cuvette (1 mm gap).
10. Electroporate at 1,250 V.
11. Add 1 ml of LB medium into the cuvette and transfer it back into a microfuge tube and incubate at 37°C for 75 min.
12. Plate the bacteria on LB agar plates containing suitable antibiotic (Amp).
13. Incubate the plates at 37°C overnight. pSC101-BAD-gbaA plasmid will be lost at 37°C.
14. Compare both the number and the size of colonies between induced and non-induced plates. You should observe more Amp^R colonies on induced plates than un-induced ones. The recombinants are sometimes smaller in size compared with background colonies.
15. Confirm the final construct with various restriction enzymes digestion.

16. If possible, create a gap in the genomic fragment using restriction enzymes and re-ligate the vector. The gap can be used to design PCR primers in the screening step of gene targeting in ES cells (see Note 10 and Subheading 3.4.2).

3.3. Construction of Replacement Targeting Vectors Using BAC Recombination

The principle of constructing a replacement vector is illustrated in Fig. 4. It requires two steps of recombination: (1) subcloning genomic fragments from BACs and (2) integration of a drug cassette into the middle of the genomic fragment retrieved.

3.3.1. Subcloning BACs into Plasmid Vectors

The first round of BAC recombination is to retrieve genomic fragments used as homologous arms in the final targeting vector from BACs. Follow procedures described in Subheadings 3.2.1–3.2.3 to generate vector 2 (Fig. 4A).

3.3.2. Integration of a Drug Cassette into a Subcloned Genomic Fragment

1. A second mini-retrieving vector, containing two homologous arms (5′Ha and 3′Ha), a prokaryotic antibiotic cassette (Zeo), and an eukaryotic drug cassette, is constructed through ligation (see Fig. 4B-a). Prepare the second donor fragments from the second retrieving vectors by restriction enzyme digestion followed by gel purification.
2. Following the steps in Subheading 3.2.2 to introduce PSC101-BAD-gbaA into DH10B. After electroporation, plate bacteria on LB agar plate with 5 μg/ml (note lower concentration) tetracycline (no Chloramphenicol!!).
3. Follow steps 1–11 (see Note 11) in Subheading 3.2.3 to electroporate both vector 2 (50 ng) and the second donor fragments (50 ng) into recombination-competent DH10B cells.
4. After incubation at 37°C, plate 1/10 of the cells on an Ampicillin agar plate which will be used to control electroporation efficiency and the rest on a Zeocin plate where only recombinants will survive.
5. The next day, pool 10–20 Zeo^R colonies into one LB liquid culture with 50 μg/ml Zeocin, shake at 37°C overnight.
6. Prepare one DNA mini-prep from the culture above and retransform them into DH5α, and plate on Zeocin agar plates.
7. Pick 10 Zeo^R single clones from the plates and prepare plasmid DNA individually. The final targeting vector will be confirmed using various restriction enzymes.

3.4. Gene Targeting in ES Cells

3.4.1. Electroporation of ES Cells

Electroporation is an efficient way of delivering DNA fragments into ES cells. Targeting vectors are linearized with restriction enzymes and purified with phenol–chloroform before electroporation (see Note 12).

A.First round recombination:

a. Preparing the 1st mini-retrieving vector

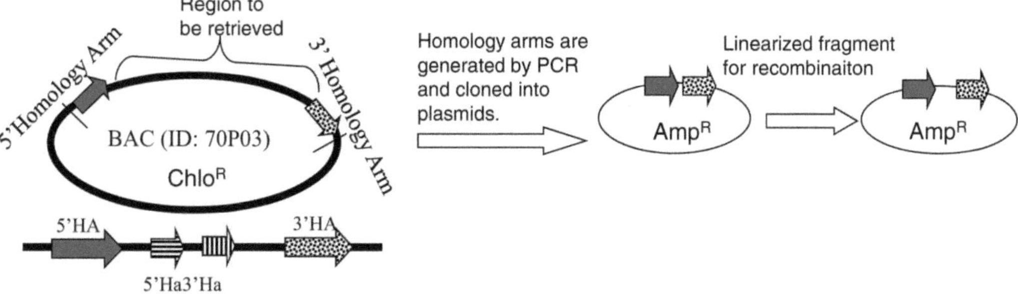

b. Preparing BAC cells for homologous recombination

c. Retrieving genomic fragments through the 1st round of Red/ET Recombination

d. Confirmation of the recombinant (vector 2)

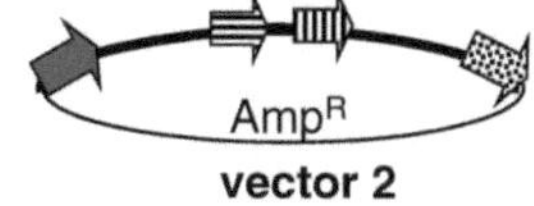

B.Second round recombination:

a. Preparing the 2nd mini-retrieving vector

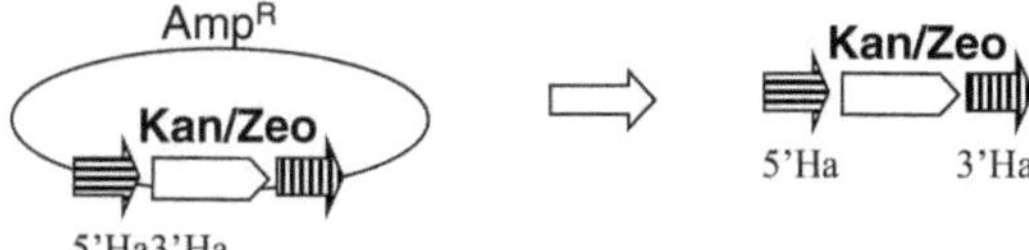

b. Preparing DH10B cells for homologous recombination

c. Inserting selectable marker into vector2 through the 2nd round Red/ET recombination.

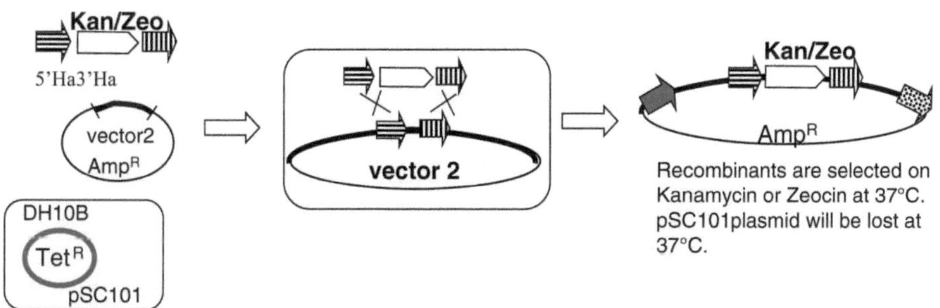

d. Retransform first, then confirmation of the recombinant (final replacement vectors).

Fig. 4. Schematic illustration of the two rounds of recombination involved in making a replacement vector. For details see Subheading 3.3.

1. Prepare one 10 cm plate of ES cells (enough cells for three to four electroporations), which should be released from any drug selection at least 2 days before electroporation.
2. Pre-feed ES cells with fresh M15 2 h before electroporation.
3. Wash cells with PBS twice to remove serum.
4. Add 3 ml of 0.25% trypsin to the 10 cm plate, incubate for 15 min in an incubator.
5. Add 3 ml of M15 to stop trypsinization and use transfer pipets to break cell clumps thoroughly by pipetting cell suspension up and down at least 15 times.
6. Spin down cells in a table-top centrifuge at 1,000 rpm for 5 min. A cell pellet should be visible at the bottom of the tubes.
7. Aspirate the supernatant and then resuspend the cell pellet with 10 ml PBS. Remove 20 μl for cell counting (dilute as necessary) using a hemocytometer.
8. Spin down cells in a table-top centrifuge at 1,000 rpm for 5 min. Aspirate the supernatant and then resuspend the cells in PBS at a concentration of 1.1×10^7/ml.
9. For each electroporation, mix 0.9 ml of cells (10^7 cells) with the DNA constructs and then transfer to an electroporation cuvette (4 mm gap). Zap at 230 V, 500 μF. The time constant should be ~6.0–8.0 ms.
10. Leave the cuvettes at room temperature for 5 min and then add 0.9 ml of M15 to each cuvette, mix with a transfer pipette for a few times. Plate electroporated ES cells at a density of 6×10^4 cells/cm^2.
11. Refeed the plate with M15 after 12–16 h.
12. Drug selection should start ~24–36 h after electroporation (see Note 13).
13. Feed the cells with fresh M15 containing appropriate drug every other day for the first 6 days until most cell debris is gone.
14. Colonies may be picked and passaged into 96-well plates ~9–10 days after electroporation.

3.4.2. Screening of Targeted ES Cell Clones

Screening of targeted ES clones is first performed by long-range genomic PCR in a multiplexed 96-well format. Positive clones are then confirmed by single-clone format. To detect gene-targeting events, two PCRs are designed to characterize the correct integration at 5′ and 3′ end, respectively. For each primer pair, one is designed within the targeting vector (usually in the drug cassette, called internal primer), the other is designed outside the targeting vector (not shared with targeting vectors, called external primer). For insertion vectors, we created gaps in the homologous arms of the targeting vectors so that the external primer can be designed within the gap to keep screening PCR products shorter (see Note 10). All positive clones identified by PCR need to be further confirmed by Southern analysis with internal/external probes (see Note 14).

1. Prepare two 96-well plates for each plate about to passage; one plate with feeders and one gelatin-treated plate without feeders. Add 100 μl M15 to each well and keep in the incubator. (Cells will reach ~80% confluency in 3–4 days after picking colonies.)
2. Wash the plates with 100 μl/well PBS.
3. Add 50 μl trypsin, incubate for 15 min at 37°C.
4. Add 50 μl M15 to neutralize trypsin, pipette the cells up and down for about 15 times to properly disaggregate cell clumps.
5. Aliquot ~45 μl of cell mixture per well to each plate and mix slightly in the well.
6. Change media the next day and then every other day until cells are ~80% confluent.
7. The plate with feeders is to be frozen down and kept as stock at −80°C (see Note 15). The other plate without feeders will be used to prepare DNA.
8. Prepare DNA from 96-well plates as described in Note 17.
9. To perform PCR screening, first take 10 μl/well DNA from each plate and pool them into 96-well plate format.
10. Set up 3′ end (see Note 18) screening PCR on the DNA pool plate first.

DNA	2 μl
10× PCR buffer with 1.5 mM $MgCl_2$	3 μl
5 mM dNTP	2 μl
10 μM Primer 1	1 μl
10 μM Primer 2	1 μl
Taq polymerase	1.5 μl
H_2O	19 μl
Total volume	30 μl

94°C 2′→(94°C 15″–60°C 15″–68°C 6′) 10 cycles→(94°C 15″–60°C 15″–68°C 6′ with 10″ increment/cycle) 22 cycles→68°C 6′→4°C ∞.

11. Run samples on agarose gel and identify well ID for the positive clones. Repeat PCR on clones with the same well ID from each plate in the pool (see Note 19).

12. After locating both plate number and well ID, the targeted clones are further confirmed using 5′ end screening primers.

3.5. Cre-Induced Recombination

After both Ago3 and Ago4 were targeted, it was critical to determine whether the two *lox*P sites are located on the same (*cis*) or different allele (*trans*) of the homologous chromosomes (Fig. 5).

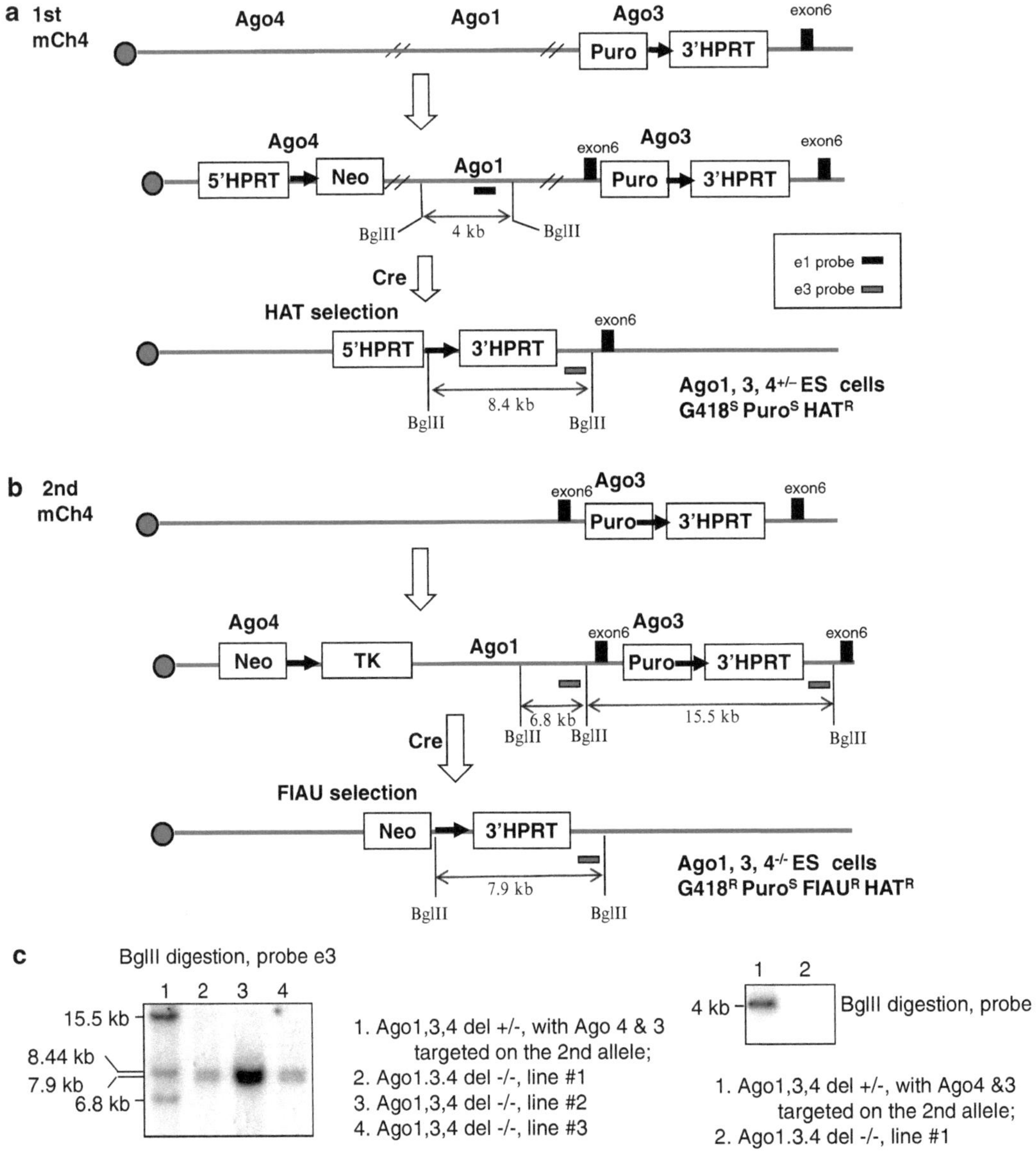

Fig. 5. Southern blot analysis of the Ago1, 3, $4^{+/-}$ and Ago1, 3, $4^{-/-}$ clones. (**a**) (**b**) Schematic of Ago1, 3, 4 genomic regions before and after Cre-mediated deletions. (**c**) *Left*: Southern blot analysis using probe e3-confirmed deletions. *Right*: Southern blot analysis using probe e1 internal to the deleted region confirmed desired chromosomal deletions.

1. We randomly select six Ago3 and Ago4 double-targeted clones and then electroporated each clone (1×10^6) with 5 μg PGKCre expression vectors to induce recombination between two *lox*P sites (follow ES electroporation procedures described in Subheading 3.4.1). Add the electroporated cells to a six-well plate and apply appropriate drug selection for recombination events.
2. When the *lox*P sites are in the *cis* configuration (see Fig. 2a), after Cre expression, at least 100 times more HAT^R clones are generated. Therefore, we can distinguish clones with *cis* configuration from *trans* based on the number of HAT^R clones after Cre expression.
3. Electroporate 10 μg of Cre expression vectors into the two *cis* clones (10^7 cells each) and then add the cells to three 10-cm plates for each electroporation. After 8–9 days, pick HAT^R colonies (see Subheading 3.4.1) into 96-well plates.
4. The candidate clones are further tested with sib-selection of Puro and G418 (see Note 16). Agol, 3, and $4^{+/-}$ cells should be $G418^SHAT^RPuro^S$.
5. Agol, 3, $4^{+/-}$ cells are further confirmed by Southern analysis using both an internal probe and an external probe (Fig. 5c, left).

3.6. Generation of Ago 1, 3, 4 Deficient Cells

The second allele of Ago3 and Ago4 in Agol, 3, $4^{+/-}$ cells are sequentially targeted as described above (Fig. 2b, right). To avoid targeting the deletion allele, the homologous arm of the new Ago4 targeting vector is designed to be within the deletion region. The same targeting vector was used for the second allele of Ago3 targeting. To determine *cis* and *trans* configurations, we tested Cre-mediated recombination as described in Subheading 3.5. We electroporated Cre expression vectors into two clones with *cis* configuration and selected for $FIAU^R$ clones. Candidate clones are expanded in triplicate for testing drug phenotype. Agol, 3, 4 deficient cells are $FIAU^RG418^RHAT^RPuro^S$. After confirming the drug phenotype, we further confirmed Agol, 3, 4 deficient cells using Southern analysis (Fig. 5c, right).

3.7. Generation of an Inducible Ago1, 2, 3, 4 Deficient ES Cell Line

Ago2 targeting vector was constructed using BAC-based recombination as described in Subheading 3.3. The first allele of Ago2 was targeted (see Note 20), whereas the second allele of Ago2 failed to be targeted after screening 400 clones. We suspected that the deletion of Ago2 together with Agol, 3, 4 may be detrimental to ES cells, thus we introduced a floxed transgene carrying an HA-tagged human Ago2 and Blasticidin gene (Fig. 1a). Both Blasticidin and hAgo2 can be excised upon Cre expression. With this transgene, the second endogenous Ago2 allele was successfully targeted. Lastly, we introduced an ERT2Cre transgene (10) into the cells where Cre is expressed in the presence of 4OH-T

(1 μg/ml). Upon the induction of Cre expression by 4OH-T treatment, the excision of hAgo2 will result in an Ago1, 2, 3, 4 deficient ES cell line.

4. Notes

1. The NM5 ES cells, SNL and SNLP feeder line can be obtained from Dr. Allan Bradley at the Wellcome Trust Sanger Institute. SNL is a STO cell line that is stably integrated with a neomycin-resistant cassette and a Leukemia Inhibitory Factor (LIF) expression vector. Therefore, no LIF supplementation is required when they are used as feeder layers for ES cells.
2. To limit the targeting of the deleted allele a second time (50% chance), the second 5′ deletion endpoint is designed to within the deletions of the first allele.
3. The BACs containing genes of interest can be located at Ensembl Mouse (http://www.ensembl.org/Mus_musculus/Info/Index) through searching by gene name or BLAST search. For example, using Ago4 as the search word, it directs to the Gene Summary page for Ago4. On the gene summary page, click on its location, which directs to a genomic map with Ago4 (also known as Eif2c4) in the center. To show BAC ID for Ago4 on the map, you need to click on "configure the page" (on the left panel of the location page). A configuration window will pop up. The option for 129S7 BAC is called "M37-129AB22" under "other DNA alignments" category. BAC clones can be requested through Sanger Genomic Center.
4. Try to use a BAC library that is derived from the same mouse strain as the ES cells you are using. The degree of sequence homology affects targeting efficiency. 129S7 BAC library is isogenic to AB2.2 cell line.
5. The tail of a restriction enzyme site on a primer is designed for constructing mini-retrieving vector and linearization in the step of BAC recombination. Avoid using large areas of repetitive sequences as a homologous arm in the retrieving vectors.
6. When selecting a genomic region as the homologous arm of an insertion targeting vector, you need to locate an enzyme that can linearize the final targeting vectors (DSBs) within the homologous region. The candidate enzyme may be found electronically based on NCBI published mouse genomic sequences. It can be achieved either by one unique enzyme cut or several cuts that create a small gap within the homologous region.

7. To eliminate Amp^R colonies derived from uncut mini-retrieving vectors, the linearized fragments usually are double gel-purified before electroporation.
8. pSC101-BAD-gbaA transformed bacteria should be used within a week. We found that the recombination efficiency drops dramatically when older ones are used.
9. Use too much mini-retrieving DNA cause high background after transformation due to the presence of undigested vectors. Generally, 50 ng DNA is enough to obtain thousands of recombinant colonies.
10. Due to the length of the homologous arm (without gaps), the expected PCR products for screening homologous recombinants usually >5 kb which makes PCR less efficient. However, because the gap will not be repaired in the events of random integration, sequences within the gaps can be used for screening gene targeting events, and make it possible to design a shorter PCR product. In addition, the same enzyme used for generating gaps can be used for the linearization of the final targeting vector (see Note 6).
11. Add only tetracycline (no chloramphenicol) to LB culture when prepare recombination-competent cells since there is no BAC in DH10B at this step.
12. Gene targeting vectors are prepared using Qiagen midi-prep kits from a 150 ml LB culture with antibiotics. For each electroporation, about 30 μg plasmid DNA is linearized with appropriate restriction enzymes in a volume of 100 μl. After confirming the complete digestion of DNA on an agarose gel, DNA is purified by phenol/chloroform extraction and ethanol precipitation. DNA pellets are washed twice with 70% ethanol and thoroughly dried before resuspension with 25 μl of 0.1× TE (sterile). DNA is then ready for electroporation.
13. To ensure efficient killing, G418 selection should start 24 h after electroporation.
14. Take advantage of the published mouse genomic sequence to design and choose your probes and enzymes for Southern analysis. Avoid using repetitive sequence as probes.
15. To freeze ES cells in 96-well format, first pre-feed cells with fresh M15 2 h before passage, then wash with PBS twice, add 50 μl of 0.25% Trypsin and incubate for 15 min, followed by adding 50 μl of 2× Freezing medium (20% DMSO, 20% FBS in knock-out DMEM). Pipette the mixture up and down 15 times to break the cell clumps. Finally, add 100 μl filter-sterilized light paraffin oil to each well. Put the plate into a styroform box and keep at –80°C overnight. The plate can be stored in cardboard boxes at –80°C for several months.
16. To perform sib-selection, the same number of cells are passaged into three plates with appropriate feeders. Each plate

is supplied with HAT or G418 or Puro, respectively, the following day.

17. Wash the plate with PBS twice. Add 50 μl cell lysis buffer (with freshly added protease) to each well and keep at 37°C incubator overnight. Add 100 μl cold ethanol with 75 mM sodium chloride and leave the plate on the bench for 15 min. Spin the plate at 3,000 rpm in a table-top centrifuge for 15 min. Carefully pour off the supernatant. Some of the transparent genomic DNA will stick to either the bottom or the side of the wells. Carefully wash the plates with 70% ethanol three times. Do not dislodge the DNA from the wells. Thoroughly dry the plates at room temperature before resuspending DNA in 100 μl ddH_2O. Make sure DNA is completely dry before adding water. Trace amount of ethanol will affect both PCR and restriction enzyme digestion. To resuspend genomic DNA well, it is recommended to first incubate the plate at 60°C for 1 h and then pipette the DNA up and down a couple of times.
18. You may screen with 5′ end primers first. Choose the primers that give better PCR result as the primary screening primers.
19. Also perform PCR using single primer on the same set of DNA. This will rule out the possibility that the PCR products are amplified from single primer.
20. The Ago2 targeting vector is designed to be a replacement vector. In addition, the drug cassette is floxed; therefore, the same targeting vector can be used to target the second allele after Cre-excision.

Acknowledgments

The authors would like to thank A. Bradley and G. Guo for providing NM5 ES cells. Funding was provided by National Institute of Health (5R21GM079528), Illinois Department of Public Health (to X.W.).

References

1. Hammond, S.M., Boettcher, S., Caudy, A.A., Kobayashi, R. and Hannon, G.J. (2001) Argonaute2, a link between genetic and biochemical analyses of RNAi. *Science*, **293**, 1146–1150.
2. Okamura, K., Ishizuka, A., Siomi, H. and Siomi, M.C. (2004) Distinct roles for Argonaute proteins in small RNA-directed RNA cleavage pathways. *Genes Dev*, **18**, 1655–1666.
3. Meister, G., Landthaler, M., Patkaniowska, A., Dorsett, Y., Teng, G. and Tuschl, T. (2004) Human Argonaute2 mediates RNA cleavage targeted by miRNAs and siRNAs. *Mol Cell*, **15**, 185–197.

4. Liu, J., Carmell, M.A., Rivas, F.V., Marsden, C.G., Thomson, J.M., Song, J.J., Hammond, S.M., Joshua-Tor, L. and Hannon, G.J. (2004) Argonaute2 is the catalytic engine of mammalian RNAi. *Science*, **305**, 1437–1441.

5. Ramirez-Solis, R., Liu, P. and Bradley, A. (1995) Chromosome engineering in mice. *Nature*, **378**, 720–724.

6. Su, H., Trombly, M.I., Chen, J. and Wang, X. (2009) Essential and overlapping functions for mammalian Argonautes in microRNA silencing. *Genes Dev*, **23**, 304–317.

7. Hasty, P., Crist, M., Grompe, M. and Brad-ley, A. (1994) Efficiency of insertion versus replacement vector targeting varies at different chromosomal loci. *Mol Cell Biol*, **14**, 8385–8390.

8. Zhang, Y., Buchholz, F., Muyrers, J.P. and Stewart, A.F. (1998) A new logic for DNA engineering using recombination in Escherichia coli. *Nat Genet*, **20**, 123–128.

9. Muyrers, J.P., Zhang, Y., Testa, G. and Stewart, A.F. (1999) Rapid modification of bacterial artificial chromosomes by ET-recombination. *Nucleic Acids Res*, **27**, 1555–1557.

10. Matsuda, T. and Cepko, C.L. (2007) Controlled expression of transgenes introduced by in vivo electroporation. *Proc Natl Acad Sci USA*, **104**, 1027–1032.

Chapter 20

Whole Cell Proteome Regulation by MicroRNAs Captured in a Pulsed SILAC Mass Spectrometry Approach

Olivia A. Ebner and Matthias Selbach

Abstract

Since gene expression is controlled on many different levels in a cell, capturing a comprehensive snapshot of all regulatory processes is a difficult task. One possibility to monitor effective changes within a cell is to directly quantify changes in protein synthesis, which reflects the accumulative impact of regulatory mechanisms on gene expression. *Pulsed stable isotope labeling by amino acids in cell culture* (pSILAC) has been shown to be a viable method to investigate de novo protein synthesis on a proteome-wide scale (Schwanhausser et al., Proteomics 9:205–209, 2009; Selbach et al., Nature 455:58–63, 2008). One application of pSILAC is to study the regulation of protein expression by microRNAs. Here, we describe how pSILAC in conjunction with shotgun mass spectrometry can assess differences in the protein profile between cells transfected with a microRNA and non-transfected cells.

Key words: Mass spectrometry, LC-MS/MS, SILAC, microRNA, Seed, Transfection, In gel digestion

1. Introduction

Measuring how gene expression changes in response to a stimulus can provide instructive insights into biological systems. Gene expression is regulated at all stages from DNA via mRNA to the protein. Most methods exclusively quantify changes in steady-state mRNA levels, neglecting posttranscriptional regulatory mechanisms. Here, we describe *pulsed stable isotope labeling by amino acids in cell culture* (pSILAC) as a method to quantify changes in protein production at a global scale. pSILAC measures the actual output of gene expression and can, therefore, reveal regulation at all levels. The method is particularly useful to study regulation at the level of translation. As an example, we show how the method can be used to quantify the effect of microRNAs on cellular protein production.

Tom C. Hobman and Thomas F. Duchaine (eds.), *Argonaute Proteins: Methods and Protocols*,
Methods in Molecular Biology, vol. 725, DOI 10.1007/978-1-61779-046-1_20,

The technology behind pSILAC is mass spectrometry-based proteomics (1). The general workflow is that proteins in a sample are digested into peptides. The resulting peptide mixture is separated by reversed phase liquid chromatography (LC). At the end of the chromatographic column, eluting peptides are directly transferred into the orifice of a mass spectrometer by a process called electrospray ionization (ESI). The mass spectrometer performs two important tasks. First, it measures the masses and intensities of the peptides in the mixture at any given time during the LC run. Second, the device fragments individual peptides and measures masses and intensities of the fragments (tandem mass spectrometry or MS/MS). The information about the masses of the nonfragmented peptides and their fragment spectra can be used to identify the peptides and hence the proteins present in the sample. In addition to identifying peptides and proteins, it is also necessary to quantify changes in their abundance. In mass spectrometry, this is most accurately achieved by stable isotope labeling (2): Incorporating heavy stable (i.e., nonradioactive) isotopes into peptides leads to a shift in mass. Differentially labeled samples can be combined and analyzed together so that all peptide peaks will occur in pairs. The ratio of peak intensities of such peptide pairs accurately reflects differences in their abundance. In *stable isotope labeling by amino acids in cell culture* (SILAC), the label is introduced metabolically. Cells are cultivated in growth medium containing heavy-stable isotope versions of essential amino acids (3). After several cell generations, all proteins have incorporated the heavy label. Mixing heavy and light cells can reveal changes in steady-state protein levels between both samples.

pSILAC is a variant of the SILAC approach (4). In contrast to standard SILAC, cells are first cultivated in growth medium with the normal light (L) amino acids. Concomitantly with differential treatment, cells are transferred to culture medium containing heavy (H) or medium-heavy (M) amino acids. All newly synthesized proteins will be made in the H or M form, respectively. Subsequently, both samples are combined and analyzed together. The abundance ratio of H versus M peptides reflects changes in protein production. Thus, pSILAC measures differences in protein synthesis integrated over incubation time of then pulse-labeling period. In principle, any essential amino acid can be used for SILAC. We prefer lysine and arginine because the protease trypsin cleaves C-terminal of these residues. Therefore, all tryptic peptides except for the protein C terminus contain a label and can be quantified. Heavy and medium-heavy lysines (Lys8 and Lys4) have a mass shift of 8 and 4 Da, respectively, compared with the normal light form (Lys0). Similarly, we use light (Arg0), medium-heavy (Arg6), and heavy (Arg10) arginine.

pSILAC is particularly useful to quantify changes in protein production induced by microRNAs (5). The protocol described here uses overexpression of short double-stranded RNAs designed to mimic endogenous microRNAs. Note that it is also possible to knock-down endogenous microRNAs, to use cells from microRNA knock-out animals, or to quantify protein production in different non-microRNA contexts. The first section describes how samples are prepared by transfecting cells with the microRNA mimic. Separation of proteins by SDS–PAGE is used as an approach to reduce sample complexity to achieve deeper proteome coverage. In the second section, mass spectrometry is described for an LTQ-Orbitrap system. In the last section, we indicate how the raw data can be analyzed to identify and quantify proteins and to investigate microRNA-mediated effects. See Fig. 1 for an overview of the described technique.

The protocol described here relies heavily on mass spectrometry-based quantitative proteomics. While the overall procedure is straight forward, the success depends on many technical details. Therefore, experience with mass spectrometry and subsequent data analysis is generally required. Alternatively, we suggest getting in touch with an expert lab while planning the project.

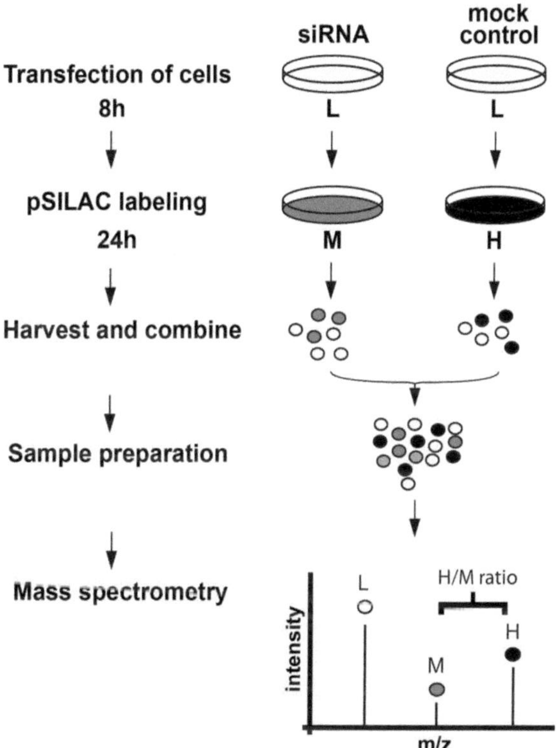

Fig. 1. Experimental setup of the method described in this chapter. Cells were cultivated in "light" (L) SILAC medium and subsequently mock- (control) or microRNA transfected. Pulse labeling was conducted after 8 h by transferring the control samples to "heavy" (H) and the microRNA-transfected samples to "medium-heavy" (M) SILAC medium. 24 h later, cells were harvested and combined. Sample preparation of both control and microRNA samples together ensures identical processing and comparability. Only M and H peaks represent newly synthesized proteins in the following mass spectrometry of the sample. Changes in protein production are reflected in the H/M ratio of peptide intensity peaks.

2. Materials

In general, HPLC and mass spectrometry require reagents and solvents of the highest available grade (HPLC grade or better). Reagent grades are indicated in the respective subsections.

2.1. pSILAC Preparation of Samples for Mass Spectrometry

2.1.1. Determination of Transfection Efficiency of BLOCK-iT™ Fluorescent Oligos

1. BLOCK-iT™ fluorescent oligo (Invitrogen).
2. 4% Paraformaldehyde (PFA).
3. Round cover slips (Menzel GmbH & Co. KG).
4. D-PBS (phosphate-buffered saline) modified without calcium chloride and magnesium chloride, sterile-filtered, liquid, cell culture tested (1×, Invitrogen).
5. Epifluorescence microscope (Leica DM-R).

2.1.2. Transfection, Labeling, and Harvest of Cells

1. HeLa cells or cell line of choice.
2. D-PBS (1×, modified without calcium chloride and magnesium chloride, sterile-filtered, cell culture tested, Invitrogen).
3. Trypsin–EDTA (0.05% trypsin with EDTA 4Na, 1×, Invitrogen).
4. SILAC amino acids: non-labeled L-lysine monohydrochloride and L-arginine monohydrochloride (Sigma-Aldrich) for “light” medium; (“Lys4”) 4,4,5,5-D_4-L-lysine monohydrochloride and (“Arg6”) L-arginine-$^{13}C_6$ monohydrochloride (Sigma-Aldrich, cat. no. 643440) for “medium-heavy” medium; and (“Lys8”) L-lysine-$^{13}C_6$ $^{15}N_2$ monohydrochloride (Sigma-Aldrich, cat. no. 608041) and (“Arg10”) L-arginine-$^{13}C_6$ $^{15}N_4$ monohydrochloride (Sigma-Aldrich, cat. no. 608033) for “heavy” medium.
5. Amino acid stock solutions: Dissolve 84 mg/ml arginine or 146 mg/ml lysine in D-PBS. Store in small aliquots at −20°C. High concentrations are necessary to avoid dilution of the culture medium. Stocks for stable isotope-labeled amino acids are prepared the same way. Sterile filtration is not necessary at this step as they will be filtered during the medium preparation.
6. Dulbecco’s Modified Eagle’s Medium (DMEM) High Glucose (4.5 g/l) w/o L-Arg, L-Lys, L-Glu (PAA, custom preparation) supplemented with 10% sterile-filtered dialyzed fetal bovine serum (dFBS, Sigma-Aldrich), and 4 mM stable Glutamine (L-alanyl-L-glutamine, PAA). Dialyzed serum is required since normal serum contains amino acids. Add dFBS and glutamine to medium, shake to mix, and fill into two 250 ml to filtration devices. Add 1:3,000 stock: medium (total amount of arginine 28 mg/l plus 48 mg/l of lysine/500 ml medium) of non-labeled lysine and arginine for “light” (L; 0/0) SILAC medium; Lys4 and Arg6 for

"medium-heavy" (M; 4/6) SILAC medium; and Lys8 and Arg10 for "heavy" (H; 8/10) SILAC medium.

Note that the given amino acid concentrations are optimized for HeLa cells. Other cell lines may require adjustments to minimize arginine to proline conversion (2, 3).

7. Synthetic microRNAs (Dharmacon): Prepare aliquot of 2 mM following the manufacturers protocol.
8. DharmaFECT 1 Transfection Reagent (Dharmacon). Store at 4°C.
9. Cell lifter (Costar).
10. Vacuum filtration with bottle, PES membrane 0.22 μm (Techno Plastic Products AG).

2.1.3. Cell Lysis and SDS–PAGE

1. NanoDrop 2000 (Thermo scientific) or alternative device to measure protein concentration.
2. RIPA buffer: 50 mM Tris–HCl pH 7.4, 150 mM NaCl, 1% Triton-X100, 1% Sodium deoxycholate, and 0.1% SDS; use a 0.22 μm filter for sterile filtration.
3. Benzonase® Nuclease-5KU (Sigma-Aldrich, optional).
4. NuPAGE Novex 4–12% gradient gels; LDS Sample buffer 4×, MES SDS Running buffer. 20× (Invitrogen).
5. 1 M DL-Dithiothreitol (DTT) in $H_2O_{bidest.}$; store 1 ml aliquots at −20°C. Harmful, prepare with caution.
6. Prestained molecular weight marker: SeeBlue Plus 2 Prestained Standard (Invitrogen).
7. Fixing solution: 20 ml $H_2O_{bidest.}$, 25 ml methanol (MeOH), 5 ml acetic acid (HAc, puriss. p.a., eluent additive for LC-MS, Sigma-Aldrich).
8. Staining Buffer A: 27.5 ml $H_2O_{bidest.}$, 10 ml MeOH, 10 ml Novex Stainer A (Invitrogen).
9. Staining Buffer B: Add 2.5 ml Novex Stainer B (Invitrogen) to Stainer A buffer in tray.

2.1.4. In Gel Protein Digestion

1. 50 mM Ammonium BiCarbonate (ABC, puriss. p.a., Sigma-Aldrich): Dissolve 40 mg ABC in 10 ml H_2O (LC-MS Chromasolv, Sigma-Aldrich). Store at RT.
2. Ethanol (EtOH, Ethanol gradient grade LiChrosolv, Merck).
3. Sequence grade-modified trypsin (Promega).
4. Trypsin solution: Dilute trypsin (0.5 μg/μl) in 50 mM ABC. Demanded Protein: Enzyme ratio = 50:1. Estimated protein amount per lane = 200 μg. Prepare immediately before use

and keep it always on ice to minimize autocatalysis. Undiluted stocks are stored in small aliquots at –80°C. Trypsin is sensitive to high urea concentrations. Concentration should be below 2 M urea/thiourea.

5. Iodacetamide solution (55 mM) in 50 mM ABC: Dissolve 10.2 mg iodacetamide in 1 ml ABC. Prepare fresh or store in small aliquots at –20°C. Keep in the dark.
6. 10 mM DL-Dithiothreitol (DTT) in 50 mM ABC. To make 1 ml dilute 10 μl of a 1 M DTT solution in 990 μl ABC and store in small aliquots at –20°C.
7. Extraction solution: 3% trifluoroacetic acid (TFA, puriss. p.a., eluent additive for LC-MS, ≥99.0% (GC), Sigma-Aldrich), 30% acetonitrile. To make 1 ml dilute 300 μl ACN (LC-MS CHROMASOLV, ≥99.9%, Sigma-Aldrich) and 30 μl TFA in 670 μl H_2O (LC-MS Chromasolv, Sigma-Aldrich). Store at RT.
8. Buffer A: 3% TFA, 5% ACN in H_2O (LC-MS Chromasolv, Sigma-Aldrich). Store at RT.

2.1.5. Desalting and Purification by C_{18} Stage Tips

1. C_{18} Empore 47 mm Disks (3 M).
2. Buffer B: 0.5% HAc, 80% ACN in H_2O (LC-MS Chromasolv, Sigma-Aldrich). Store at RT.

2.2. Mass Spectrometry of the Samples

1. Nanoflow HPLC system.
2. For HPLC columns: Pack your own columns with ReproSil-Pur 120 C18-AQ, 3 μm beads HPLC bulk packing material (Dr. Maisch GmbH, Germany) and a 360 mm OD, 75 mm ID pulled capillary column. Columns can also be bought ready-to-use from several companies.
3. High performance mass spectrometer. Ideally, this should be an instrument with high resolving power, high dynamic range, high speed, and high sensitivity. We describe the protocol for an LTQ-Orbitrap system (Thermo Fisher).

2.3. Processing of Mass Spectrometry Data

2.3.1. Processing of Raw Data

We use the freely available software package MaxQuant for peptide identification, protein assembly, and quantification (6). Alternatively, MSQuant (http://msquant.sourceforge.net/) or other software packages capable of SILAC-based quantification can be used.

2.3.2. Possibilities of Data Analysis

Among the open or commercial (programming) tools for data analysis, we found R (Bioconductor package) and PERL extremely useful for handling big datasets. Other tools with one main focus are, e.g., Sylamer, miReduce, Cytoscape, and STRING.

3. Methods

A common effect in microRNA overexpression experiments is the down-regulation of many (i.e., hundreds) of proteins. Yet many of these proteins are only mildly regulated (5). To be able to discern the effect and signature of the microRNA in a dataset, a high transfection efficiency of the microRNA is required. Thus previous to the transfection of cells for proteomics, a test of transfection efficiency via BLOCK-iT™ Fluorescent Oligos should be performed. To accurately determine the effect of the transfected microRNA, it is also necessary to treat both microRNA transfected cells and mock-transfected controls in the very same way. The cultures should have similar confluence of cells and the transfection procedure of microRNA samples and controls should be the same with the one exception of omitting the microRNA in the control. After transfection controls are usually plated into heavy (Lys8/Arg10) medium, whereas microRNA samples are plated into medium-heavy SILAC medium (Lys4/Arg6). Identical treatment for mass spectrometry sample preparation is assured by combining both control and microRNA sample upon harvest (see Fig. 1). Separation of lysed proteins is achieved via SDS–PAGE and single slices of the SDS-gel are subsequently subjected to reduction, alkylation, and digestion into peptides with trypsin. Prior to the digestion step, clean handling of the samples is essential to avoid contamination with contaminants such as keratin. Following the extraction of the peptides from the gel slices, the samples are desalted on reverse-phase C_{18} STAGE-tips. This step also provides further filtering and concentration of the samples. For mass spectrometry analysis, samples are eluted from the columns and further separated by HPLC. At the end of the HPLC column, the peptides are ionized (electrospray ionization, ESI) and analyzed by tandem mass spectrometry (LTQ-Orbitrap). Raw data files are analyzed with the MaxQuant software package, which is able to detect peaks and SILAC-labeled peptide triplets, identify peptides by database searching and deduce proteins from the latter. For further practical and background information about SILAC and related methods, consult also the available *Nature Protocols* (3, 6–8).

3.1. pSILAC Preparation of Samples for Mass Spectrometry

3.1.1. Determination of Transfection Efficiency of BLOCK-iT™ Fluorescent Oligos

1. As a nontargeted dsRNA oligomer, the BLOCK-iT fluorescent oligo resembles a microRNA and the efficiency of its transfection can be measured via fluorescence microscopy. Conduct the transfection of cells (on sterile coverslips) for oligos as well as for controls. In this case, however, six-well culture plates with area-adjusted amounts of reagents are employed (see Note 1).

2. Wash the cells after 8 h of transfection with D-PBS once and fix them with 4% paraformaldehyde (PFA) in D-PBS. To determine the transfection efficiency, compare the fluorescence of oligo-transfected with non-transfected cells under the epifluorescence microscope.

3.1.2. Transfection of Cells, Labeling, and Harvest

The transfection is carried out using DharmaFECT1 according to the manufacturer's protocol. In parallel, control transfections of HeLa cells for each microRNA sample are done under the same conditions expect for serum-free DMEM replacing the synthetic microRNA. Using another short dsRNA as a control (for example with scrambled sequence) is generally not recommended because every short dsRNA will have off-target effects on protein production making result interpretation more difficult.

1. Prior to the transfection, cells are cultured in "light" SILAC medium for a week to give them time to adjust to these growth conditions. Split the cells every 3 days and cultivate 37°C with 5% CO_2.
2. Plate the cells 1 day before transfection on 10 cm^2 dishes in "light" SILAC medium and incubate at 37°C with 5% CO_2 overnight. Confluence at the time of transfection should be 60–70% (see Note 2).
3. Add 300 μl microRNA (2 μM; final plating concentration of 100 nM) to 300 μl serum-free DMEM and incubate for 5 min at RT. Mix gently.
4. Add 16 μl DharmaFECT1 to 584 μl serum-free DMEM and incubate for 5 min at RT. Mix gently.
5. Combine both, mix by slow pipetting and incubate for 20 min at RT. Be careful not to disrupt the lipid–microRNA complexes from this point on – only very slow pipetting!
6. Remove medium from cells, rinse with D-PBS once, and then add 4.8 ml "light" SILAC medium.
7. Slowly pipet the transfection mix (1.2 ml in total) onto the cells and gently rock the plate to ensure equal distribution.
8. At 8 h posttransfection, wash cells twice with D-PBS and change the SILAC medium of microRNA samples to "medium-heavy" (Lys4/Arg6) and of the controls to "heavy" (Lys8/Arg10).
9. Wash once again with D-PBS 24 h later, scrape cells off in 6 ml ice-cold D-PBS, combine the microRNA sample and control, and centrifuge for 10 min at 600 × *g* and 4°C. Pellets can be stored at −20°C.

3.1.3. Lysis and SDS–PAGE

1. Resuspend cell pellets in 250–400 μl RIPA buffer and lyse for 20 min on ice with occasional vortexing. After centrifugation (14,000 rpm for 10 min in a table-top centrifuge at 4°C),

transfer the supernatant to a fresh microtube. If the supernatant appears to be viscous, this is due to DNA. To reduce viscosity, employ Benzonase.

2. Measure protein concentration of the whole-cell lysate via the NanoDrop 2000 ProteinA280 method or other suitable methods.
3. For SDS–PAGE, you can use NuPAGE Novex 4–12% gradient gels with MES as running buffer according to the manufacturer's protocol. However, other denaturing gel electrophoresis systems work as well. Note that precast gels are generally preferable because acrylamide in self-made gels can introduce protein modifications. Optimal load of protein is approximately 200 μg per lane although in some cases, it is possible to use as little as 50 μg of sample.
4. Mix the desired amount of lysate with 1:20 of DTT (1 M) and 1:4 of NuPAGE buffer and incubate at 75°C for 10 min. Subsequently, load the samples onto the gel together with 7.5 μl of the protein marker and run 180 V for approximately 50 min (i.e., until the marker reaches the lower boundary of the gel).
5. Treat the gel first with fixing solution and then staining buffer A for 10 min each. Next, incubate gel with staining buffer B for 30–60 min (or until bands are visible). Wash the gel with ddH_2O once or twice until the solution of the gel appears to be clear. This step functions mainly to make sure that the gel ran smoothly and enough protein has been loaded.

3.1.4. In Gel Protein Digestion

All steps before the trypsin digestion step should be performed under a flow hood. Use extreme care not to contaminate samples with keratin from skin or other contaminants (see Note 3). The samples should always be incubated using a microtube shaker. Prior to the gel extraction step, the supernatants are discarded at each step. Use enough solution to cover the gel slices. Washing can be done with an excess of solution (around 500 μl); however, extraction solutions generally should not exceed 300 μl to contain sample volume for evaporation. Unless stated otherwise all step are done at RT.

1. Cut the gel into 12 slices on a clean plastic foil. Chop each slice into smaller pieces (approx. 1 × 1 mm) and place the pieces of one slice in a clean 1.5 ml microtube (see Note 4). Cutting and transferring slices is easier with a gel that is neither too dry nor too wet. When possible, it is best to cut out individually stained bands. However, for regions of low protein content where individual proteins are not clearly discernable, one can combine two to three slices (see Note 5).
2. Wash gel pieces with 1:1 ABC/EtOH for 20 min. Gel pieces should be clear before proceeding to the trypsin digestion

step. If they are not clear, wash again with ABC for 20 min and then ABC/EtOH for 20 min.

3. Dehydrate the gel pieces by incubating for 10 min in absolute EtOH.
4. Dry the samples in a Speed-Vac for 10–15 min until the gel pieces are bouncing in the tube (see Note 6). Samples can be stored at 4°C for several days at this point.
5. Rehydrate the gel pieces and reduce the proteins by incubating for 45–60 min in DTT solution at 56°C. Discard all the liquid afterwards.
6. Block-free sulphydryl groups by incubating for 45 min in iodacetamide at RT in the dark.
7. Wash gel pieces once with ABC for 20 min at RT.
8. Dehydrate the gel pieces by incubating for 10 min in absolute EtOH.
9. Remove remaining ethanol from gel pieces by vacuum centrifugation. Samples can be stored at 4°C for several days at this point.
10. Add enough trypsin solution at 4°C to cover the dehydrated gel pieces and place tubes on ice. Make sure that the gel slices are fully covered with trypsin solution (otherwise add ABC) after swelling as much as possible (approx. 20 min). Place in microtube shaker at 37°C over night.
11. Add 2 μl TFA to stop the digestion and quickly finger-vortex the solution. Spin down the gel pieces at low speed in a microfuge and transfer the liquid to a fresh tube.
12. Extract the gel pieces by adding Extraction solution to cover the gel. Shake the mixture vigorously for 10 min at RT. Remove the liquid and combine with that from step 11.
13. Dehydrate gel pieces in 100% ACN for 10 min at RT. Spin down the gel pieces, recover the supernatant, and combine with supernatant from steps 11 to 12.
14. Dry the samples in a Speed-Vac until 10–20% original volume to remove ACN. Adjust the samples to a low pH with buffer A by adding approx. 50 μl to the samples. The resulting pH should be <2.5.

3.1.5. Desalting and Purification by C_{18} Stage Tips

1. Prepare as many desalting columns (Stage Tips) as necessary by punching out small disks of C_{18} Empore Filter and eject the disks (three per tip) into a P200 pipette tip using a blunt-ended syringe needle (Fig. 2). Ensure that the disks are securely wedged in the bottom of the tip.
2. Condition tips by adding 50 μl methanol to the Empore disk. Use this step to check whether the Stage Tips are leaky.

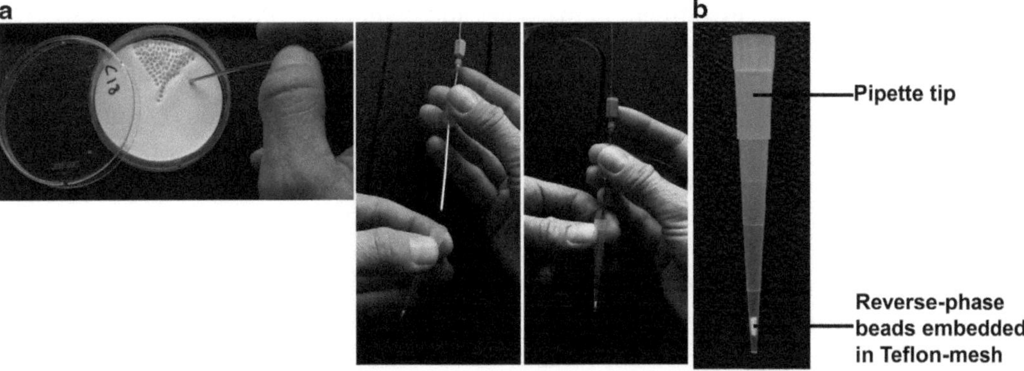

Fig. 2. (**a**) Preparation of Stage Tips. Small disks of C_{18} Empore Filter are excelled with a blunt-ended syringe needle and inserted into the tip. Disks should be pushed to the end of the tip and wedged securely but not too tightly (impedes flow-through) to the *bottom* of the tip. (**b**) Final Stage Tip. The C_{18} Empore Filter consists of chromatographic beads embedded in a Teflon mesh and is ready-to-use for micro-purification. To activate, wash and to load the sample, liquid is always added from the top and passes through the column by either centrifugation or pressure from an air-filled plastic syringe.

For centrifugation (5 min at 5,000 rpm), a standard table-top centrifuge can be used. Punch a hole into the lid of 2 ml microtubes and place the Stage Tips into them. Discard flow through once it reaches the end of the tip, and always make sure that the Stage Tips are empty before adding the next solution.

3. Remove any remaining organic solvent in the column by adding 100 μl buffer A to the disk and centrifuge for 5 min at 5,000 rpm.
4. Force the acidified peptide sample (From 3.4, Nr.14) through the Stage Tip at 5,000 rpm for 5 min.
5. Wash the column with 120 μl buffer A for another 5 min at 5,000 rpm. Samples can be stored at this point at 4°C for up to 1 year. When using tips that have been stored for a longer period, wash them again with washing buffer.
6. Elute the peptides from the C_{18} material using 50 μl buffer B. Elute directly into microtube the autosampler plate of your HPLC system (see Note 7).
7. Dry the autosampler plates in the Speed-Vac at 45°C until all acetonitrile has evaporated (~3 μl remaining final volume). (Do not over dry because otherwise you will lose most of the sample.) For two MS measurements (each from samples of 6 μl), add sample buffer to a final volume of approx. 15 μl (see Note 8).

3.2. Mass Spectrometry of the Samples

While sample preparation for shotgun mass spectrometry can be done in almost every wet lab, setting up the resources for both mass spectrometry itself and subsequent data analysis requires investment in machines and computing power as well as expertise in handling both. The following protocol can thus only provide the rough outlines of sample processing after in gel digestion. The analysis is performed using liquid chromatography coupled to tandem mass spectrometry (LC-MS/MS). Peptides are separated on a nanoflow HPLC system. At the end of the column, they are directly ionized and transferred to the mass spectrometer. The mass spectrometer determines masses and intensities of the eluting peptides during the gradient (MS). In addition, it isolates individual peptides for fragmentation and records their fragment spectra. Ideally, the machine should be able to fragment all peptides eluting from the column. However, due to limited dynamic range, sensitivity, and speed, only a fraction of peptides are picked in each run. Measuring every sample twice alleviates undersampling.

1. Analyze the eluted samples via LC-MS/MS on a high performance mass spectrometer (LTQ-Orbitrap). Coupled to the LTQ-Orbitrap, an HPLC nanoflow system (for example Eksigent) separates peptides by reversed phase chromatography on C_{18} columns directly connected to the electrospray ion source.
2. Inject each sample twice with an injection volume of 6 μl and a 150 min flow gradient ranging from 10 to 60% acetonitrile in 0.5% acetic acid at a flow rate of 250 nl/min. There are several alternative gradients that can be used at this point. Supervision by a scientist experienced in LC-MS/MS is recommended.
3. The Orbitrap performs precursor ion scan/survey MS spectra (m/z range of 300–1,700; resolution $R = 60{,}000$; target value of 1×10^6; profile mode). For MS/MS, isolate five to ten of the most intense peaks (monoisotopic precursor selection enabled; charge state 2; target value 5,000), fragment via collision-induced dissociation (normalized collision energy 35%; wideband activation enabled; centroid mode) in the LTQ part of the LTQ-Orbitrap and exclusion-selected peptides using dynamic exclusion (duration 60 s, exclusion list size 500). For an example of a full MS scan showing a SILAC triplet see Fig. 3.

3.3. Processing of Mass Spectrometry Data

3.3.1. Preprocessing of Raw Data

While there are different ways to process the obtained spectra, we employ the MaxQuant software package. This package consists of several modules with adjustable parameters. In most cases, the default parameter settings can be used. More information about MaxQuant can be found online (http://www.maxquant.org).

1. Settings for Quant are Arg6 and Lys4 as “medium-heavy,” Arg10 and Lys8 as “heavy” labels with maximum of three

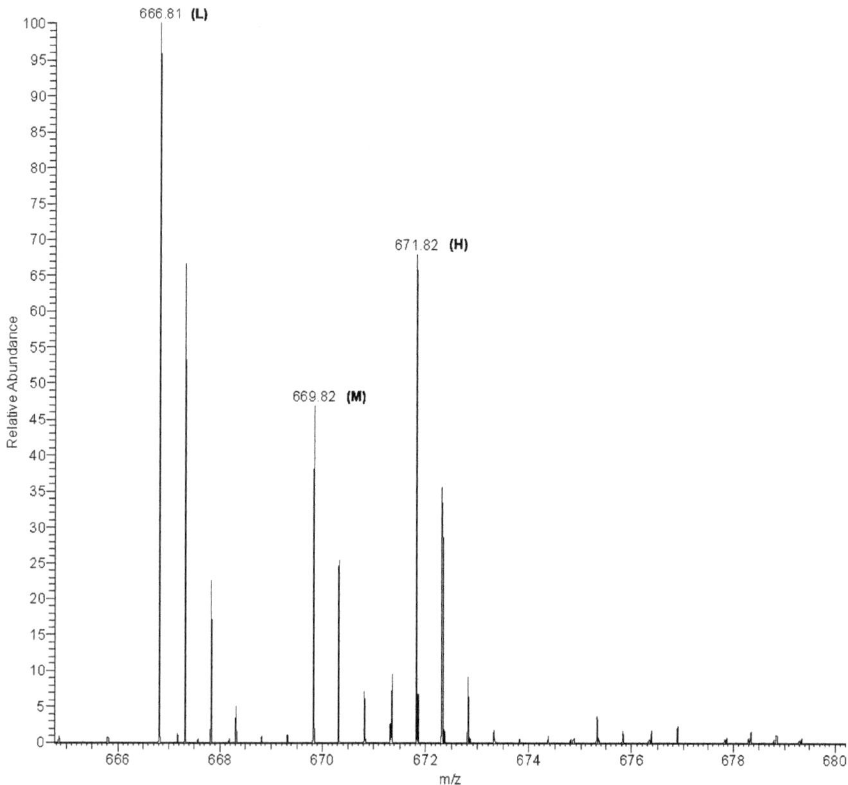

Fig. 3. Peptide mass spectra displaying pulsed SILAC triplets of Destrin (ASGVQVADEVCR). The full MS scan of the miR-34-transfected HeLa sample shows the light (L), medium-heavy (M), and heavy (H) monoisotopic peaks of Destrin. The doubly charged peptide contains labeled arginine which leads to a shift of 3 Da for the M label and 5 Da for the H label (Arg6 and Arg10, due to the double charge, the shift is cut into halves). The H/M ratio of Destrin in this sample is 1.12, indicating a reduction of Destrin in the miR-34-transfected sample of −0.05 log2-fold change.

labeled amino acids per peptide. Top six MS/MS peaks per 100 Da are used and polymer detection is enabled. For the database search, use carbamidomethylation of cysteines as fixed modification. Acetylation of the protein N terminus and oxidation of methionine are used as variable modifications. A maximum of two missed cleavages and specificity of cleavage by trypsin should be required. Set the mass tolerance to 0.5 Da for low-resolution fragment ions and mass accuracy cut-off to 7 ppm for the high resolution parent ion. Note that this is just the initial value: MaxQuant will determine the optimal cut-off value by target-decoy database searching.

2. Set up a suitable protein database for the search. In general, an organism-specific database should be used (for example from http://www.uniprot.org). To this database, you should add common contaminants (human/sheep keratins, trypsin, bovine serum albumin, etc.).

3. For peptide and protein identification, we normally require at least two identified peptides per protein group in total, one of them being unique in the database. The minimum peptide length is set to six amino acids. Quantification is done only for protein groups with at least two ratio counts. A more conservative analysis uses only proteins with three ratio counts as the median of three peptide ratios is outlier-robust. With the protocol described here, it should be possible to identify at least 4,000–5,000 and to quantify 3,000–4,000 proteins per sample.

3.3.2. Possibilities of Data Analysis

Besides the preprocessing via MaxQuant, the data may still need further steps of filtering before being ready for downstream analysis. Recheck your list for contaminants such as keratins that might have been missed because they were not in the contaminant part of the database. Proteins with very high/low ratios should be examined carefully. As a rule of thumb, proteins quantified based on many counts are reliable despite a high variability of individual peptide ratios. For proteins represented by only a few peptide counts, it may be necessary to check individual cases. For example, you can check the posterior error probability (PEP) of proteins and peptides to see how confident individual peptides/proteins were identified. If you have experience with mass spectrometry, you can also look at individual fragment spectra and try to assign all peaks to expected fragments.

For data analysis, it is useful to transform measured ratios into logarithmic fold changes; we use the –log2 of the heavy to medium-heavy normalized ratios of the MaxQuant output (H/M normalized) to visualize the data. The two main subsets that can be distinguished in microRNA-related datasets are proteins that possess a seed site in the 3′UTRs of their respective mRNAs versus proteins without such seed sites. Even though the effect of microRNAs on many targets may not be explained by 3′UTR seed-mediated regulation, this is one of the most feasible ways to identify likely targets of a microRNA. To make sure that the overexpression of the microRNA worked, monitor whether the seed signature of the microRNA is seen in the data. Sylamer is a tool to assess nucleotide sequences over- or underrepresented in data sorted list of genes. When searching for enriched sequences of 6 nt length, the microRNA seed sequence should be enriched mainly in the downregulated fraction (Fig. 4). Furthermore, the comparison of cumulative fractions of all proteins versus seed-containing proteins should have a bias of the latter to negative fold changes. Another way to validate data is to investigate how many known targets of the microRNA are found to be downregulated in the mass spectrometry data. Network relationships can be obtained using tools such as Cytoscape, PathVisio, or STRING. To assess functional implications

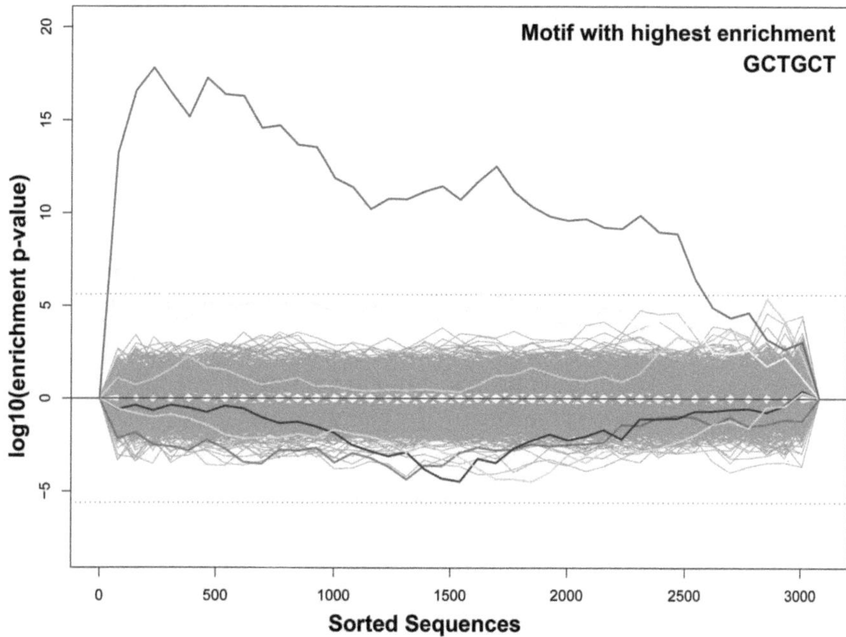

Fig. 4. Sylamer analysis of HeLa cells transfected with miR-16. Proteins quantified in the sample are sorted in bins from most downregulated (*left*) to most upregulated (*right*) on the *X*-axis. The nucleotide (nt) sequences of the 3 UTRs of these proteins were analyzed for enrichment of 6 nt long sequences. The hyper-geometric significance for each leading bin of all detected 6 nt sequences is shown on the *Y*-axis scaled as log10 (*P*-value). Enrichment is represented on the positive *Y*-axis, whereas depletion in negative values. The *E*-value threshold of 0.01 (Bonferroni-corrected) in indicated as *horizontal dotted line*. Significance *curves* for all sequence motifs are drawn along the *X*-axis. Most of the sequences are not significantly enriched; however one sticks out of the *gray* shaded background. This *curve* represents the miR-16 seed sequence of GCTGCT. The profile shows a clear peak in the region of downregulated proteins as signature of the microRNA with lesser but significant enrichment in the whole protein fraction.

of the microRNA targets, especially the seed-containing ones, Gene Ontology analysis in conjunction with hierarchical clustering is a commonly used method.

4. Notes

1. The transfection protocol has been optimized for HeLa cells, but can be adapted to other cell lines as well. If you choose a different cell line, make sure that DharmaFECT1 has high transfection efficiency in the chosen cell line. Otherwise you might need to try different lipids for transfection. The same is true for transfection and labeling times; depending on the subject of investigation shorter or longer times until the harvest of the cells might be required.
2. Make sure the confluence of cells does not exceed 70% at the time of transfection; to sufficiently SILAC label long-lived

proteins the cells need to be able to further grow after transfection to incorporate the labeled amino acids.

3. While in gel digestion is not very error prone, it is crucial to avoid contaminants because they will be predominant in the mass spectrometry spectra. Therefore, work under the hood and make sure all surfaces are clean. Do not work over open microtubes.
4. We use the inner lining of small autoclaving waste bags for cutting the gel but any clean plastic foil will do.
5. A good rule for cutting the gel is to keep regions with a thick distinct protein band in a single slice, while pooling regions of lower protein content where you do not see clearly distinguished protein bands. In general, you can cut the gel into as many pieces as you like but keep in mind that each slice prolongs measuring time. Accordingly, you may not necessarily get better results from increasing the number of gel slices.
6. Be careful when handling the gel cubes in the microtubes; especially after Speed-Vac evaporation they are prone to "jump" from the tube.
7. For elution from stage tips, we recommend to use a 20 or 50 ml syringe. Push out the sample by hand by sticking the tip on top of the syringe; do not elute too fast but rather in slow drops to elute most peptides on the column.
8. Be careful when diluting the final sample in the autosampler plate. You will actually need a few more microliter of sample than what is injected. Because the sample is contained in a very small volume, even adding a single micorliter of buffer A will lead to a significant dilution of the sample. For good MS results, you should keep the sample as concentrated as possible.

Acknowledgments

The authors would like to thank Björn Schwanhäusser for providing data and templates for the shown figures.

References

1. Aebersold, R., and Mann, M. (2003) Mass spectrometry-based proteomics, *Nature 422*, 198–207.
2. Bantscheff, M., Schirle, M., Sweetman, G., Rick, J., and Kuster, B. (2007) Quantitative mass spectrometry in proteomics: a critical review, *Anal Bioanal Chem 389*, 1017–1031.
3. Ong, S. E., and Mann, M. (2006) A practical recipe for stable isotope labeling by amino acids in cell culture (SILAC), *Nat Protoc 1*, 2650–2660.
4. Schwanhausser, B., Gossen, M., Dittmar, G., and Selbach, M. (2009) Global analysis of cellular protein translation by pulsed SILAC, *Proteomics 9*, 205–209.

5. Selbach, M., Schwanhausser, B., Thierfelder, N., Fang, Z., Khanin, R., and Rajewsky, N. (2008) Widespread changes in protein synthesis induced by microRNAs, *Nature* *455*, 58–63.

6. Cox, J., Matic, I., Hilger, M., Nagaraj, N., Selbach, M., Olsen, J. V., and Mann, M. (2009) A practical guide to the MaxQuant computational platform for SILAC-based quantitative proteomics, *Nat Protoc* *4*, 698–705.

7. Shevchenko, A., Tomas, H., Havlis, J., Olsen, J. V., and Mann, M. (2006) In-gel digestion for mass spectrometric characterization of proteins and proteomes, *Nat Protoc* *1*, 2856–2860.

8. Rappsilber, J., Mann, M., and Ishihama, Y. (2007) Protocol for micro-purification, enrichment, pre-fractionation and storage of peptides for proteomics using StageTips, *Nat Protoc* *2*, 1896–1906.

Index

Tom C. Hobman and Thomas F. Duchaine (eds.), *Argonaute Proteins: Methods and Protocols*,
Methods in Molecular Biology, vol. 725, DOI 10.1007/978-1-61779-046-1,

D

E

M

N

O

P

Q

R

S

MIX
Papier aus verantwortungsvollen Quellen
Paper from responsible sources
FSC® C105338

If you have any concerns about our products,
you can contact us on
ProductSafety@springernature.com

In case Publisher is established outside the EU,
the EU authorized representative is:
Springer Nature Customer Service Center GmbH
Europaplatz 3, 69115 Heidelberg, Germany

Printed by Libri Plureos GmbH
in Hamburg, Germany